Crystal Clear

Vintage American Crystal Sets, Crystal Detectors, and Crystals

Crystal Clear

Vintage American Crystal Sets, Crystal Detectors, and Crystals

Maurice L. Sievers

The Vestal Press, Ltd.

Library of Congress Cataloging-in-Publication Data

Sievers, Maurice L., 1922-
Crystal clear : vintage American crystal sets, crystal detectors, and crystals / Maurice L. Sievers.
p. cm.
Includes index.
ISBN 0-911572-86-4 (pbk. : alk. paper) :
1. Crystal sets (Radio)—History. 2. Crystal sets (Radio)—Collectors and collecting. 1. Title.
TK6563.S44 1990
621.384'18—dc20 90-12863
CIP

First Printing 1991
Printed in the United States of America

Contents

RADIO NEWS

REG. U.S. PAT. OFF.

25 Cents
August
1923
Over 200 Illustrations

Edited by H. GERNSBACK

THE 100% WIRELESS MAGAZINE

Preface

This is a book that I had hoped someone else would write. For 15 to 20 years, I waited in vain for such a tabular, descriptive, and illustrated publication. During those years, I needed authoritative information to satisfy my fascination with crystal sets, detector stands, and crystals, so I began compiling reference cards from material found in many early radio magazines and catalogs, and from observations of various crystal set items that I acquired or viewed in other collections.

About five or six years ago, after completing hundreds of these reference cards, I felt that perhaps I should consider writing the book that I needed. But it seemed a gargantuan task. I wanted a comprehensive and accurate volume, yet I knew that I could never compile a complete list of *all* early commercial crystal sets, detectors, and crystals marketed in the United States. In fact, even for some items that I had seen, information was elusive. Obviously, I could not discover all manufacturers, trade names, and details of lesser-known crystal sets and detectors.

Eventually, I concluded that no one could write the ideal book but that, nonetheless, the information that I had assembled was useful to me and might be of interest to others. Such a book, I knew, must be based on a thorough search of most major early radio magazines and catalogs. To accomplish this literature review, I had the generous assistance of some early-radio-literature collectors who made their reference materials available to me. Many persons also supplied information about items in their collections, and several collectors gave helpful suggestions and encouragement.

In this regard, I am deeply indebted to Lester Rayner and David McKenzie for providing access to their numerous references; to Gerald "Jerry" Granat for supplying invaluable details about the

THE GALENA PUZZLE

Editor, RADIO NEWS:

Referring to long distance galena records, I wish you would print something about the possibility of the seeming long distance signals coming, in fact, by relay from a nearby tube set. I got St. Louis recently, a distance of 980 miles, and a friend with a tube set is trying to explain it as above. I referred him to the bit you wrote in your January editorial about the galena "puzzle" but he declines to be puzzled. Quite illogically, it seems to me, he grants that I can get KDKA at 450 miles and my lesser "gets." He got these himself in the crystal stage of his experience.

Now, if it be true that each tube set is a transmitter, not merely of squeals but of good radiophone, it raises hob with all records of distance whether by galena or tube set, for if galena can get these local relays, the tube sets can also, with all the more reason. This leaves it free for the owner of the best set to claim that the weaker sets are not really getting Mexico and San Francisco, but are merely hearing him get them. This ought to stir things up. Important, if true. Also it ought to be easily proved. If Jones with the big set goes to bed, no more fine "gets" that night, etc., etc.

And what about the thousands of crystal sets in cities, mixed in promiscuously with tube set relays? Why does it not make them all good? Why the exceptional set anyway? Did not galena make fine records before tubes were invented? How about records in vessels at sea with no relays near? What, why, how?

JUNIUS T. HANCHETT,
Antrim, N. H.

(Yes, Why? "We want to know." Watch for our "Galena Record Prize Contest."—EDITOR.)

Radio News (Apr. 1923), p. 1842

Philmore Manufacturing Company; and to Al Reymann, Ed Sharpe, and Don Iverson for being the source of many descriptions and facts enhancing the tables and text, and of several photographs that appear in this book. I am also pleased to acknowledge the important contributions of information or suggestions by Woodrow Beshore, Tom Burgess, Jim Clark, Ray Claude, Dexter Deeley, Floyd Engels, Ralph Ernstein, Frank Falkner, Greg Farmer, Jack Farrell, George H. Fathauer, Jerry Finamore, Mike Feher, Paul Giganti, King Harrison, Joe Horvath, Bob Kerns, Bob Lane, Guy Martin, E. F. Schmidt, Mike Schiffer, Russ Schoen, Paul Thompson, William W. "Woody" Wilson, and Rich Wolven. Finally, to Harvey Roehl, Grace Houghton, Lucie Washburn, and Katha Fauty of The Vestal Press; to my wife, "Bud;" and to many unnamed friends whose patient, helpful encouragement gave direction and inspiration, I am extremely grateful.

The processes required to develop this book produced much joy and a little anguish. Perhaps most painful was the need to exclude available material because of space constraints. Selecting from the multitude of wonderful old ads was a formidable task. I hope the readers enjoy the ads included in *Crystal Clear*; but I must say, as in the fisherman's tale, "You should have seen the ones that got away!"

Maurice L. (Maury) Sievers

Pacific & Atlantic

A FINGER-RING RECEIVER

This is claimed to be the smallest set made. Do our readers know of one smaller?

Popular Radio (Nov. 1922), p. 221

COMMON ERROR

Mr. P. Fish, Rustling Palms, Alaska, says:

Q. 1. Following the directions in your issue of February, 1889, I wound my tuner on a Quaker Oats Corn Meal carton, but I hear only a fluttering noise when it is connected.

A. 1. You are hearing what is known as "mush". Rewind on a table-salt box, and everything will come in clear as crystal.

Radio News (Mar. 1923), p. 1660

The sun-dodgers of the Chicago Board of Trade station WDAP are a bunch of regular fellows. They've chipped in and bought over 500 crystal sets for shut-ins and the Boy Scouts lent their aid by going around and installing them.

Radio Topics (May 1923), p. 40

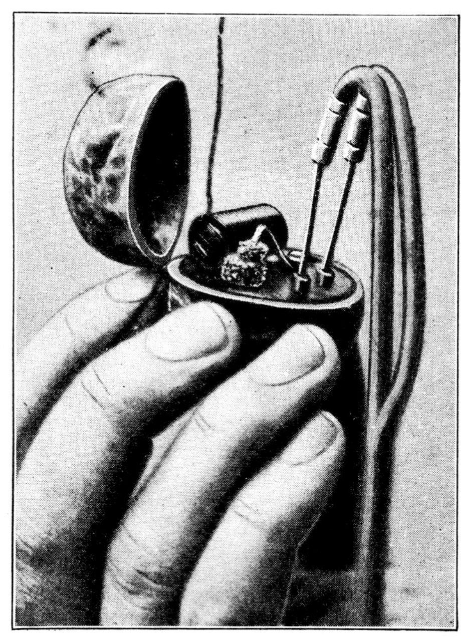

Radio in a Nutshell.

Radio News (Feb. 1924), p. 1142

Ready July 1st "Musio" Ready July 1st.

TRADE MARK REG. U.S. PAT. OFF.

A NEW RADIO RECEIVING SET

With Patented Automatic Crystal Detector

Every Set Guaranteed!

Every "MUSIO" is guaranteed to give as good or better results than any other crystal set on the market, regardless of price.

$12.50

Newest Invention in Radio

"Musio" Automatic Crystal Detector

A slight turn of the knob picks out another "sensitive spot"—always at the correct pressure, insuring the sharp and clear reception of music, lectures and news.

The crystal detector is located inside the cabinet. It cannot collect dust. It cannot be tampered with nor damaged.

"Musio"

RECEIVING OUTFIT
Includes the "MUSIO"
and
Copperweld Antenna Wire
Copperweld rubber insulated Lead-in and Ground Wire
Antenna Insulators
Entrance Tube
Ground Clamp and Splicer

All packed in a token box with explicit directions for installing and wiring correctly in accordance with Underwriters' Rules.

Without Phones Complete for $12.50

RADIO DEALERS
"Musio" offers you a big opportunity. Write today for particulars.

THE most efficient, low-priced receiving set ever devised. *Easily operated by a child.* Designed by the best engineering talent. The very highest quality of materials used throughout.

Beautifully built.

Finished in mahogany.

Musio Radio Co.
Westinghouse Building
Garrison Way and French St.
PITTSBURGH, PENN.

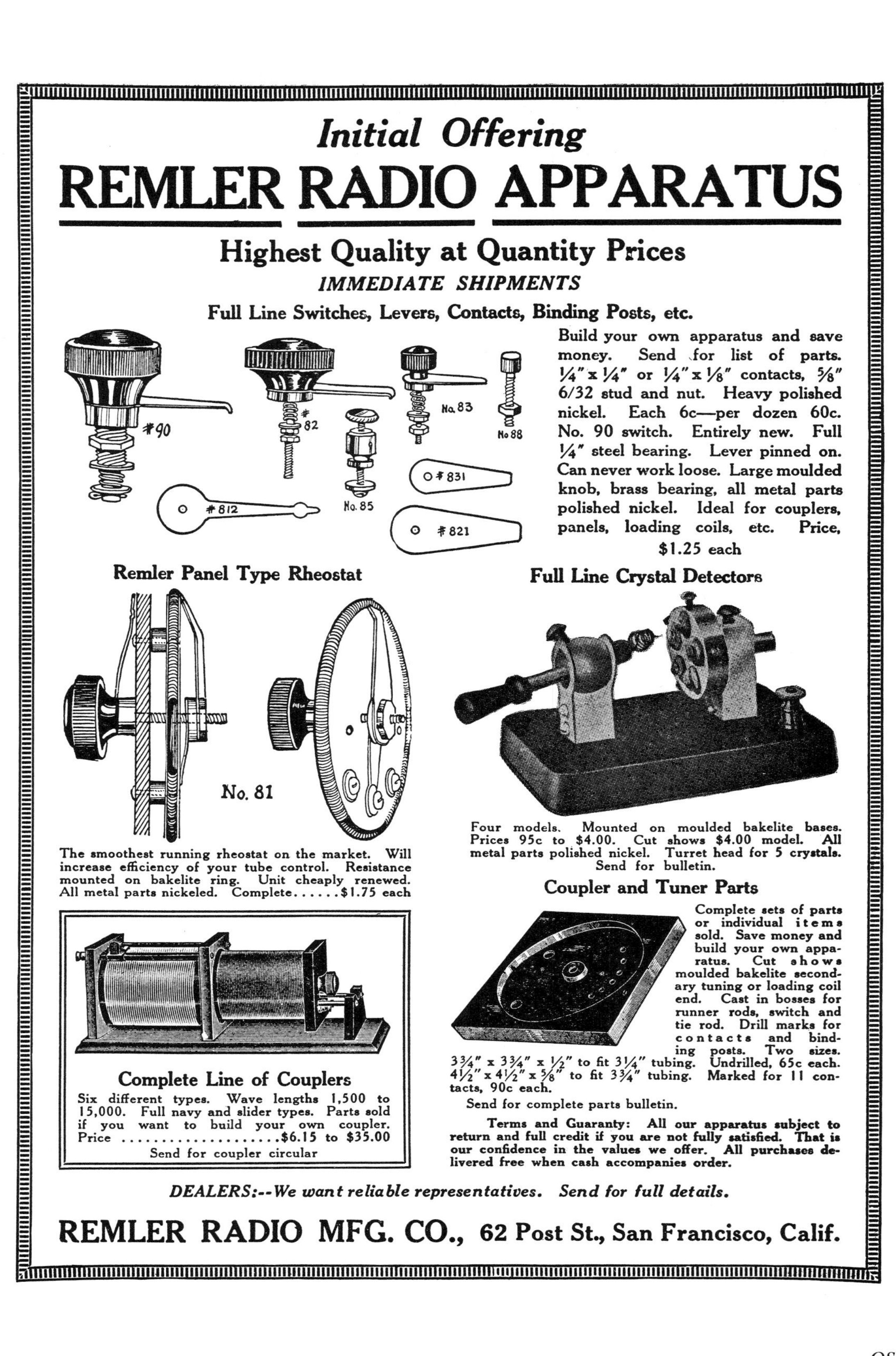
Initial Offering
REMLER RADIO APPARATUS
Highest Quality at Quantity Prices
IMMEDIATE SHIPMENTS
Full Line Switches, Levers, Contacts, Binding Posts, etc.
#90
#82
No. 83
No 88
No. 85
#831
#812
#821
Build your own apparatus and save money. Send for list of parts. 1/4" x 1/4" or 1/4" x 1/8" contacts, 5/8" 6/32 stud and nut. Heavy polished nickel. Each 6c—per dozen 60c. No. 90 switch. Entirely new. Full 1/4" steel bearing. Lever pinned on. Can never work loose. Large moulded knob, brass bearing, all metal parts polished nickel. Ideal for couplers, panels, loading coils, etc. Price, $1.25 each
Remler Panel Type Rheostat
Full Line Crystal Detectors
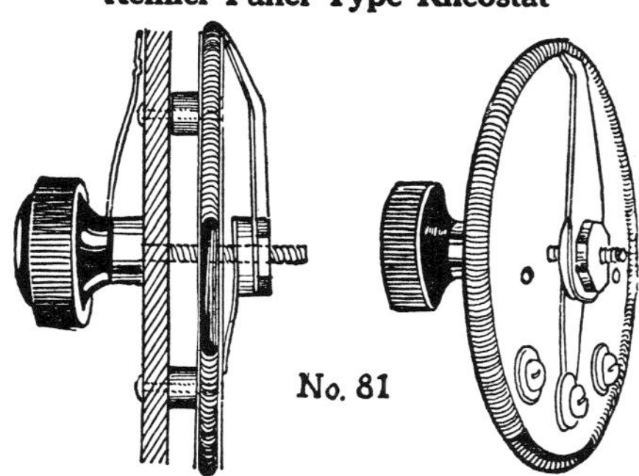
No. 81
The smoothest running rheostat on the market. Will increase efficiency of your tube control. Resistance mounted on bakelite ring. Unit cheaply renewed. All metal parts nickeled. Complete......$1.75 each

Four models. Mounted on moulded bakelite bases. Prices 95c to $4.00. Cut shows $4.00 model. All metal parts polished nickel. Turret head for 5 crystals. Send for bulletin.
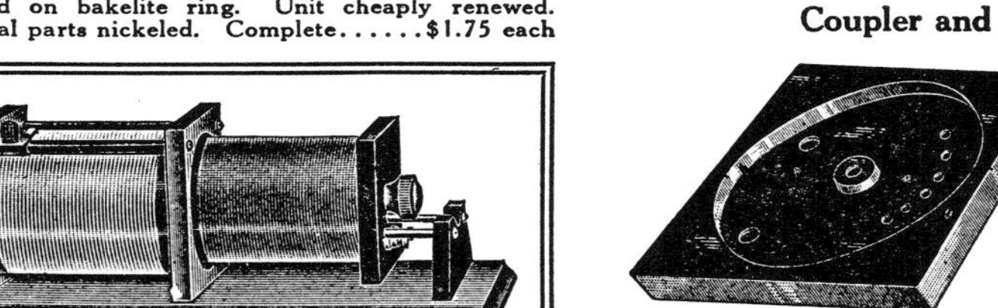
Complete Line of Couplers
Six different types. Wave lengths 1,500 to 15,000. Full navy and slider types. Parts sold if you want to build your own coupler. Price ....$6.15 to $35.00
Send for coupler circular
Coupler and Tuner Parts
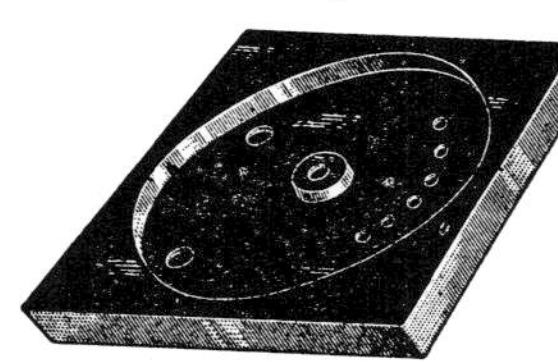
Complete sets of parts or individual items sold. Save money and build your own apparatus. Cut shows moulded bakelite secondary tuning or loading coil end. Cast in bosses for runner rods, switch and tie rod. Drill marks for contacts and binding posts. Two sizes. 3 3/4" x 3 3/4" x 1/2" to fit 3 1/4" tubing. Undrilled, 65c each. 4 1/2" x 4 1/2" x 5/8" to fit 3 3/4" tubing. Marked for 11 contacts, 90c each.
Send for complete parts bulletin.
Terms and Guaranty: All our apparatus subject to return and full credit if you are not fully satisfied. That is our confidence in the values we offer. All purchases delivered free when cash accompanies order.
DEALERS:--We want reliable representatives. Send for full details.
REMLER RADIO MFG. CO., 62 Post St., San Francisco, Calif.

Crystal Clear

Vintage American Crystal Sets, Crystal Detectors, and Crystals

Crystal-Clear Detection

The importance of the crystal to undistorted detection of radio signals cannot be overstated. From the early years of the wireless telegraph when the crystal detector supplanted the coherer and supplemented the electrolytic detector to present-day radio and television receivers where transistors and integrated circuits have replaced thermionic vacuum tubes, the crystal has provided inexpensive clarity of reception.

The principle of crystal detection is, indeed, simple—it acts as a "rectifier" to convert high frequency radio waves from alternating current (A. C.) to direct current (D. C.). Certain natural crystals and synthetic materials allow the passage of A. C. electrical impulses or oscillations in one direction but inhibit their return surge (i.e., act as a one-way valve). The resultant "half wave" is, in effect, D. C. For *amplitude*-modulated (A. M.) radio signals, this half wave has an imaginary "envelope" that follows the contour of the varying heights (the *amplitude*) of the carrier radio waves. This undulating "envelope" effectively changes the high radio-frequency (R. F.) waves into the much lower audio-frequency (A. F.) waves required to produce an audible response for headphones and loud speakers by drivers. The A. F. range, generally from 15 to 20,000 hertz, is heard by the human ear as a reproduction of the sound originally presented to the microphone at the transmitting station.

As discussed later in the section on the Wireless Specialty Apparatus Company, Greenleaf Whittier Pickard discovered the principle of the crystal detector in 1902. His discovery was based on the research of Professor Karl Ferdinand Braun of Germany, who had found that certain minerals, such as galena and pyrites, have the property of "unidirectional" conductivity. In Pickard's studies, he discovered more than 250 minerals capable of detecting radio signals when used together with a metal contact or in combination with a different mineral. His crystal detector was first produced and marketed in 1906; the first patent was granted January 21, 1908. Pickard discussed his landmark discovery in an article, "How I Invented the Crystal Detector," published in the August 1919 *Electrical Experimenter*. General H. H. C. Dunwoody of the U. S. Army is usually considered the first experimenter to put the crystal detector to practical use. He also produced the carborundum detector in 1906.

The crystal detector was the first really sensitive and stable detector. Its sensitivity equalled or exceeded that of the Fleming valve (diode) or the DeForest Audion (triode). Crystal detectors and receivers, however, do not amplify signals.

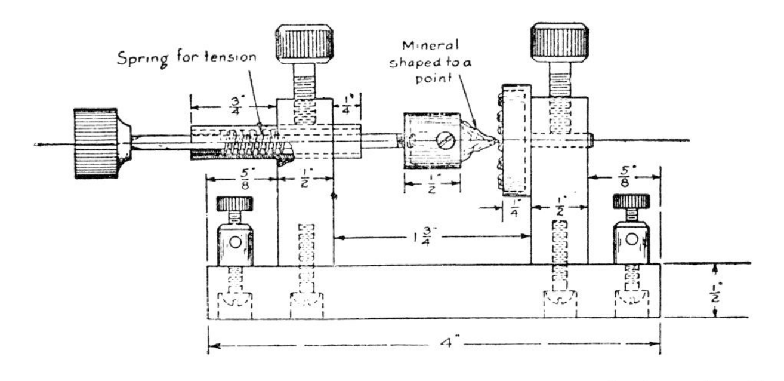

Fig. A. Detector for bornite and zincite crystals ("Perikon" detector). See next page for source of Figures A- D.

They require high-impedance (usually 2,000 ohms or greater) headphones. If a vacuum-tube A. F. amplifier is added to a crystal set, a loud speaker may be used for strong signals.

Two essential components of a crystal detector are a holder for the mineral—usually a crystal cup or clamp—and a device for probing the mineral by making electrical contact—usually a thin spring-metal wire called a "catwhisker" (although rigid, needle-type contact and crystal-to-crystal contact, "Perikon," are also used). Ideal pressure for the contact with the mineral varies from light (e.g., galena) to moderate (e.g., iron pyrites) to heavy (e.g., carborundum). In fixed crystal detectors, the electrical contact is retained at the pre-set point, and pressure is established by the manufacturer. Adjustable detectors may be either open or enclosed, but fixed and semi-fixed detectors are always enclosed. For many fixed detectors, this enclosure is sealed. (By the mid-1920's, fixed crystal detectors were used not only in crystal sets, but also for the detectors of several vacuum-tube receivers with the reflex or inverse duplex circuit.) Examples of four types of early, adjustable crystal detector stands are shown in the line drawings of Figures A–D.

Early wireless experimenters often purchased or made crystal-set components (detector, slide tuner or loose coupler, variometer or variable capacitor, fixed condenser, binding posts) and assembled the parts on their own (i.e., homebrew sets). Individual parts and kits were readily available. Many early factory-built crystal receivers had elaborate tuners with several dials or tap-switch controls and large cases. Later factory crystal sets were generally much smaller and usually had simplified tuners.

Examples of two common crystal-set circuits are shown in the schematic diagrams of Figures E and F. The selective crystal receiver (Fig. F) has three circuits. Its primary (or antenna and ground) circuit achieves the necessary electrical properties for tuning to the desired frequency by variable inductance (coil), variable capacity (condenser or capacitor), or both. In the secondary circuit, variation of inductance (by tapping the coil at various turns, by a sliding contact with the coil, or by sliding or rotating one coil within another) or capacity (by rotating a series of mobile metal plates—the rotor—within a parallel set of

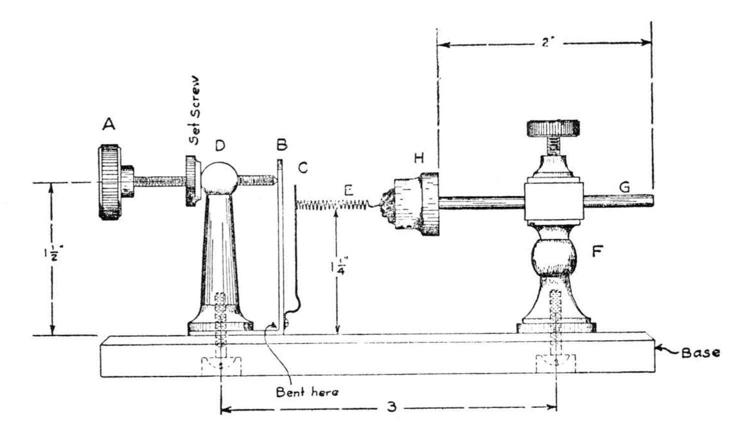

Fig. B. Detector for crystals requiring light pressure (e.g., galena).

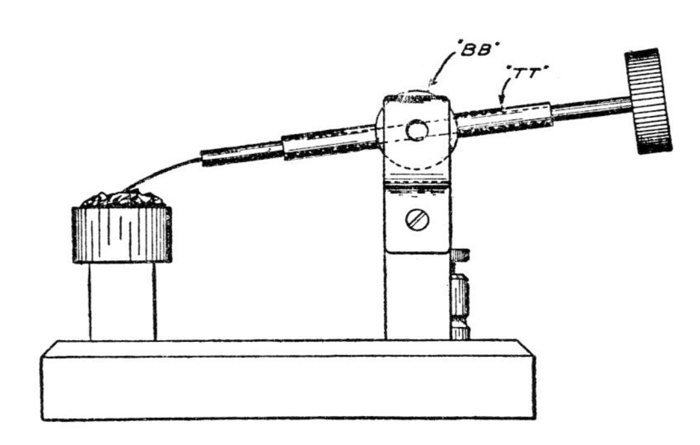

Fig. C. Detector for universal adjustment (ball-and-socket joint).

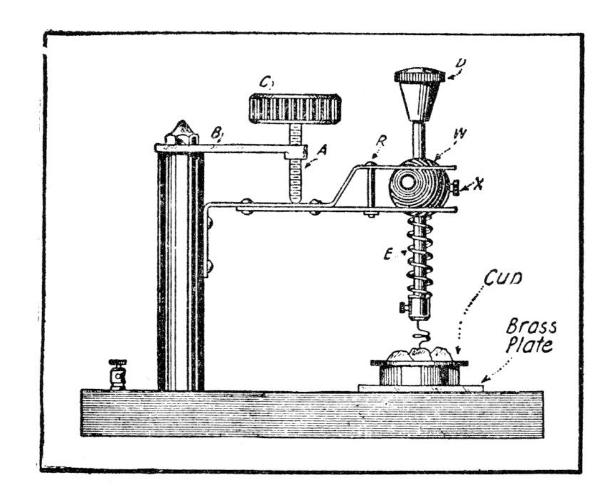

Fig. D. Vertical detector stand that permits accurate adjustment of contact pressure.

Figs. A-D: Bucher, E. E., *The Wireless Experimenter's Manual* (New York: Wireless Press, 1920), pp. 192-193

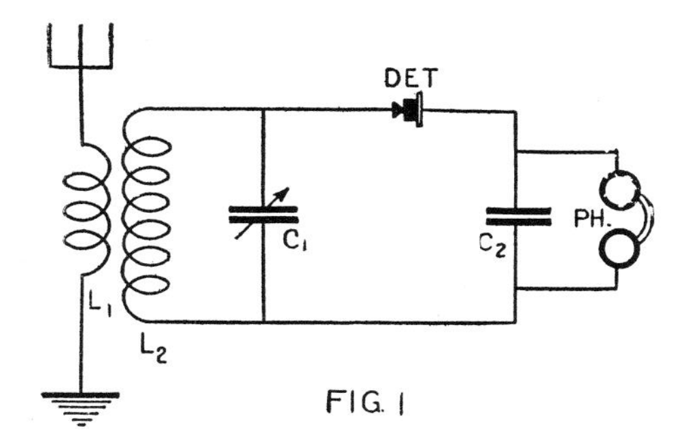

Radio News (Nov. 1925), p. 628

Fig. E. A simple crystal set that is not very selective.

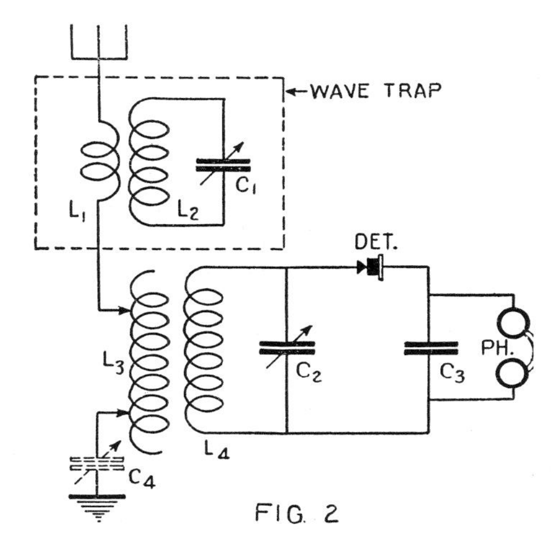

Radio News (Nov. 1925), p. 628

Fig. F. A much more selective crystal set.

fixed plates—the stator) results in tuning. Energy from the secondary circuit is received by the third circuit—the crystal detector and headphones. A fixed shunting condenser (often called a phone condenser, usually .001 mfd. capacity) is in a circuit parallel with the headphones; it allows the R. F., but not the A. F., signals to bypass the phones.

Long after the vacuum tube had generally replaced the crystal detector, radio rooms aboard ships continued to have back-up receivers using carborundum crystal detectors. In case of power or tube failure, the reliable carborundum detector provided continued reception of radio signals. Carborundum remains unscathed by a heavy burst of "static" that would knock out a galena detector. The heavy contact required by the carborundum detector gives it much greater stability than the galena detector, which needs a light contact. Galena crystals are more sensitive than carborundum, but they require more critical adjustments. Carborundum, produced at high temperatures in electric furnaces, can withstand great heat. By contrast, galena crystals are adversely affected by heat; to mount them, an alloy that melts at a low temperature (i.e., Wood's metal) is needed. Carborundum can be used in high-resistance circuits, but galena is effective only in low-resistance circuits.

Radio News (Apr.-May 1922), p. 920

The Catwhisker and the California Connection

The catwhisker (or cat whisker/cats whisker) has appeared in a variety of shapes. In its simplest form, it is only a direct, unbent, spring-wire extension of the detector arm. A variation of this style has a right angle bend at a point above the crystal cup, as in the Jove detector (Fig. 249) and the open detector of the 1921 Quaker Oats crystal set (Figs. 124A and B). Most styles of catwhiskers, however, have some sort of coiled spring near the detector arm with a straight portion of the wire preceding the tip that contacts and probes the crystal.

A few types of the coiled catwhiskers have distinctive shapes so that their manufacturers are readily apparent. Two of these recognizable forms are the loosely coiled, vertically suspended, inverted-cone type made by Kilbourne & Clark (Figs. 68A and B, 253) and the tightly coiled, mildly tapered, horizontally directed type made by the Philmore Manufacturing Company (Figs. 106, 107A and B, 206, 268, 269). Indeed, the Philmore catwhisker is so well known that any catwhisker with a similar appearance (e.g., "Elkay Cats Whisker") is often described as being the Philmore type, although other manufacturers had actually used this style earlier (e.g., in 1916—the early-version Murdock No. 324 crystal detector).

By the early 1920's, only a minority of adjustable crystal detectors had catwhiskers without a coiled portion. Among the exceptions, however, were the detectors of some crystal sets made in California. They used an arched (or "tipped-over C") type of spring-wire catwhisker. The Lee Electric & Manufacturing Company was the largest (but not the earliest) user of this style of catwhisker. It appeared on both versions of the LEMCO No. 340 crystal set (Figs. 71A and B, 72A and B). This catwhisker style is often called

CRYSTAL RADIO RECEIVER $1.00

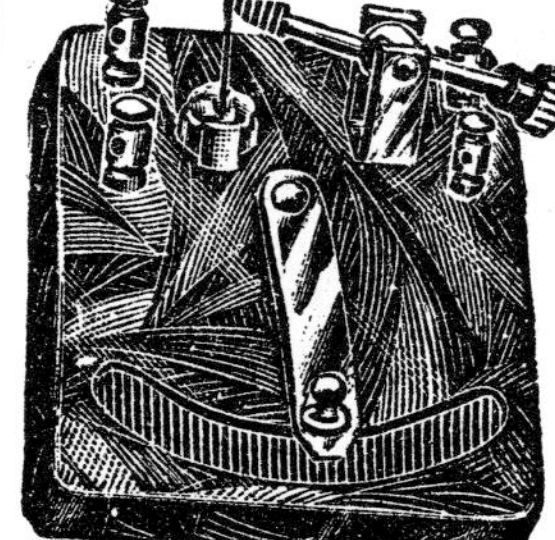

Say what you will about Electrical and Battery Sets, a CRYSTAL set has them all beat for simplicity and clearness of tone. There is absolutely no noise and no distortion—no batteries to buy—no tubes to wear out—no maintenance expense whatsoever. Cost nothing to operate and will last indefinitely. Reception is loud and clear. This Receiver is guaranteed to work equal to the most expensive Crystal Set you can buy. Has a receiving radius of over 25 miles, or under favorable conditions up to 100 miles. Constructed of the finest materials throughout, including supersensitive crystal, assuring quick results. PRICE of Receiver only $1.00 postpaid. Or complete with Ear Phone and Aerial Kit—everything all complete—nothing more to buy. $2.95 postpaid. Johnson Smith & Co., Dept.

Johnson Smith & Co., *Supplementary Catalogue* (1936), p. 16

Early Philmore Little Wonder crystal set with the Philmore catwhisker that was commonly used on other crystal sets.

No. 2615F.

W. M. Welch Scientific Co., *Wireless Apparatus Catalog* (1919), p. 3

Early Murdock No. 324 detector.

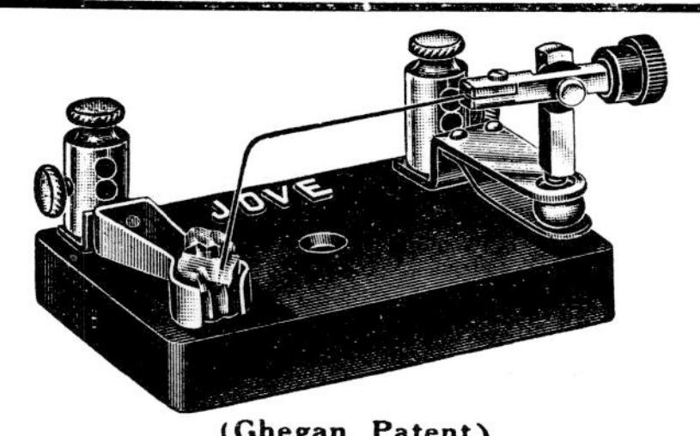

BUNNELL *Instruments Always Reliable*

THE JOVE DETECTOR

HANDIEST, HANDSOMEST AND BEST
Sample by Mail, $2.00
Tested Galena Crystal 25c.

(Ghegan Patent)

Bunnell Keys, Transformers, Condensers, Spark-Gaps, Receivers etc. are High Grade but Inexpensive.
Distributors of Standard Electric Novelty Company Type B "Cyclone" Audion Batteries. Also DeForest, and all other makes of High Class Wireless Apparatus.
Send Stamp for New Edition, 42 Q Catalog.

J. H. BUNNELL & CO. **32 Park Place, New York**

QST (May 1920), p. 86

THE WILLIAM B. DUCK COMPANY, TOLEDO, OHIO. 203

TURRET CUP DETECTOR.

A cat whisker detector—low in price—of pleasing design and easily adjusted. The turret cup detector is mounted on a composition base, nickel plated throughout. It is arranged so that five separate crystals can be brought consecutively into service, this feature making it really five detectors mounted on one base, doing away with the inconvenience of switching separate detectors into circuit. Every portion of each crystal can be reached either by rotating the multiple cup or by moving the sliding shaft.

This detector can be furnished with a fixed condenser in base or not, as desired.

No. A512 Without condenser in base, shipping weight 8 oz., reduced to......$1.75
No. A512CC with condenser in base, shipping weight 8 oz., reduced to...... 2.00

Note.—See raw material section for phosphor bronze wire for Cat Whisker Detectors.

TYPE "G" CAT WHISKER DETECTOR.

Formerly Sold at $5.00. .Now Only $2.75

Galena as a sensitive detector mineral is well known among all amateurs. For the best results Galena requires a very light contact made by means of a fine point. The detector shown above is one of the most practical and sensitive detector stands that has ever been put on the market. It is intended for advanced experimenters and commercial work. A study of the half tone will reveal the many exceptionally excellent features of this detector.

The cup is mounted on a hard rubber disk with knurled edges. This disk in turn sits on a ball and socket joint having a universal motion in all directions. The mineral is clamped in the cup by three screws. It can be moved around with greatest ease by means of the knurled rubber disk allowing any portion of the mineral to be brought under the contact spring. The

Wm. B. Duck Co., *Catalog No. 16* (1921), p.203

the "LEMCO type," although this style had been used previously on some detectors sold by the Edgcomb-Pyle Co., Wm. B. Duck Co., Electro Importing Co., A. H. Grebe & Co., Manhattan Electrical Supply Co., Parkin Manufacturing Co. (of San Rafael CA), and others. The LEMCO type catwhisker was used on at least three other crystal sets made in California—the Brownie Airphone (Fig. 23A), ROMCO (Fig. 23B), and Excello Galena (Fig. 49). These sets were made in San Francisco (or neighboring Berkeley, for the Excello), the city where the LEMCO factory was located (see "Crystal Set Differences, Details, and Dilemmas" for further information about these three crystal sets). Although production of this arched type of catwhisker in the 1920's was largely a regional phenomenon, it also appeared on the Sypher detector made in Toledo, Ohio.

NO. CEK8888

The "Electro" Radiocite Detector With Gold Catwhisker **$3.50**

POSITIVELY THE MOST SENSITIVE CRYSTAL DETECTOR MADE

FEATURES

Gold Catwhisker	Ultra-Sensitive
Bakelite Base	3/4 in. Felt Sub-base
Non jar-out	Adjustment Lock
Quadruple adjusting Range	Non Surface-leaking
Long distance tested	Rotary Detector Cup

Electro Importing Co., *Wireless* Catalogue (1918),p. 41

Please All the Family This Christmas with a

MAIL ORDERS
filled promptly. Send in your Christmas order NOW, and avoid the rush.

No. 340 Crystal Receiver

$6.00 Delivered

Provides clear, distinct, loud reception of both voice and music within 50 miles radius. Thousands in use attest their popularity. For clarity and purity of tone nothing has yet been developed that equals a crystal.

LEMCO RADIO PARTS

offer more real value for the money than any made. Note the low prices.

No. 250 Reflex Coils $2.00
No. 275 Reflex Units, with 17-plate condensers 8.00
No. 300 Fixed Crystal ..... .60
No. 401 Audio Transformer.. 3.50
No. 100 Broadcast Tuner ..$5.00
No. 340-B Crystal Receiver 6.00
No. 55 Knockdown kit (2-tube Harkness reflex) ..25.00
No. 50 Neutroflex Receiver 50.00

If your dealer can't supply you—send direct. We'll deliver C. O. D. anywhere.

LEE ELECTRIC & MFG. CO.
220 EIGHTH STREET SAN FRANCISCO, CALIF.

Write for Booklet R-81

Radio News (Dec. 1924), p. 1028

RADIO

DETECTORS

Size 3¼ x 2″

An improved detector of real merit. Nickel-plated posts.

Price 75 cents, post paid

Dealers get our prices on Panel Switches and Binding Posts.

SYPHER MFG. CO.
Dept. R., TOLEDO, OHIO

Radio News (July 1922), p. 116

USE LAST SUMMER'S

WHY NOT ?

Radio News (Mar. 1923), p. 1660

Wireless MESCO Manual

Single Detector Stand

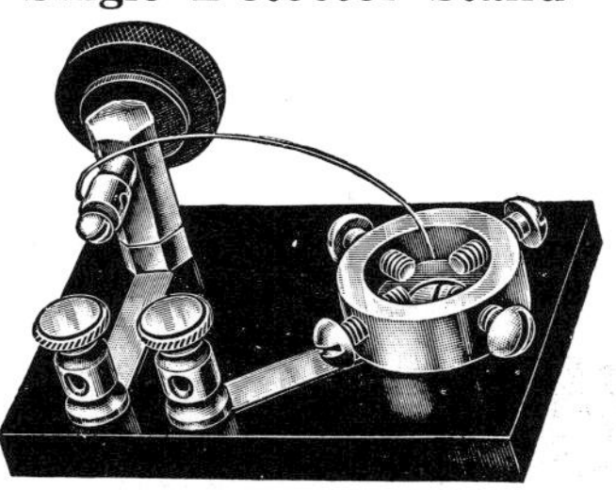

This Single Detector Stand has several distinctive features not found in any other one detector. The stand is mounted on a polished genuine hard rubber base. All metal parts are nickel plated.

The cup is readily adjustable to materials of any size or shape within wide limits, and it is free to revolve so as to bring various parts of the surface under the point of the special composition wire. The standard with the adjustment revolves, the pressure of the wire on the sensitive material is regulated by the hard rubber knob, so that with these three adjustments, a sensitive spot can be found.

List No. 323 Single Detector Stand .... Price. $6.00

Manhattan Electrical Supply Co. (MESCO), *Manual of Wireless Telegraphy and Catalog of Radio Apparatus* (1916), p. 105

April, 1920 RADIO AMATEUR NEWS 529

Efficiency of the Crystaloi Detector

From Nebraska an amateur writes us a letter about the Crystaloi Detector, which says:

I am greatly interested in the Crystaloi Detector as I use one on my wireless set. It might be of interest to you to know that I copy NAA, NAR and NAT without any trouble on the type O Crystaloi Detector.

(Name on Request)

Stations named above are:

NAA—Washington, D. C. (Arlington)
NAR—Key West, Florida.
NAT—New Orleans, Louisiana.

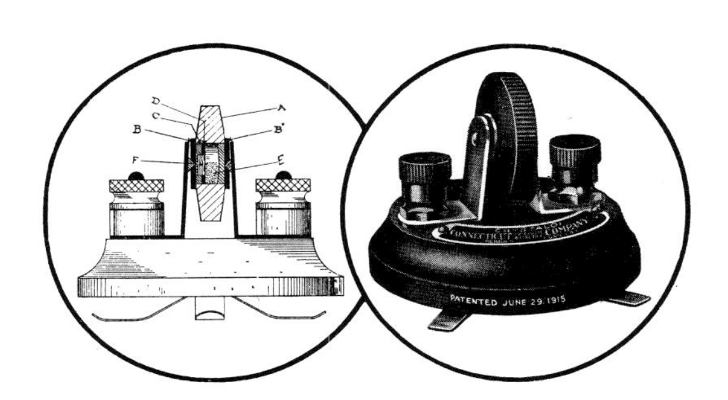

Your results can be as good; you can copy the big Stations, with a Crystaloi Detector. Send for our booklet.

A few crystal detectors, e.g., Crystaloi, did not use a catwhisker. (Eugene T. Turney Co. was the original manufacturer of the Crystaloi; Connecticut Telephone & Electric Company was the successor to Eugene T. Turney Company.)

Terminal Posts to Bind or Clip

Terminal posts provide a ready means for connecting a crystal set or crystal detector with the antenna, ground, headphones, or another component. Such terminals usually are either binding posts (a nut-and-bolt device) or Fahnestock clips. Binding posts, in a variety of styles, were used to make electrical connections long before wireless items appeared. The type of terminal posts on a crystal set or crystal detector provides some indication of the item's vintage.

U.S. Patent No. 1,163,371 for the Fahnestock clip was issued December 7, 1915, to Fahnestock Electric Company of New York (John Schade, Jr., Assignor) more than 3½ years after the filing date (May 22, 1912). Almost eight more years elapsed, however, before this device was first advertised in *Radio News* (November 1923, p. 637). "Fahnestock connectors" were listed in the 1922 catalog of the Manhattan Electrical Supply Company (MESCO) for "experimenters who are particularly interested in the construction of rough laboratory circuits" (p. 182). By 1922, Fahnestock clips began to appear on commercial crystal sets and crystal detectors, but binding posts continued to be used widely for several more years.

After the mid-1920's, crystal sets of some manufacturers were marketed in two versions, differing only in the type of terminals used (e.g., Wecco Gem—see Figs. 183A and B). Usually the binding-post version antedated the Fahnestock-clip sets. Collectors may assume that a factory-made crystal set or detector with Fahnestock-clip terminals was not marketed before 1922.

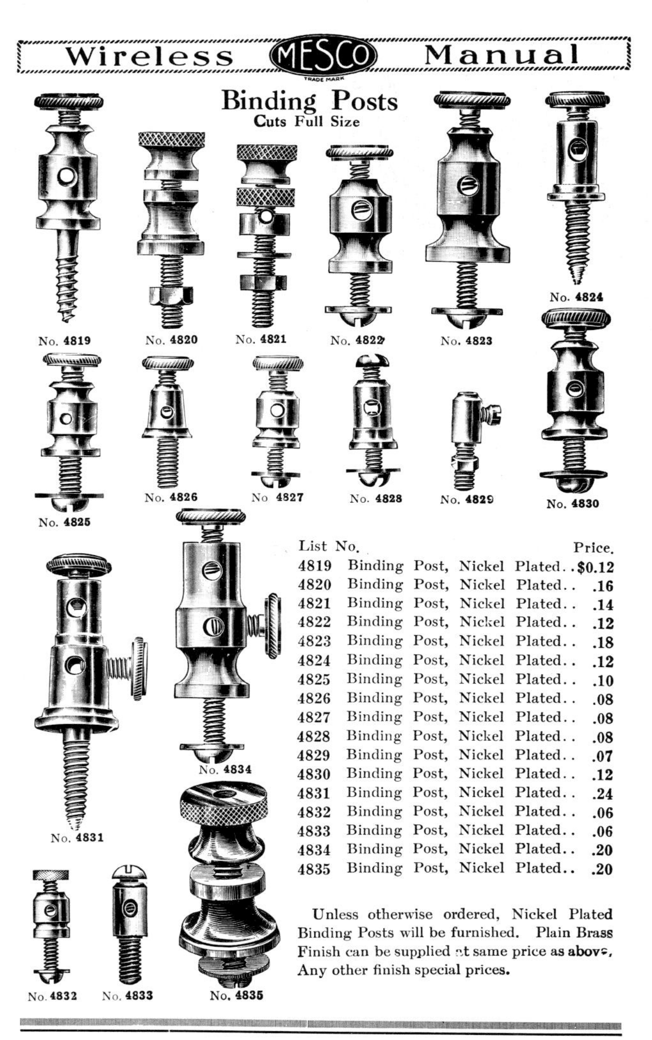

Wireless MESCO Manual

Binding Posts
Cuts Full Size

No. 4819 No. 4820 No. 4821 No. 4822 No. 4823 No. 4824 No. 4825 No. 4826 No. 4827 No. 4828 No. 4829 No. 4830 No. 4831 No. 4834 No. 4832 No. 4833 No. 4835

List No.		Price.
4819	Binding Post, Nickel Plated..	$0.12
4820	Binding Post, Nickel Plated..	.16
4821	Binding Post, Nickel Plated..	.14
4822	Binding Post, Nickel Plated..	.12
4823	Binding Post, Nickel Plated..	.18
4824	Binding Post, Nickel Plated..	.12
4825	Binding Post, Nickel Plated..	.10
4826	Binding Post, Nickel Plated..	.08
4827	Binding Post, Nickel Plated..	.08
4828	Binding Post, Nickel Plated..	.08
4829	Binding Post, Nickel Plated..	.07
4830	Binding Post, Nickel Plated..	.12
4831	Binding Post, Nickel Plated..	.24
4832	Binding Post, Nickel Plated..	.06
4833	Binding Post, Nickel Plated..	.06
4834	Binding Post, Nickel Plated..	.20
4835	Binding Post, Nickel Plated..	.20

Unless otherwise ordered, Nickel Plated Binding Posts will be furnished. Plain Brass Finish can be supplied at same price as above. Any other finish special prices.

Manhattan Electrical Supply Co. (MESCO), *Manual of Wireless Telegraphy* (1916), p. 79

Fahnestock Connectors

This type of Binding Post or Connector is at the same time cheap and efficient. It has enjoyed wide popularity among those experimenters who are particularly interested in the construction of rough laboratory circuits. It is extremely simple to mount and the positive nature of the contact made with the wire particularly recommends it. May be obtained in various shapes and sizes, the most popular of which is shown below.

List No.		Price
13605	**Type No. 3 Fahnestock Connector**	**$0.05**

MESCO, *Wireless Manual* (1922), p. 182

POCKET RECEIVER **$5.00**

Compact Well Made Efficient

THE HELIPHONE is one of the most remarkable receivers developed in late years. It is complete in every detail—permits sharp tuning.

This is just the instrument for your vacation—can be used aboard train, steamboat, while camping or fishing—in fact, everywhere you go the HELIPHONE will bring the outside world to you.

Order this and other RADISCO products from your dealer

RADIO DISTRIBUTING COMPANY

"If It's Radio—We Have It"

NEWARK, NEW JERSEY

Radio News (July 1922), p. 118

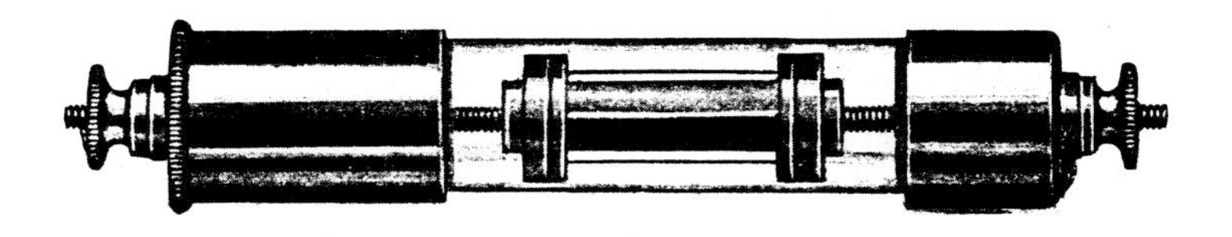

Pickard Carbon Steel Detector, 1902

Wireless Apparatus Co. *Wireless Telephone & Telegraph Equipment* (1919), p. 35

SILICON DETECTOR, TYPE I-P-200

THE principal advantage of the Silicon Detector is the permanence of its adjustment under the influence of high voltages induced across it by static or local transmitters.

The blunt contact point is of antimony. The crystal is of special furnace silicon. The silicon is embedded in soft alloy held in a cup by a set screw. The contact pressure between the silicon and the antimony is obtained by the variation of a spring holding the antimony point and controlled by a hard-rubber-capped thumb screw. The point of contact on the crystal is varied by the motion of two micrometer screws, which slide the face of the silicon crystal under the antimony point.

This unit is mounted on a substantial bakelite dilecto base, which is supported by four rubber feet.

Size, 5½" x 4½" x 2½" Weight, 10 ounces

SILICON DETECTOR

GALENA DETECTOR, TYPE I-P-201

THE Galena Detector is especially sensitive and requires an extremely fine regulation of pressure between a metal point and the galena crystal, together with a rigid method of locking the sensitive adjustment. When a proper contact has been secured, the Galena Detector is the most sensitive of all crystal detecting devices. This type of detector, as shown in the photograph, is provided with a compound spring adjustment that has proven highly successful in practice.

Size, 5" x 2⅜" x 2⅛" Weight, 10 ounces

Both of these detectors were priced at $10.00.

Wireless Specialty Apparatus Co., *Catalog* (1919), p. 36

COMBINED SILICON AND GALENA DETECTOR, TYPE I-P-202

A SINGLE-STAND combination of the silicon detector and the galena detector is provided for securing reliability of operation. This combination insures successful reception of signals through heavy static, by means of the silicon, and long-range operations in cases of light atmospherics, by means of the galena.

The detectors are mounted on a single bakelite dilecto base, with a three-point switch, allowing operation of either detector at the operator's will, or the short-circuiting of both detectors, during transmission. The base is mounted on four rubber feet, to eliminate vibration.

This detector is standard with The Great White Fleet.

Size, 7 3/8" x 5 1/2" x 2 1/2"
Weight, 2 pounds

Combined Silicon and Galena Detectors

TRIPLE CRYSTAL DETECTOR, TYPE I-P-203

A TRIPLE detector combination has been designed to give the maximum possible freedom from both electrical and mechanical disturbances.

The detectors are all adjusted by our compound plunger spring contact mechanism. The degree of contact pressure and the permanency of adjustments are insured by simple clamping screws. A four-point switch permits the connection of any desired detector in the circuit on receiving, or the open circuiting of all the detectors on transmission. This arrangement protects the detector during transmission and makes readjustment after transmission unnecessary.

The United States Navy uses this type exclusively.

Size, 5 3/4" x 5 3/4" x 2 5/8" Weight, 2 1/2 pounds

Triple Crystal Detector

Wireless Specialty Apparatus Co., *Catalog* (1919), p. 37

The Combined Silicon and Galena detector was priced at $25.00, and the Triple Crystal detector was priced at $35.00.

May 1, 1922

Warning to Patent Infringers

Various types of crystal detectors, renewals therefor, and crystal detector radiophone receiving sets now being offered for sale employ the inventions of one or several of the following United States patents (commonly referred to as the Pickard patents) the property of the Wireless Specialty Apparatus Company.

836,531	904,222	924,827
886,154	912,613	1,104,073
888,191	912,726	1,137,714
13,798 (reissue of 877,451)	963,173	1,225,852
933,263	1,104,065	1,257,526
1,213,250	1,118,228	1,136,044
1,136,045	1,136,046	1,136,047

The above patents cover, among other things, the most efficient circuit arrangement of apparatus commonly used in crystal detector radiophone sets, various kinds of crystal members, means for mounting the crystals and holding the mounting, special forms of contacting conductors for the crystals, and mechanism permitting the user's selection of contact points of the contacting conductor on the crystals.

Authorized crystal detectors now are available through the distributors of the Wireless Specialty Apparatus Company, also renewals therefore, and complete crystal detector radiophone receiving sets, all in large quantities, which are sold under the various above-mentioned patents.

The Wireless Specialty Apparatus Company purposes to prosecute, vigorously, all infringers of its patents, and therefore, those manufacturers, distributors, jobbers and dealers who have not been authorized as yet are warned to cease the manufacture or the sale or distribution of crystal detectors, renewals therefor, or crystal detector radiophone receiving sets or any other radio devices which infringe these patents.

Unauthorized distributing or selling, wholly independent of manufacturing, is just as much an infringement as the manufacturing itself, and any seller is separately liable to suits for accounting for damages or profits in addition to injunction.

For their own protection, the distributors, jobbers and dealers who yet may be offering for sale unauthorized crystal detectors, renewals therefor, or complete crystal detector radiophone receiving sets, should demand a guarantee from the manufacturer from whom they purchase radio equipment holding them harmless in case of damage suits arising through their distribution and sale of radio apparatus which infringes the above-mentioned patents.

Crystal detectors, renewals therefor, or crystal detector radiophone receiving sets made and sold with the authorization of the Wireless Specialty Apparatus Company can be readily identified by the data of the above patents and restriction notices prominently marked on the apparatus.

Wireless Specialty Apparatus Company

BOSTON, MASS. Established 1907 U.S.A.

ALWAYS MENTION QST WHEN WRITING TO ADVERTISERS 111

QST (June 1922), p. 111

Wireless Specialty

Cleartone Crystal Detector

This supersensitive Galena Detector, made by the Wireless Specialty Apparatus Company, represents the highest development of the crystal detector art, as the manufacturers have designed and built the

Cleartone Detector

"solid rectifier" type of detector for fifteen years.

The very reasonable price of $1.60 is possible only by extremely large quantity production.

This Cleartone Detector is licensed for amateur, experimental or entertainment purposes only under U. S. patents, the property of the Wireless Specialty Apparatus Company.

Price **$1.60**

Robertson-Cataract *Radio Equipment & Supplies* (1922), p.71

Key Companies

Wireless Specialty Apparatus Company (WSA)

The Wireless Specialty Apparatus Company (WSA) was prominent in many aspects of crystal detectors and receivers. This pioneer company was formed in New York in 1906 as a partnership between Greenleaf Whittier Pickard and Philip Farnsworth. Pickard is generally credited with the invention of the crystal detector in 1902.

With the addition of John Firth to WSA in 1907, the company became incorporated. Early that year, the Wireless Specialty Apparatus Company received its first order—for 35 silicon detectors—from the Signal Corps of the U. S. Army. Firth left the organization in 1912 when WSA moved to Boston; he later produced and sold crystal detectors and receivers bearing his name.

About that time, the expansion of the company accelerated when the United Fruit Company decided to purchase radio apparatus designed and produced by the Wireless Specialty Apparatus Company to equip all of its ships and shore stations in the West Indies, and Central and South America. Thereafter, WSA became the major supplier of crystal detectors and receivers for the vast majority of other commercial ships and shore stations. During World War I, the Wireless Specialty Apparatus Company developed and built apparatus for the U. S. Navy. By the end of the war, WSA claimed that it had become the second largest radio engineering and manufacturing organization in the Western Hemisphere, producing many other items besides crystal detectors and receivers.

Although Pickard discovered the principle of the crystal detector in 1902 while working with a carbon steel detector he had designed, he was not ready for production and marketing until 1906; his first patent date was January 21, 1908. After his 1902 discovery, he investigated a multitude of minerals and furnace products to determine which materials would best serve as rectifiers in radio receiving circuits. About 250 of these substances proved to be satisfactory detectors when used with metal contacts or in combination with another mineral. Among the more successful pairs of minerals were bornite and zincite, silicon and antimony, and silicon and galena. The word "Perikon," a registered trade mark of Pickard and WSA (said to be derived from *PER*fect p*ICK* c*ON*tact), is generally used to designate these mineral-to-mineral detectors.

Crystal detectors of the Wireless Specialty Apparatus Company are now prized collector's items because of the excellence of their design and construction; the distinctiveness of their attached labels; and the historical significance of this early company and its co-founder, Greenleaf Whittier Pickard. Among these detectors are Type IP-200, Silicon Detector (with an antimony contact point); Type IP-201, Galena Detector (with a metal point contact); Type IP-202, Combined Silicon and Galena Detector (both detectors mounted on a single base and having a three-point switch); and Type IP-203, Triple Crystal Detector (three detectors mounted on one base with a four-point switch).

In 1907, WSA marketed its first crystal receiver. The set designation, Type IP-76, was retained for subsequently improved and vastly enlarged models in 1909 and 1914 (the latter was a "double-deck" receiver). Sometime before 1919, the Type IP-77 crystal receiver was introduced; as with most early sets, it had the broad wave-length tuning range (from 200 to 3500 meters) used by commercial ships and shore stations.

By 1919 or 1920, the Wireless Specialty Apparatus Company was producing crystal detectors for other companies with the names of both the manufacturer (WSA) and the contract company listed on the attached label. Among these contract companies were Crosley Manufacturing Company, MESCO (Manhattan Electrical Supply Co.), Precision Equipment Company, SORSINC (Ship Owners Radio Service, Inc.), and Westmore-Savage Company (Figs. 222, 323–325). The WSA "Cleartone" detector, marketed in 1919 for $1.60, was sold by a mail order company (without a label) as late as 1932—for $0.29.

In May 1922, the Wireless Specialty Apparatus Company began to warn other manufacturers of crystal detectors—with varied results—of their infringement on the Pickard patents. Earlier, in 1920, WSA had entered into an agreement (along with the General Electric Company, Westinghouse Company, American Telephone & Telegraph Company, United Fruit Company, and Western Electric Company) with the newly formed Radio Corporation of America for RCA to be the exclusive selling agency for all broadcast receiving sets manufactured for public sale by WSA and the other companies. After this agreement, RCA marketed the Radio Concert Crystal Receiver (Radiola), Models AR-1375 and AR-1382, both made by the Wireless Specialty Apparatus Company. The labels of these receivers not only listed the Radio Corporation of America but also gave the name and address of WSA, the manufacturer.

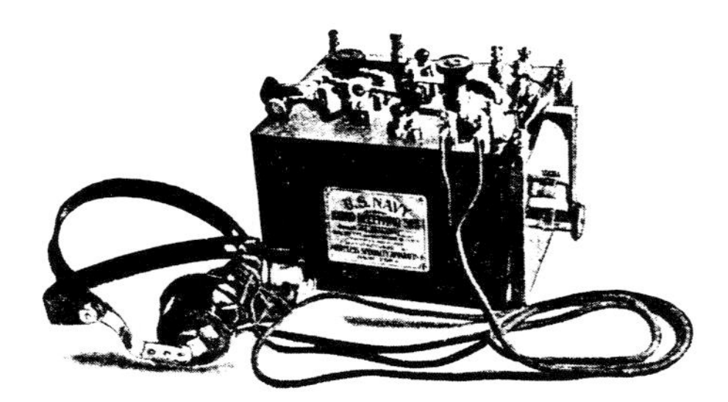

1907 Type Receiver

1908 Type Receiver

1914 "Double-Deck" I-P-76 Receiver
Originally designed for use in the United Fruit Company's ship and shore stations.

Wireless Specialty Apparatus Co., *Radio Telegraph & Telephone Equipment* (1919), p. 29

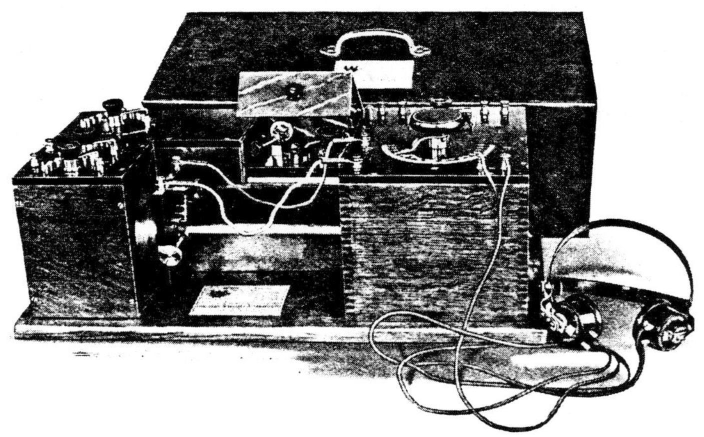

Early Form of the I-P-76 Receiver

Wireless Specialty Apparatus Co., *Radio Telegraph &Telephone Equipment*, (1919), p. 6

RECEIVER, TYPE I-P-77

Size: 19 3/4″ x 14 3/4″ x 9 1/2″. Price: $396.00.

Wireless Specialty Apparatus Co., *Radio Telegraph & Telephone Equipment* (1919), p. 34

RECEIVER, TYPE I-P-500

(WITH AMPLIFIER CONTROL BOX, TYPE TRIODE A)

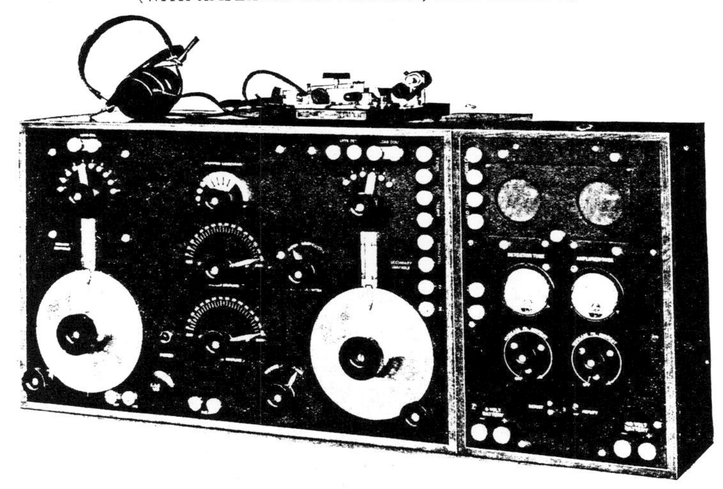

Overall dimensions, Receiver Type I-P-500: 23″ x 11″ x 14 1/2″. Overall dimensions, Amplifier Control Box Type Triode A: 14 1/2″ x 9 1/2″ x 6 1/4″. Price: $425.00

Wireless Specialty Apparatus Co., *Radio Telegraph & Telephone Equipment* (1919), pp. 30, 32

RECEIVER, TYPE I-P-501

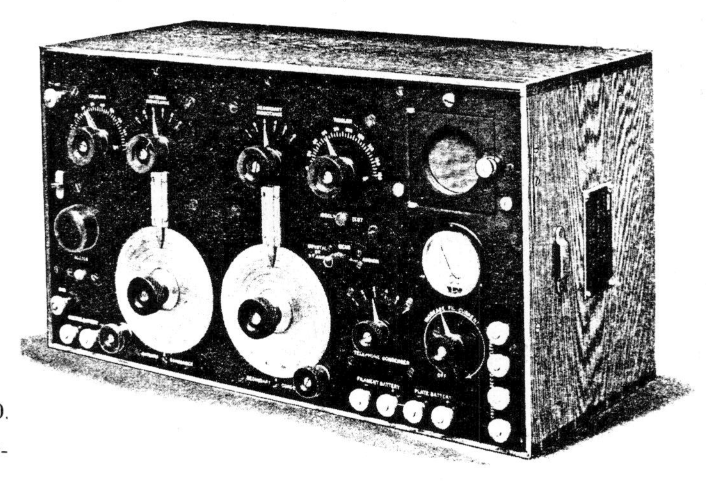

Overall size: 20″ x 11″ x 9″. Price: $348.00.

Wireless Specialty Apparatus Co., *Radio Telegraph & Telephone Equipment* (1919), p. 33

RADIO EQUIPMENT & SUPPLIES

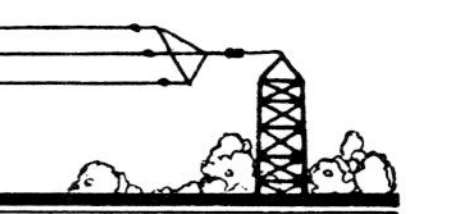

Radio Concert Receiver, Model AR-1375

This receiver was designed by the Wireless Specialty Apparatus Co. to fill the need for a high-class crystal receiver covering a wave-length range of 170 to 2650 meters, thus permitting the reception of broadcasted concerts as well as daily time signals sent by Arlington (Radio, Va.) on a wavelength of 2500 meters. The entire unit is built in an artistically finished metal case, having a bakelite dilecto front panel. The set is sold complete with a pair of highly sensitive telephone receivers.

Model AR-1375

Radiola AR-1375 is portable, rugged and remarkably sensitive. The ideal Receiver for all around work.

Wave Changing Switch

A wave change switch having three positions is mounted on the left hand side of the panel, providing three distinct wavelength ranges; 170-410, 350-965, 925-2650 meters. Variations between the lower and upper portions of these three ranges can easily be obtained by manipulating the tuning knob found in the center of the front panel.

The tuning knob is provided with an indicator which moves over a graduated dial engraved directly upon the front panel itself. This knob is used to bring in desired, and to cut out undesired stations.

The crystal detector employed with this outfit is mounted directly on the front panel and is of the "catwhisker" type, provided with a very sensitive crystal.

Binding Post Feature

The binding posts on this outfit are of unique design. To connect external wires, it is merely necessary to push down on the top, insert the end of a wire and then release the top. The wire is then automatically held in place by a strong tension spring. This type of post is the simplest and most effective brought out to date.

The receiver is provided with a metal cover which is held in place by two snap catches. One end of the receiver case is removable and forms a suitable receptacle for the telephone receivers when the set is not in use, or when it is being carried about.

An added feature of Receiver AR-1375 is that provisions are made to connect a vacuum tube amplifier unit for loud speaker operation.

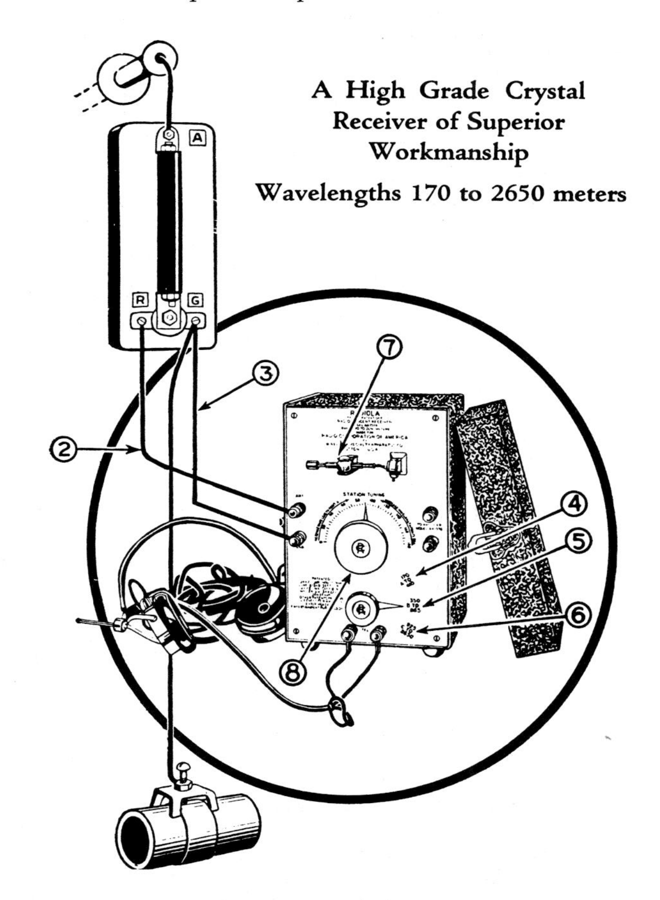

A High Grade Crystal Receiver of Superior Workmanship

Wavelengths 170 to 2650 meters

Robertson-Cataract Elec. Co., *Radio Equipment & Supplies* Catalogue No. 22 (1922), p. 11

Operating Instructions for Model AR-1375

Text numbers correspond to diagram.

No. 1. First, refer to accompanying sketch, then erect antenna and place protective device in position as described under antenna outfits.

No. 2. Connect a wire leading from terminal marked R on protective device to binding post marked ANT.

No. 3. Connect a wire leading from terminal G on protective device to terminal marked GND on receiver. Connect telephone cord tips to terminals marked TEL.

No. 4. For wavelengths between 170-410 meters, place wave change switch at point A. Most broadcasting reception will be heard with the switch in this position.

No. 5. For stations between 350 and 965 meters, turn wave change switch to point B. This range includes commercial stations, radio compass stations and many Naval stations.

No. 6. For wavelengths between 925 and 2650 meters turn wave change switch to point C. This range includes Naval stations and special commercial stations, as well as Arlington (Radio, Va.).

No. 7. After wave change switch has been set for the desired wavelength range, adjust detector by pulling movable spring-point away from crystal and then allowing it to come in contact again at various points. While making this adjustment, rotate tuning knob (8) slowly over the scale, listening until sound is heard in the telephone receivers. Temporarily stop adjustment of detector and manipulate tuning knob until maximum sound is obtained. Leave tuning knob in this position and readjust detector. After a short time the operator will become skillful in finding delicate adjustments on this crysal detector. Once the detector is properly set various stations may be heard by simply rotating the tuning handle over scale.

Note: A black deposit sometimes forms on detector crystals, decreasing the sensitivity of the set. This deposit may be scraped off lightly with a penknife.

Complete Radio Concert Receiver, Model AR-1375, 170-2650 Meters, with Head Telephone Receivers, Spare Crystals, Antenna Equipment and Full Instructions **$47.50**

Radio Concert Receiver, Model AR-1375, as above, less Antenna Equipment **$40.00**

Dimensions: 9¾ in. x 7 in. x 7 in.

Weights: Net, 7 lbs.; Shipping, 12 lbs.; with Antenna Equipment, 18 lbs.

Note: For prices of other Complete Receiver Combinations see other pages.

Model AR-1382 Receiver

This receiver is a two-tuned circuit design, equipped with crystal detector. The antenna circuit comprises a stepwise variable inductance and a variable condenser. The secondary circuit likewise comprises a stepwise variable inductance and a variable condenser. A very wide range of coupling is provided between the primary and secondary and the design is so arranged that the coupling may be swept through a zero value at ten degrees on the coupling scale. This wide range of coupling is for the purpose of permitting the separate tuning of antenna and secondary circuits or a so-called "untuned secondary" arrangement. The receiver has a range of 250-750 meters.

If the secondary condenser is placed at a zero value and the coupling handle thrown to maximum, tuning can be accomplished by the operation of the primary knob alone. If great selectivity is desired, the coupling can then be set to a low value, for example, a setting of 15 degrees and the secondary then brought into resonance with the antenna circuit. The degree of selectivity then depends upon the value of coupling between the primary and secondary circuits.

This receiver is particularly adapted for use with a radio frequency amplifier. In this case the cat whisker is removed from the surface of the crystal, opening the crystal circuit.

Binding posts marked "to amplifier" are arranged for the connection of a radio frequency amplifier, or a tube detector and audio frequency amplifier. The very wide range of coupling included in this set will give a very delicate control of the stability and selectivity of the receiving system.

If, however, it is desired to connect the receiver to an audio frequency amplifier not provided with a tube detector, the amplifier may be connected across the telephone binding posts of the receiver.

Model AR-1382

Model AR-1382 Wireless Specialty Receiver **$70.00**

Robertson-Cataract Elec. Co., *Radio Equipment & Supplies* Catalogue No. 22 (1922), p. 12

Radio Corporation of America (RCA)

The Radio Corporation of America (RCA) was chartered and organized in October, 1919, after a merger agreement with the Marconi Wireless Telegraph Company of America. General Electric Company was involved in the formation of the corporation. By 1920, through cross-licensing agreements, RCA had acquired exclusive selling rights for all broadcasting receiving sets (including crystal sets) manufactured for sale to the public by General Electric Company (GE), Westinghouse Electric Company, American Telephone & Telegraph Company, Western Electric Company, United Fruit Company, and Wireless Specialty Apparatus Company (WSA). Each of these companies owned important patents for radio apparatus. Three of them (GE, Westinghouse, and WSA) were involved in the manufacture of crystal receivers and detectors.

Labels of crystal sets and detectors sold by RCA generally gave the name of the manufacturer; RCA was also listed on most items (but not on the Aeriola Jr. or the ER-753 crystal sets, or the DB detector). The Aeriola Jr. Crystal Receiver/Model RE (Fig. 5), first marketed in 1921, was made by Westinghouse. Radio Broadcast Receiver Model ER-753 (Fig. 54), Model ER-753A/Radiola I (Fig. 55), and Radio Receiver Model AR-1300 (Figs. 131A and B) were manufactured by GE. Radio Concert Receiver (Radiola) Models AR-1375 and AR-1382 were produced by WSA. RCA also sold the Type DB Crystal Detector (Figs. 321A and B), made by Westinghouse, and the Cleartone Radio Concert Detector Model UD-1432 (Fig. 221), manufactured by WSA.

The Aeriola Jr. was by far the most successful of the crystal receivers sold by RCA. Its sales also exceeded those of the contemporary DeForest Everyman DT-600 (both were priced at $25 with headphones), as well as all of the other *early* commercial crystal receivers. No other crystal sets sold in greater numbers until the inexpensive, small, widely marketed sets of the Philmore Manufacturing Company achieved that distinction in the 1930's and beyond.

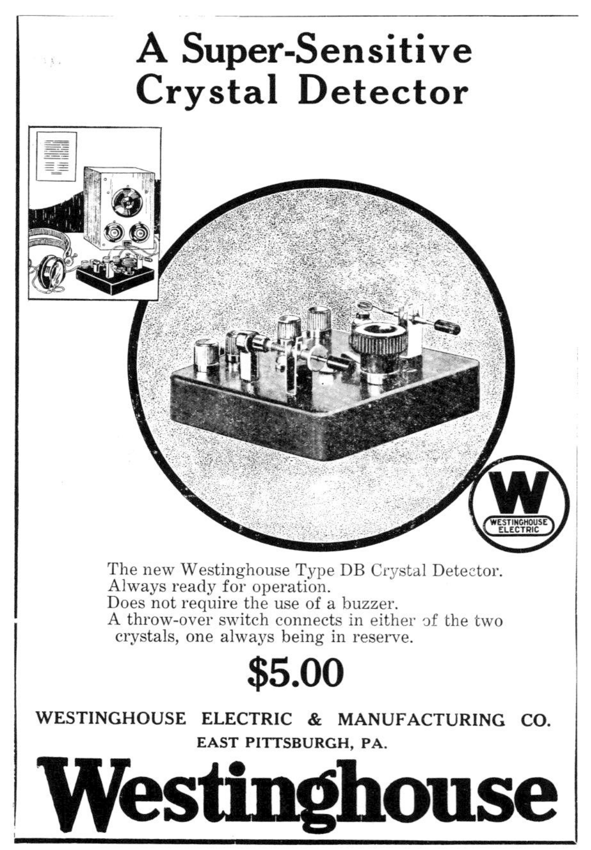

QST (July 1921), p. 71

The New Westinghouse Radio Receiver

A single-circuit tuner; a supersensitive crystal detector; a fixed condenser; and a set of head telephones, all in one case.

Just the outfit to take on camping trips or automobile tours to keep in touch with your home town.

Weighs only 5 lbs. Wave-length range 190 to 500 meters. Send for Folder 4465.

PRICE 25.00

WESTINGHOUSE ELECTRIC & MANUFACTURING CO.
PITTSBURGH, PA.

Westinghouse

QST (Aug. 1921), p. 97

Radio News (June 1922), p. 1137

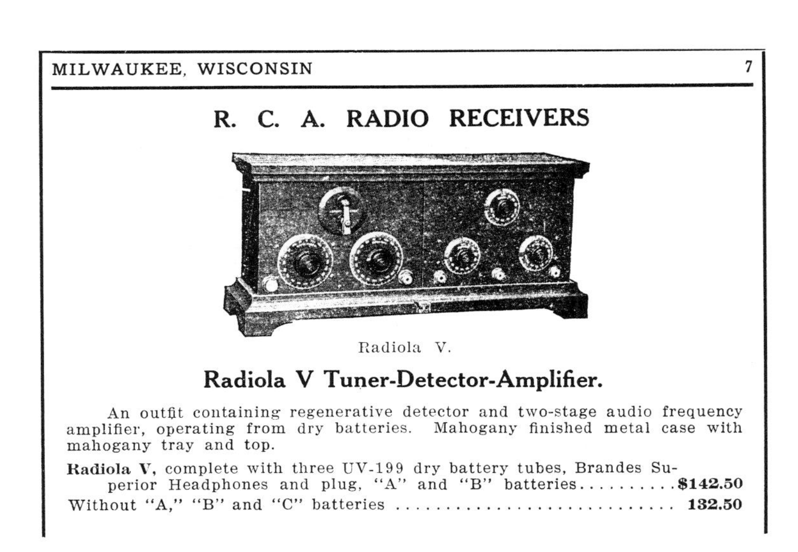

MILWAUKEE, WISCONSIN 7

R. C. A. RADIO RECEIVERS

Radiola V.

Radiola V Tuner-Detector-Amplifier.

An outfit containing regenerative detector and two-stage audio frequency amplifier, operating from dry batteries. Mahogany finished metal case with mahogany tray and top.

Radiola V, complete with three UV-199 dry battery tubes, Brandes Superior Headphones and plug, "A" and "B" batteries **$142.50**
Without "A," "B" and "C" batteries **132.50**

Julius Andrae & Sons Co., *Radio Catalog No. R-6* (1923), p. 7

Radiola V was the "unified" successor to the two-set combination of Model AR-1300 and Model AR-1400.

Low in Cost
Simple to manipulate

THE radio enthusiast who lives within ten to twenty miles of a broadcasting station has exactly what he wants in Radiola I (ER 753-A)—low cost, compactness, portability, and simplicity of manipulation.

Open the walnut cabinet, and on the front panel you find the tuning control, the crystal detector and the binding posts. In the body of the cabinet are the head-telephones. Tuck away the telephones, close the front panel, and you can carry the whole set as you would a satchel.

Opened like a book

Radiola I at your dealers, $25.00

This symbol of quality is your protection.

The Book That Brings Radio Into the Home. For 35 cents you can obtain from your dealer or from us a copy of the book "Radio Enters the Home." It explains the principles, the fascination of radio in plain English. It describes Radiolas and their accessories. It contains the most valuable wiring diagrams ever published.

Carried like a satchel

Sales Department, Suite 2068
233 Broadway,
New York, N. Y.

District Office
10 South La Salle St.,
Chicago, Ill.

QST (Nov. 1922), p. 149

Crystal or Vacuum Tube Detection with the same set

Model AA-1400

These two sets (radio receiver Model AR-1300 and Detector Amplifier Model AA-1400) meet the demand of the novice who wishes to start with a simple crystal detector and later to pass on to vacuum tube detection and amplification at minimum cost.

Radio receiver Model AR-1300 is a new tuner for the broadcast enthusiast. Used as a crystal detector it is a complete receiver. Used with Model AA-1400, here shown, the crystal detector is switched off and amplification is controlled by regeneration.

Detector Amplifier Model AA-1400 consists of a vacuum tube detector and two stages of audio-frequency amplification. It is especially adapted for use with receiver Model AR-1300 to increase the strength of broadcasted concerts. The individual filament control permits close regulation of the received energy. Distortion of broadcasted music is avoided by a special high-frequency resistance across the secondaries. Three telephone jacks insure ideal selectiveness ranging from simple tube detection to two stages of amplification.

PRICES (NOT INCLUDING ANTENNA, TUBES, AND BATTERIES)

Radio receiver Model AR-1300	*$50.00*
Detector Amplifier Model AA-1400 . .	*75.00*
Total for Combination	*$125.00*

See these New G. E. Products at Your Nearest Dealer

Sales Department, Suite 1803

233 Broadway, New York City

QST (June 1922), p. 93

Philmore Manufacturing Company, Inc.

Formed in 1921 by Philip Schwartz and Morris (M. L.) Granat as Ajax Products Company, the Philmore Manufacturing Company was incorporated under its new name in 1925. "Philmore" was a combination of the first names of the two founders. The company, originally located in New York City, has moved several times; its sequence of known addresses will be given later.

Although production figures are not available, there seems no doubt that Philmore has marketed far more crystal sets, crystal detectors, crystals, and components than any other manufacturer. This company was not one of the earliest producers of crystal sets, but it has been the most enduring, continuing this activity into the early 1980's.

Philmore began operation in 1921, but it probably did not make complete crystal sets until the mid-1920's (personal communications to author, 1985 and 1988, from Gerald "Jerry" Granat, son of one of the co-founders and President of Philmore until transfer of company ownership in December, 1988). The August 1925 *McGraw-Hill Radio Trade Catalog* listed (pp. 115–116) a crystal set ("Ajax") for Philmore, but a September 1926 catalog of the Philmore Manufacturing Company announcing the introduction of "the new Super Tone Crystal Radio Receiving Set" is the earliest known source of published details about Philmore crystal-set products (actually, still called Ajax Products at that time). By 1931, the catalogs of numerous retail, wholesale, and mail order companies listed Philmore crystal sets and components with illustrations, descriptions, and prices.

The sales prices were, indeed, attractive. Perhaps the influence of these small compact items on crystal-set customers might be compared to the effect of the Model-T Ford on the automobile market. Both Ford and Philmore products were inexpensive, simple, and mass-produced, as well as being attractive, effective, and durable.

Of the four models of Philmore crystal sets in early production, the Selective (Figs. 109, 110) was at the top in sales price and performance, and the Little Wonder (Figs. 106, 107A) was at the bottom. Not unexpectedly, the other two models—the Supertone (Figs. 112–116) and the Blackbird (Fig. 105)—performed indistinguishably because both were the same basic receiver, differing only in cases and controls. These early Philmore crystal sets continued to be produced for many years. Indeed, the Little Wonder crystal set remained available into the early 1980's (by then, a kit set called the "Sky Rover").

In each item that Philmore made for many years, however, some changes did occur. For example, early set cases of the Little Wonder and the Supertone were metal; later sets were Bakelite. Terminals shifted from binding posts to Fahnestock clips with an overlapping interval when both types appeared. The early Blackbird had four binding-post terminals on the slant-front panel, but later models of this set had no terminals on the panel. Instead, two wires emerged at the rear for the antenna and ground connections, and a phone-tip jack was located at the right side of the set. Cases for the Selective and the Blackbird were black leatherette in early versions, but later cases were painted metal. Along with this change in material, the case of the Selective was also altered in form—from a slant-front to a vertical panel with a recessed domed detector, and from a flat-top case to a cathedral-type cabinet. (A slightly larger version of this cathedral metal case was used by Philmore previously for its 1-, 2-, and 3-tube battery radios [Fig. 111].)

The early-type Selective was marketed with at least two label variations of trade name—"Peerless" and "Luxor." One Philmore crystal set was called by four different names—"Super Tone," "Supertone," "Super," and "Little Giant." But even more remarkable is the fact that the name "Selective" was used both for an early cabinet-model crystal receiver made during the 1930's (Figs. 109, 110) and, by the 1950's, for an entirely different, breadboard-type crystal set with a diode detector—Model VC-1000 (Figs. 205A and B). Philmore altered its catalog numbers from the early 3 digits (in the 300 series for crystal sets) to the later 4 digits (in the 7000 series); cartons changed hues and arrangements of colors; and the dome for the enclosed detector for the Supertone shifted from round-top glass to flat-top clear plastic.

The following sequence of known company addresses (as found on cartons, instruction labels, and pamphlets) may be useful to collectors and historians interested in determining the vintage of Philmore items:

1921 or later: 105 (also 105–107) Chambers St., New York NY

By 1926 or earlier: 106 Seventh Avenue, New York NY

By early 1930's: 106–110 Seventh Avenue, New York NY

By late 1930's: 113–115 University Place, New York NY (zone numbers added in 1943)

In 1943–1959: 113 University Place, New York 3, NY

1959 to 1964 or later: 130–01 Jamaica Avenue, Richmond Hill 18, NY (changed from zones to zip codes in 1963)

1964 or later through 1988: 40 Inip Drive, Inwood NY 11696

1989 to present: 2240 15th Street, Rockford IL 61104

The Philmore crystal sets manufactured for the longest time and in the greatest numbers are the Little Wonder and the Supertone. These two models will be discussed in detail.

The Little Wonder

The earliest type of the Little Wonder—introduced in 1926, but first documented in 1931 publications—had an unpainted metal case; the word "Philmore" did not appear on the set (Fig. 106, left). Its four binding-post terminals had identifying metal ring labels beneath them. The open-type detector had a horizontal arm with a plastic handle knob; the ball inserted on the arm had its socket in a fitted concavity formed in the two vertical supports. A crystal cup with a set screw held the mineral. Philmore assigned catalog number 337 to the Little Wonder (next in sequence to the Super Tone, marketed in 1926 or earlier).

By about 1937—certainly before 1939—the metal case, still lacking the Philmore name on it, had acquired an overlapping sunray-burst crackle paint finish (Fig. 106, right). The binding-post terminals, which were smaller than before, were labelled by impressed lettering in the metal case. A winged, U-shaped, spring-metal crystal holder replaced the crystal cup, but the open-type detector assembly was, otherwise, unaltered.

Prior to 1947, the case (Fig. 107A) was changed to "genuine BAKELITE in 4 brilliant colored pastel shades of GREEN—ORANGE—RED and ROSE" (other colors appeared later). At the same time, the detector arm acquired a different type of knob handle—black painted wood—and a different type of socket for the ball of the arm—two small fitted holes near the upper ends of the vertical supports. Then, for the first time, in the space between the detector assembly and the far upper edge of the case, the word "Philmore" appeared. This word and the label for each binding-post terminal were in raised letters. As with the earlier metal case, the first Bakelite sets had bases of unpainted metal (later sets had black cardboard bases). Despite these changes, this set continued to be identified by catalog number 337. By 1955, however, four-digit catalog numbers had been adopted by Philmore, and the Little Wonder was assigned No. 7000 (No. 7000K for the kit set). At that time, Fahnestock clips were being used for terminals, although the binding posts continued to appear for several years. Prior to 1969, this crystal set acquired a new name—"Sky Rover"—and a new model number—400. Marketed as a kit and unchanged in appearance from the previous Bakelite sets, it was available in a coral-colored plastic case with an open-type detector and Fahnestock-clip terminals until the early 1980's.

The Super Tone/Supertone (Super; Little Giant)

Introduced in 1926 as the "Super Tone," this Philmore crystal radio receiving set was subsequently called by the one-word name, "Supertone." As with the Little Wonder, the Supertone first had a metal case and base. Unlike the smaller set, however, the Supertone, in its earliest-known version (Fig. 112), had a painted case; the sunray-burst crackle paint finish was the same as that used several years later for the Little Wonder (see above). Binding-post terminals of the Supertone were labelled by lettering impressed into the metal case. Unlike the early Little Wonder, "Philmore Mfg. Co., New York" was stamped into the metal panel. The semi-fixed detector was enclosed by a round-top glass dome (a flat-top, clear plastic enclosure was used much later).

Philmore assigned catalog number 336 to the Supertone. Sporadically—in 1933, 1934, 1936, and 1939, at least—this model was called "Little Giant" in some retail catalogs while remaining "Supertone" in others. Although a few of the available cartons have the name "Supertone" on them, many others (probably later) used only "Super."

Before 1939, much earlier than the Little Wonder, the case of the Supertone was changed to Bakelite (Fig. 115). At first, these Bakelite cases had metal bases, but later sets used black cardboard bases. A single word—"Philmore"—replaced the two-line name and address of the manufacturer on the case; both these words and the labels for the binding-post terminals were in raised letters.

By 1955, Philmore was using a four-digit catalog number (7001) for the Supertone. Fahnestock-clip terminals appeared on some of these sets (including kits), although binding posts also continued to be used. The manufacture of the Supertone beyond the late 1950's or early 1960's has not been documented.

Listen to Music, News, Sports, With A

Crystal Radio Receiving Set

No Batteries! No Tubes! No Distortion! No Noise! No Expense to Maintain!

The world's best entertainment and educational features are to be had for the listening in. The very low prices asked for these really good and practical sets puts them within the reach of every man, woman, boy and girl. "If you really wish to learn all about radio," advises the editor of Radio News, "start with a good crystal set." A crystal set has many advantages, but the main ones are simplicity and absolute clearness. There is no distortion and no noise. There are no batteries to buy. No tubes to wear out, in fact there is no maintenance expense whatsoever. They cost absolutely nothing to operate and they will give you constant service for an indefinite period of time. The reception is loud and clear. **Prices below do not include aerial or ear phones.**

Little Wonder Crystal Radio
Price $1.00

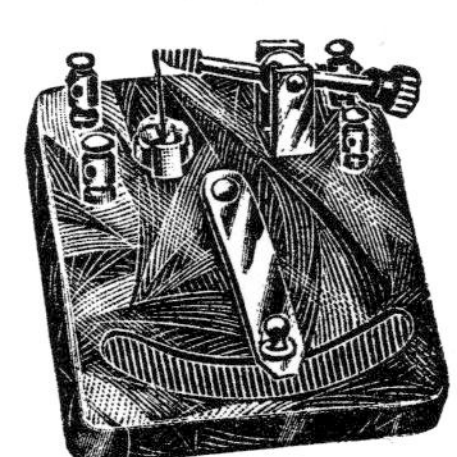

Compact in size but big in results. This set will receive broadcasting within 25 miles of a station or even greater distances when conditions are favorable. The open type detector permits adjustments to be made to the finest degree. The set includes the Philmore supersensitive crystal which assures quick results when "looking" for a station, because the entire surface of the crystal is sensitive.
No. 6581. Little Wonder Crystal Radio. Price Postpaid. $1.00

SUPERTONE CRYSTAL RADIO

A rigidly constructed radio receiving set which will give everlasting service. Has a range of 200 to 600 meters with a receiving radius of over 25 miles. Under favorable climatic conditions, reception may be had within 100 miles from a broadcasting station. Reception is loud and clear without distortion of any kind.
No. 6582. Supertone Crystal Set $1.50

BLACKBIRD CRYSTAL RADIO

The "Blackbird" is a most efficient crystal radio receiving set. Will receive broadcasting from stations up to 25 miles (greater distances under favorable conditions) loud and clear that will bring joy and happiness to the listener. This radio is equipped with a super-sensitive crystal enclosed in a glass housing, thus effectually keeping out dust, etc. No tubes or batteries required. Price does not include Aerial Kit or Phones. **SHIPPED POSTPAID.**
No. 6583. Blackbird Crystal Radio $2.50

SELECTIVE CRYSTAL RADIO

Our best crystal set. Posssses unusual selectivity and designed to give most efficient results. Built with a tapped coil matched with a .00035 Mfd. variable condenser. On the panel is a dial for tuning and logging stations, two double posts to take two pairs of headphones, and a dust-proof glass enclosed crystal detector which includes a supersensitive crystal. With it you can hear messages broadcast anywhere within 25 mles, or even up to 100 miles under favorable conditions.
No. 6584. Selective Crystal Radio $3.95

Johnson Smith & Co., *Supplementary Catalogue* (1936), p. 506

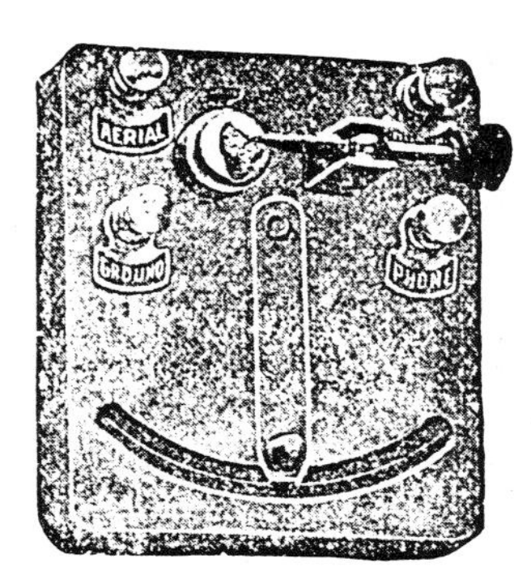

S. S. *Kresge 1931-1932 Radio and Television Catalog*, p. 4

Early version of the Little Wonder crystal set.

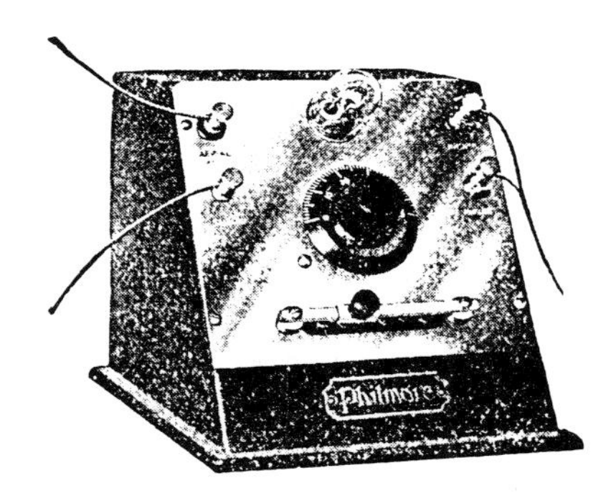

S. S. *Kresge 1931-1932 Radio and Television Catalog*, p. 4

Early version of the Selective crystal set.

A Wonderful Achievement!

PATENTED
Sept. 4, 1917; Oct. 12, 1920; Pat. Pend.
Jan. 21, 1908; Sept. 7, 1909; Nov. 24, 1914
Nov. 17, 1908; July 21, 1914; April 27, 1915
June 15, 1909; Sept. 8, 1914; Jan. 23, 1917

Licensed for amateur, experimental or entertainment purposes only. Any other use will constitute infringement.

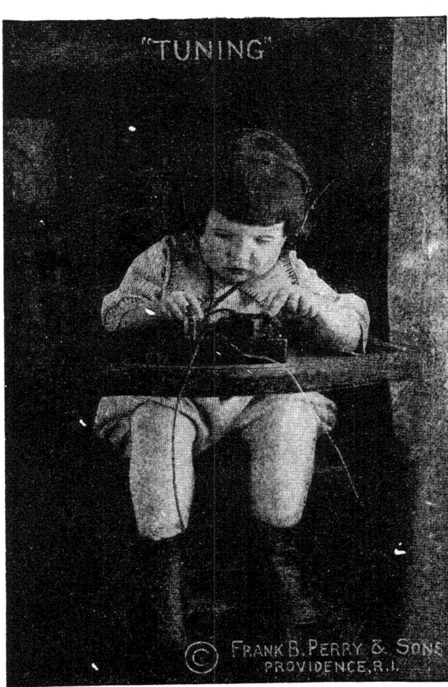

So Simple a Child May Operate It!

RECEIVING SET

The smallest practical receiving set manufactured

Size, $5\frac{5}{16}$" x $3\frac{11}{16}$" x $3\frac{1}{2}$"—Net Weight, 1 lb.

THE THREE IN ONE SET

(1) A practical, efficient, self-contained Wireless Receiving Set for Radio Phone Broadcasting and Commercial Stations approximately 150 to 800 meters wave length.
A newly designed dustproof crystal detector, extremely sensitive. Can be quickly adjusted by means of buzzer test and locked in position.

(2) A miniature Sending Set within limits of buzzer radiation.

(3) A Practice Set for code work.
Ready for phones, aerial and ground connections.
Sold on merit with our usual guarantee.

Real Results With the Radio Blinker Receiving Set!

EASTERN RADIO INSTITUTE

TEL. BACK BAY 5964

899 Boylston St.
Boston — Mass.
RADIO DEPARTMENT

May 23,1922

Frank B. Perry & Sons,
513 Hospital Trust Bldg.,
Providence, R. I.

Gentlemen;

After giving your Radio Blinker Receiving Set a very rigid test here at the Institute we take pleasure in writing you concerning the results that we obtained during this test.

The outstanding feature of your receiving set is the simplified adjustment of the crystal detector which we found very sensitive on all signals.

Moreover the purchaser of this crystal receiver is obtaining a combined receiving and buzzer practice set which is excellent for one to learn the code on.

Wishing you success with your little set, we are,

Very truly yours,
EASTERN RADIO INSTITUTE
A. L. D. Moulton
Superintendent

adm/m

Equipment No. 1

1 Radio Blinker Receiving Set
1 Pair Western Electric Co. Head Phones of best quality.

Equipment No. 2

1 Radio Blinker Receiving Set
2 Pair Murdock 3000 ohm Head Phones.
150′ Antenna Wire.

Price, for either equipment as above, \$25.00. Ask your nearest dealer and if he cannot supply you write us for further information.

Manufactured Solely by

Frank B. Perry & Sons

Radio News (Aug. 1922), p. 313

Double- or Triple-Item Companies

For obvious reasons, many crystal set manufacturers also made crystal detectors. Likewise, several set or detector companies produced (or sold under their own labels) crystals with trade names. These double- or triple-item companies included in the tables of this book are listed below.

CRYSTAL SETS	CRYSTAL DETECTORS	CRYSTALS
Adams-Morgan Co. (AMCO)	Yes	Yes
Aerovox Wireless Corp.	Yes	Yes
Ajax Elec. Co.	Yes	---
Ajax Elec. Specialty Co.	Yes	---
Allen, Alva F.	Yes	Yes
American Electro Technical Appliance Co.	Yes	---
American Radio & Research Corp. (AMRAD)	Yes	---
American Specialty Mfg. Co.	Yes	---
---	Andrae & Sons Co., Julius (not mfr.)	Yes
---	Argentite Radio Corp.	Yes
ARJO Radio Products Co.	---	Yes
Atwater Kent Mfg. Co.	Yes	---
---	Barawik Co. (not mfr.)	Yes
---	B-Metal Refining Co.	Yes
Bowman & Co., A. W.	Yes	---
---	Brach Mfg. Co.	Yes
Bronx Radio Equipment Co.	Yes	Yes
Brooklyn Metal Stamping Corp.	Yes	---
---	Brownlie & Co., Roland	Yes
---	Bunnell & Co., J. H.	Yes
Buscher Co., C. A. (not mfr.)	Yes	Yes
Case Toy & Sales Co.	---	Yes
---	Celerundum Radio Products Co.	Yes
Chambers & Co., F. B.	Yes	---
Champion Radio Products Co.	Yes	---
Cheever Co., Wm. E.	Yes	Yes

CRYSTAL SETS	CRYSTAL DETECTORS	CRYSTALS
Cherington Radio Labs.	---	Yes
Clapp-Eastham Co.	Yes	---
---	Clearco Crystal Co.	Yes
Connecticut Tel. & Elec. Co.	Yes	---
Crosley Mfg. Co.	Yes	---
DeForest Radio Tel. & Tel. Co.	Yes	---
Detroit Radio Co.	Yes	---
Doron Bros. Electrical Co.	Yes	---
---	Dubilier Condenser & Radio Co.	Yes
Duck Co., J. J. (not mfr.)	Yes	---
Duck Co., Wm. B. (not mfr.)	Yes	Yes
Durban Elec. Co.	---	Yes
Edelman, P. E.	---	Yes
Edgcomb-Pyle Wireless Mfg. Co.	Yes	---
Electric City Novelty & Mfg. Co.	Yes	---
Electro Importing Co. (EICO)	Yes	Yes
---	Electro-Set Co.	Yes
Essex Specialty Co.	Yes	---
Everett Radio Co.	---	Yes
Federal Tel. & Tel. Co.	Yes	---
Firth & Co., John (FIRCO)	Yes	Yes
Flash Radio Corp.	Yes	---
---	Foote Mineral Co./Foote Radio Corp.	Yes
Ford Co., K. N.	Yes	Yes
Forman & Co.	Yes	Yes
Freed-Eisemann Radio Corp.	Yes	Yes
---	Freshman Co., Chas.	Yes
Gehman & Weinert	Yes	Yes
Gibbons-Dustin Radio Mfg. Co.	---	Yes
Gilbert Co., A. C.	Yes	---
Gordon Radio & Elec. Mfg. Co.	Yes	---
---	Great Lakes Radio Co.	Yes
---	Grewol Mfg. Co.	Yes
Haller, W. B.	Yes	Yes
Hamburg Bros.	---	Yes
Hargraves, C. E. & H. T.	Yes	Yes
---	Harris Lab.	Yes
Hearwell Elec. Co.	Yes	Yes
Hyman & Co., Henry	Yes	Yes
J. K. Corp.	Yes	---
---	Keystone Products Co.	Yes
Kilbourne & Clark Mfg. Co.	Yes	---
Kresge Co., S. S. (not mfr.)	Yes	---
Lamb Co., F. Jos.	Yes	---
Lee Elec. & Mfg. Co.	Yes	Yes
---	Lenzite Crystal Corp.	Yes
Los Angeles Radio Supply Co. (not mfr.)	---	Yes

CRYSTAL SETS	CRYSTAL DETECTORS	CRYSTALS
Magnus Elec. Co.	Yes	Yes
Manhattan Electrical Supply Co. (MESCO)	Yes	---
---	Martin-Copeland Co.	Yes
Melodian Labs.	Yes	Yes
Miller Radio Co., A. H.	Yes	Yes
Modern Radio Labs.	Yes	Yes
Montgomery Ward & Co. (not mfr.)	Yes	Yes
Moore Mfg. Co.	Yes	---
---	M. P. M. Sales Co.	Yes
Murdock Co., Wm. J.	Yes	---
National Airphone Corp.	Yes	---
---	Newman-Stern Co.	Yes
Nichols Elec. Co.	Yes	---
Non-Skid Crystal Mfg. Co.	---	Yes
Novelty Radio Mfg. Co.	Yes	Yes
Pacent Elec. Co.	Yes	---
---	Pacific Radio Specialty Co.	Yes
Parkin Mfg. Co.	Yes	Yes
Philmore Mfg. Co.	Yes	Yes
Powertone Radio Products Co.	Yes	Yes
Precision Equipment Co.	Yes	---
Premier Dental Mfg. Co.	Yes	---
Pyramid Products Co.	Yes	---
Radio Apparatus Co. (Pittsburgh PA)	Yes	---
Radio Distributing Co. (RADISCO)	Yes	---
---	Radio Engineering Co.	Yes
---	Radio-Ore Labs	Yes
Radio Receptor Co.	Yes	---
Radio Service & Mfg. Co.	Yes	Yes
---	Radio Shop of Newark	Yes
Radio Specialties Co.	---	Yes
Radio Supply Co. (not mfr.)	Yes	Yes
Radio Surplus Corp. (not mfr.)	Yes	---
---	Radio Trading Co.	Yes
RAY-DI-CO	---	Yes
Remler Radio Mfg. Co.	Yes	---
Reynolds Radio Specialty Co.	---	Yes
Roll-O-Crystal Co./Roll-O Radio Corp.	Yes	Yes
---	R-U-F Products Co.	Yes
---	Rusonite Products Corp.	Yes
R-W Mfg. Co.	Yes	---
Schenectady Radio Corp.	Yes	---
Sears, Roebuck & Co. (not mfr.)	Yes	Yes
---	Shamrock Mfg. Co.	Yes
Signal Elec. Mfg. Co.	Yes	---
Stafford Radio Co.	Yes	---
---	Star Crystal Co.	Yes

CRYSTAL SETS	CRYSTAL DETECTORS	CRYSTALS
---	Star-King Co.	Yes
Star Mfg. Co.	Yes	---
Steinite Labs.	Yes	Yes
Steinmetz Wireless Mfg. Co.	Yes	Yes
Tanner Radio Co., C. D.	Yes	---
Taylor Elec. Co.	---	Yes
Telephone Maintenance Co.	Yes	Yes
Teleradio Engineering Corp.	---	Yes
Tel-Radion Co.	Yes	---
Towner Radio Mfg. Co.	Yes	---
Tri-City Radio Elec. Supply Co. (TRESCO)	---	Yes
Turney Co., Eugene T.	Yes	---
Tuska Co., C. D.	Yes	---
U. C. Battery & Elec. Co.	---	Yes
United Metal Stamping & Radio Co.	Yes	Yes
---	Walthom Radio Products Co.	Yes
Westinghouse Elec. & Mfg. Co.	Yes	Yes
Westwyre Co.	Yes	---
Winn Radio & Elec. Mfg. Co.	Yes	---
Wireless Improvement Co.	Yes	---
Wireless Specialty Appar. Co.	Yes	Yes
Wolverine Radio Co.	---	Yes
---	Wonder State Crystal Co.	Yes

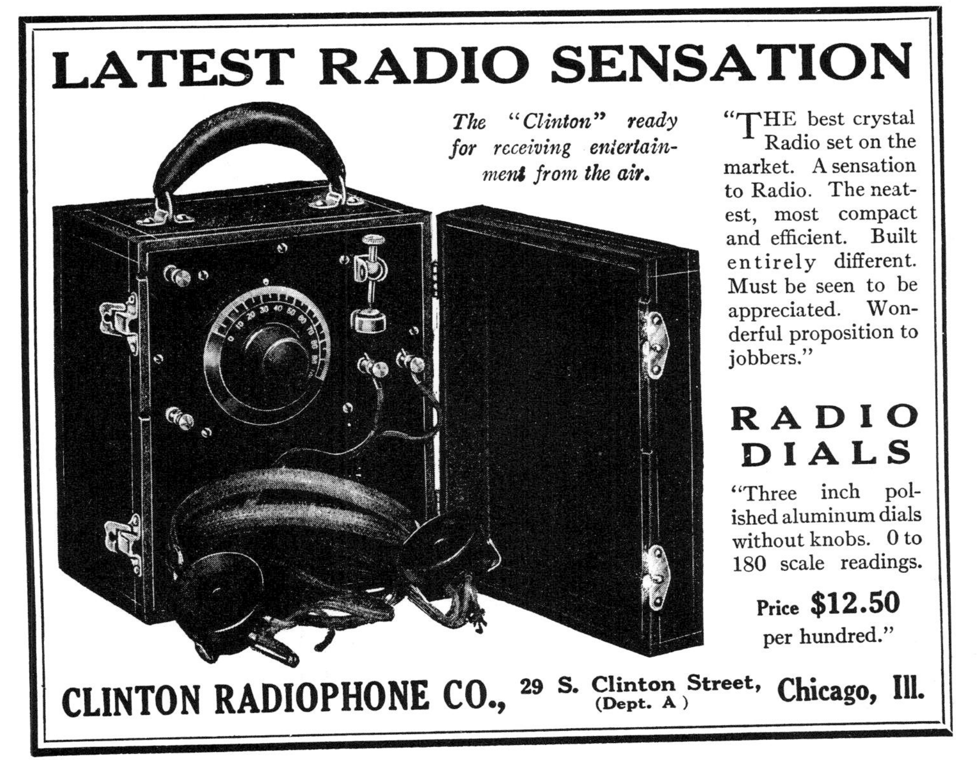

Radio News (Aug. 1922), p. 278

Crystal Set Differences, Details, and Dilemmas

Many models of factory-made crystal sets were marketed with variations in components or appearance. Some events related to certain sets were unusual or remarkable. This section describes differences and special features of several sets, and discusses several noteworthy occurrences.

Aeriola X / Crain Craft Jr.

Crain Bros. Radio Shoppe of Oakland, California, began advertising their inexpensive ($6.00 without accessories) Aeriola X (or "Aeriola 10") crystal set (Fig. 6) in *Radio News*, September 1925. This trade name was an obvious infringement on that of the Aeriola Jr., which had been on the market for more than four years. Of course, RCA lost no time in demanding that the name of the newly introduced set be changed. Crain Bros. responded promptly as shown by their ad in the January 1926 *Radio News* announcing, "The Aeriola X Crystal Set will be known in the future as the Crain Craft Jr. Crystal Set." Perhaps some minor victory was achieved by its losing the first word but gaining the second part of the prestigious name "Aeriola *Jr.*" In this regard, many other

Aeriola X Crystal Set

$12 *Complete with*

100 ft. aerial wire, 35 ft. ground and lead-in wire, aerial insulators, percelain tube, lead-in strip, ground clamp, nail-tite knobs, set head phones. (Without accessories, $6.00.)

MONEY BACK GUARANTEE
Send for FREE
Descriptive Booklet.

CRAIN BROS. RADIO SHOPPE
2304 Telegraph Avenue : : Oakland, Calif.

Radio News (Oct. 1925), p. 480

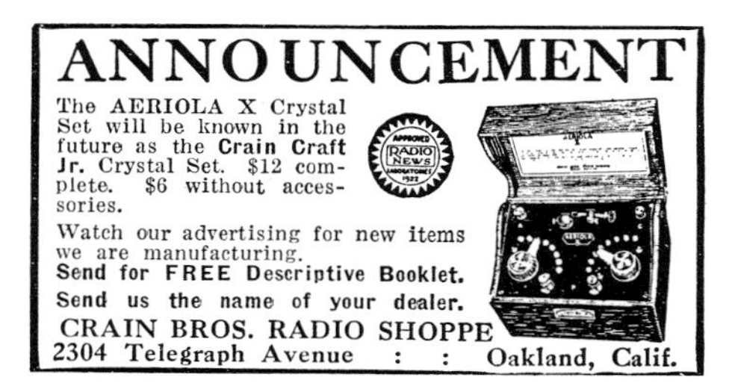

Radio News (Jan. 1926), p. 1060

Radio Broadcast (Dec. 1922), advertiser section

crystals sets also include "Jr." in their trade names.

Audiola; Splendid

The Audiola Model A No. 1 (Figs. 15A and B) and Splendid (Fig. 16) are almost identical crystal sets except for the engraved label on the metal dial-indicator plate. In addition, terminal labels of the Audiola are engraved in the panel, but the Splendid has metal tags inserted beneath the binding posts. For attachment to the case, the Audiola has six screwholes at the panel edges—one hole at each end and two at each side (away from the corners); the Splendid has only four screw holes—one at each corner of the panel. Both crystal sets have the same distinctive triangular vertical support frame for the detector. Both manufacturers—Audiola Radio Co. and Splendid Radio Co.—were located in Chicago. Whether the two sets were marketed at the same time or sequentially is unknown, but evidence from ads supports the earlier appearance of the Audiola.

Beaver Baby Grand

The very small Beaver Baby Grand Vest Pocket crystal set appeared on the U. S. market in 1923. The label on its carton (Fig. 20A) states, "*Licensee* and Mfr. Beaver Machine & Tool Co., Inc / Sales Office. . . . N.Y. City," (metal tag on case has ". . . Laboratories / Forest Hill, N.J."); it does not, however, list the original *licenser*. In the classic British book *Vintage Crystal Sets 1922–1927* by Gordon Bussey, an ad of the Reliance Radio Service Co. (London, England), reproduced from January 1923, has an illustration of the Reliance Bijou No. 2, "Manufactured by one of the oldest firms in the trade. . . . The smallest and most efficient crystal set ever offered. . . . Size of base 3¾ x 2 x ¾." This British crystal set is identical in size and appearance (except for a different detector handle knob) to the Beaver Baby Grand model with a wood case. It seems likely that the Reliance Radio Service Co. in London or its supplier was the source of the license by Beaver Machine & Tool Co. The U. S. collectors' revered book, *Vintage Radio*, has a photograph of the Beaver Baby Grand on page 66; its description ("one of the smallest crystal sets made") is similar to that in the British ad for the Reliance Bijou No. 2. The usual version of the Baby Grand has an oak case with a metal label on front of the case (Fig. 20A). A less common variation (Fig. 20B) has a black, molded, hard-rubber case; on its panel, in raised letters appear "*B M & T Co—Licensee*" ("B M & T Co" is Beaver Machine & Tool Co.) and the three-line "*BEAVER / TRADE MARK / BABY GRAND.*" The sequence of appearance of these two versions of the Baby Grand is unknown.

Brownie Airphone; ROMCO

Despite the color implication of its name, the *Brownie* Airphone has a *black* metal case. Unexpectedly, each of the phone terminals clearly has the painted letter "R" (not "P") on the panel by the two binding posts. Presumably, the "R" means "receivers," a term commonly used in the pre-1920 period for "wireless head set" (later called "headphones" or "phones").

The case, detector stand, and internal components of the Brownie (Fig. 23A), made by Brownie Mfg. Co., and the ROMCO crystal set (Fig. 23B), made by R & O Mfg. Co., appear to be identical. Both manufacturers were located in San Francisco. Incidentally, the ROMCO case *is* brown.

Radio News (Jan. 1924), p. 1006

BUSCO Special; BUSCO Jr.; Radioceptor

The Kansas City Radio Mfg. Co. introduced its Radioceptor (Fig. 30) with an ad in *Radio News*, July 1922. This set has a distinctive, red, basketball-type variometer; the word "Radioceptor" in cursive script appears on the upper center of its panel. The C. A. Buscher Co., also located in Kansas City, Missouri, marketed the BUSCO Special (Figs. 26, 27) and BUSCO Jr. (Fig. 28) in 1923 as shown in the illustrated catalog of this company. The BUSCO Special has the same general features as the Radioceptor, including the variometer (but not the panel logo). The BUSCO Jr. has an identical case and some other similarities. Both BUSCO crystal sets were almost certainly made by the Kansas City Radio Mfg. Co. The BUSCO Jr. differs from the Radioceptor or BUSCO Special in its: (1) detector stand (crystal cup on panel, not mounted on end of detector stand); (2) tuning dial (larger with etched numbers on dial, instead of pointer type with numbers on panel); and (3) variometer (cylindrical, not round basketball type).

California Diamond Jubilee; Radiowunder; Receptor Radionola

The California Diamond Jubilee (Figs. 32A and B) is representative of some small, "fan-coil-tuned," often book-type crystal sets sold during the latter half of the 1920's. It was distributed by Ben Cooper Sales Dept., San Francisco, but the manufacturer is unknown. Its commemorative purpose and the production date (1925) are apparent. This crystal set is quite similar to the Radiowunder (Fig. 31), which was made in Boston in 1925, and to the Receptor Radionola (Figs. 34A and B). U. S. Patent Office records indicate that Herbert M. Hill filed a patent application for this basic crystal receiver on April 20, 1925. The patent was not issued until March 19, 1929. During 1929, the Spencer Pocket Radio crystal receiver (see below and Figs. 140A and B, 141A and B) was marketed. Its components (e.g., "fan-coil" tuner) are also included in the Hill patent.

Crain Craft Jr. (see *Aeriola X*)

The Crystal-Dyne (see *World*)

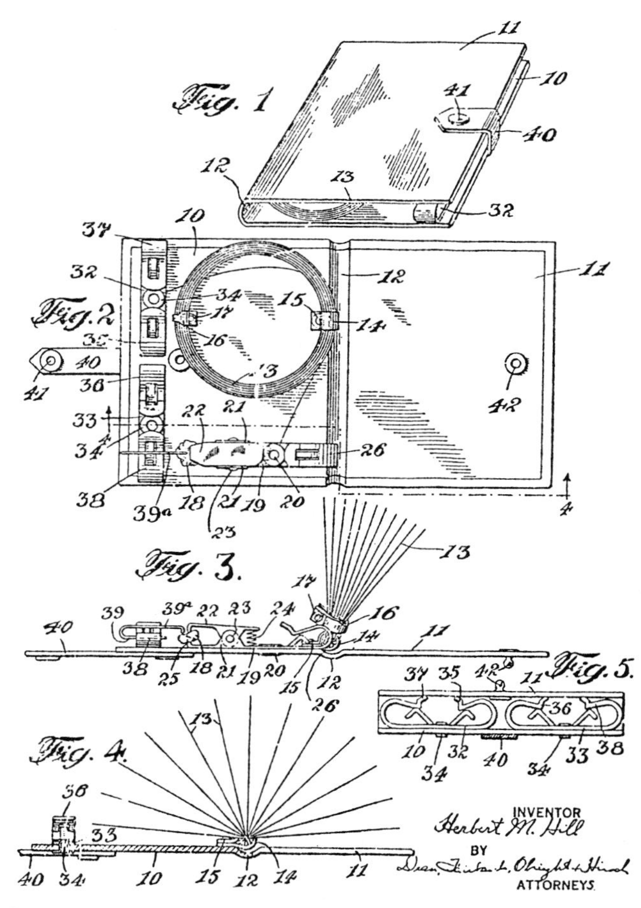

U. S. Patent Office. Applicant Herbert M. Hill of Leonia, NJ

The Hill patent was used in the California Golden Jubilee, Radiowunder, Spencer Pocket Radio crystal sets, and probably others.

Cunningham's Radio Receiving Outfit; Carter

The Cunningham (Cunningham, Detroit) and Carter (Carter Radio Co., Chicago) are almost identical crystal sets (Figs. 36, 41A and B). These two sets appear *similar* to the DeForest Everyman DT-600 (Fig. 42). Each of these three has a lidded wood case, headphone compartment, horizontal enclosed detector, plug-in inductance coil provisions, and two pointer-type knobs that control a set of wiper blades on the interior tuning coil. Surprisingly, the Corbin lid latch on the case of both the Cunningham and Carter crystal set is mounted *upside down* (Fig 41B). Probably the same manufacturer made these two sets.

Federal Jr.

In its presumed early version, the Federal Jr. has a brass case painted black. The metal cases of apparently later sets are not brass. A list of nine patent dates (between January 21, 1908, and January 23, 1917) appears on one end of some sets (Fig. 51B), but this feature is not found on most existing Federal Jr. crystal sets (Fig. 51A).

Radio News (Mar. 1922), p. 903

Dick Tracy Two-Way Wrist Radio (see *Pee Wee*)

Flivver; Moore

The Flivver (a term commonly used in the 1920's for a small cheap automobile), by its name, clearly indicates the small size and price of this crystal set. Its aluminum case is unpainted in most surviving sets (Fig. 52), but brown painted cases are not uncommon. The Lightrite Co. (Bloomfield NJ) probably preceded the Moore Mfg. Co. (Bloomfield; later, Nutley NJ) in the manufacture of this set; each company used its own label. The Flivver and Moore are alike except for the slide-tuner knob (metal on the Flivver; black wood on the Moore) and the enamelled wire wound on the tuning coil (tarnished-brass color on the Flivver; red on the Moore).

General Electric ER-753; ER-753A, Radiola I

The General Electric ER-753 (Fig. 54) and the ER-753A, Radiola I (Fig. 55) are similar sets, differing only in their cases and tuning systems (pancake coil, ER-753; variometer, ER-753A). A removable panel on the slant-front, triangular metal case of the ER-753 provides access to a storage space for the headphone. Although the unlidded metal-case set was very durable and functional, the Radiola I (ER-753A) with its attractive lidded wooden case (and rear compartment for the headphone) was introduced three months after the ER-753 and soon displaced it from the market. (See "RCA," pp. 20–23.)

A. C. Gilbert No. 4016; Tuska No. 4007; Tuska

The A. C. Gilbert No. 4016 crystal set is known to exist with features varying from those shown in Fig. 57C including: no lid-label insert, headphone space in a deeper lid, detector of the type on the Radiotector (Fig. 276), and swing-type lid latches on both sides of case (Figs. 57A and B). On most sets of both versions, the panels have a black celluloid facing on a fiber-sheet backing, but some sets have black painted wood panels. Radio pioneer Clarence D. Tuska (co-founder of ARRL, first editor of *QST*, founder and president of C. D. Tuska Co.) was closely associated with the early wireless activities of the A. C. Gilbert Co. This interaction is apparent in the similarities of crystal sets by the two companies. Both the A. C. Gilbert No. 4016 (in the more common version) and the Tuska No. 4007 use the Radiotector detector (Fig. 276). The unlidded Tuska set (Fig. 153) has the same distinctively styled domed, unplated brass taps and tap-switch tuning blade as the A. C. Gilbert No. 4016.

Howe

A metal plate attached to the front of the Howe case has a square upper part containing scale marks and numbers for the tuning dial; a rectangular lower area has the two-line trade name "HOWE / RADIO RECEIVER." In small letters immediately above the line separating the upper and lower sections of the plate, the words "PATENTS PENDING" appear on many sets but are absent on several others. In other ways, the two versions are identical (Figs. 63A and B) except for the type of knob on the detector arm handle. An ovoid, black-painted, smooth wood knob (Fig. 63A, left) is present on most existing sets *that do not have the "Patents Pending" on the plate.* Howe crystal sets with (and a few without) the "Patents Pending" (Fig. 63A, right) have black plastic knobs with a knurled rim, similar to terminal thumb nuts. Early ads and the Howe store display box (Fig. 63B) show the ovoid knob on sets with plates lacking the "PATENTS PENDING"; these features, therefore, seem to identify sets that were marketed earlier than the other version.

HOWE RADIO RECEIVER
This complete radio receiving set is of very small size and is entirely enclosed in a metal case. Tuning is effected with a knob attached to a slider that makes contact with the tuning coil inside. An adjustable crystal detector is mounted on the

top. It works very well with the average size aerial for local broadcast reception. Manufactured by the Howe Auto Products Co., Chicago, Ill.
AWARDED THE RADIO NEWS LABORATORIES CERTIFICATE OF MERIT NO. 697.

Radio News (Apr. 1925), p. 1907

Kitcraft Ready Built Radio (see *Pee Wee*)

Lambert Long Distance;
Leon Lambert Long Distance

Leon Lambert Radio Co. marketed two types of assembled "Long Distance" crystal sets. One type (Fig. 75) has an unlidded, sealed, wood-box cabinet with a two-line decal on the panel ("Genuine Long Distance Crystal Radio / Leon Lambert"); it was marketed in versions with either binding posts (earlier) or Fahnestock clips as terminals. The other Lambert Long Distance crystal set (Fig. 76A) has a wood base with a vertical single-slide coil covered with green paper upon which the label appears. Both versions have a free-swinging, weighted detector arm and a bundled, multiple-contact ("horse-tail" type) catwhisker. The patent number, issued in 1926, is shown on the label covering the coil of the open set (Fahnestock-clip terminals) but does not appear on the set with a wood-box case (Fig. 75; version with binding posts pictured). Apparently, the enclosed Lambert set was marketed earlier (possibly in 1925) than the open set.

Except for their cases, however, both sets are basically the same (Fig. 76B). Instead of the slide-tuner of the open set, the vertical coil of the set in the wood case uses a wiper blade controlled by a tuning knob on the panel. The chassis of the enclosed set is not readily accessible. Its panel is glued to a base board that slides into the cabinet. At the mid-point of the underside of the case, a recessed screw holds the base board of the chassis in the case; the screw is securely sealed with embossed sealing wax. Apparently, the manufacturer intended to prevent tampering.

Between September, 1923, and April, 1925, before he marketed his commercial crystal sets, Leon Lambert advertised—mostly in *Radio News* (*RN*)—his copyrighted "Long Distance Crystal Hookup Plans" for $1.00. These plans, he claimed, would enable a purchaser to build or remodel a crystal set to get distances of 400 to 1,000 miles. An *RN* ad in February, 1925, even proclaimed, "Yes, we heard England." In addition to the hookup plans, his March 1925 ad in *RN* included an offer for parts to build the Long Distance crystal set ($5.00).

LEMCO: No. 340; No. 340-A; No. 340-B

The popular small LEMCO crystal set No. 340 (Figs. 71A and B, 72A and B) apparently was marketed in 1923 as suggested by the advertisements in early 1924 radio magazines. In these early ads, the set was listed, but not described or shown. Later, during June through September, 1924, each monthly issue of *Radio News* had an identical ad for the LEMCO No. 340 showing a set having a tuning dial (with variometer tuning). In October, 1924, an ad stated for the first time that sets No. 340-A and No. 340-B were available. During the following two months, ads contained an illustration of a LEMCO crystal set with two sets of tuning tap switches for variocoupler tuning. Although the large-print heading of these ads is "LEMCO *No. 340*," the accompanying price list entry is "LEMCO *No. 340-B*, $6.00."

Thus, a dilemma was created. Although logic might suggest that the tap-switch/variocoupler style appeared first (the No. 340?), the earliest known illustrated ad for the No. 340 pictures the tuning-dial/variometer model. The earliest found illustrated ad listing No. 340-B shows the tap-switch version. No ads with illustrations of a set said to be No. 340-A have been found. Perhaps only two variations exist; if so, after the second

Radio News (June 1924), p. 1863

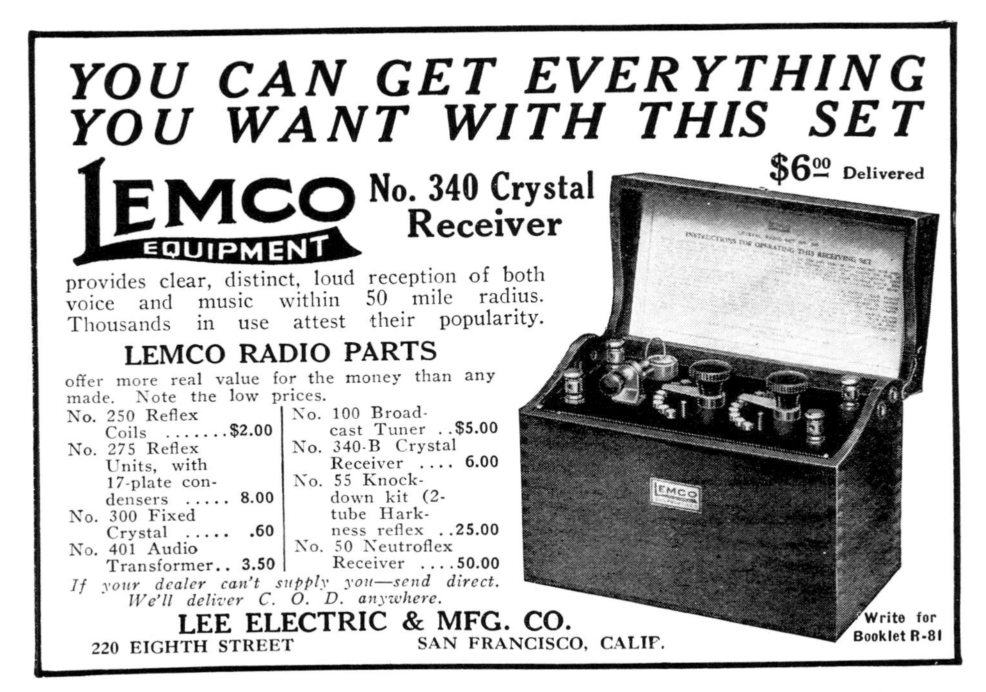

Radio News (Nov. 1924), p. 766

style (No. 340-B) was introduced, the original type (No. 340) probably was designated set No. 340-A.

Comparison of the lid instruction sheets for the two LEMCO versions fails to resolve the confusion. The lid sheets of both types list the same set designation—"No. 340." Some of the wording for the tap-switch style, however, suggests its earlier status as in these examples (quotations listed after "vs." are for the dial-tuning model):

(1) "Radio Telephone Crystal Set No. 340"
vs.
"No. 340 Crystal Receiving Set"

(2) "To hear radio signals or music . . ."
vs.
"To hear radio broadcasting . . ."

(3) ". . . will respond to wave lengths up to 900 meters"
vs.
". . . will respond to all broadcast wave lengths."

In essence, whether the LEMCO No. 340 crystal set appeared in two or three versions is unknown. Only two styles have been seen (Figs. 71A and B, 72A and B), but the sequence of appearance and their designations (No. 340, 340-A, 340-B) is unclear.

The Little Giant

Marketed in 1922 by the Metropolitan Radio Corporation, The Little Giant (Fig. 80A) differs from most lidded-case, American crystal sets by lacking either a tuning dial or tap switches on the panel. Instead, a tuning handle emerges from the left side of the cabinet. To use this handle for tuning, the lid instruction sheet states, "slide in or out while 'Listening In' with head set. . . ." Within the case is a loose coupler. The tuning handle, outside the cabinet, is used to vary the position of the movable inner coil relative to the fixed outer coil of the loose coupler.

An aerial tuner intended for use with The Little Giant was marketed in a matching case (Fig. 80B). This rare item was not mentioned in early ads for the crystal set. It has a tuning dial on the panel; a honeycomb inductance coil and variable condenser, in series, are within the cabinet.

The LITTLE GIANT Outfit

Trade Mark

The Gift That Everybody's Sure To Like!

GIVE someone the endless pleasure of radio entertainment with an outfit that anyone, even if they know nothing of radio, can operate ten minutes after unpacking. Give them the Little Giant that brings in the entertainment so marvelously clear and natural. Costs nothing to maintain and never gets out of order.

The attractive Holly Box contains everything needed to receive the radio program, the Little Giant Set, Metro Headphones, aerial wire, insulator, ground clamp, ground wire and all accessories. The Little Giant is wonderfully compact and finished as beautifully as a high-class phonograph. Every outfit is unconditionally guaranteed to operate perfectly. Can you think of anything you can give that would be more acceptable? It is the ideal gift.

Complete Outfit List $15

Your dealer should have his stock of Little Giant outfits by this time. In case he has not, send us your order, mentioning your dealer's name.

DEALERS: The Best Value For Them--The Most Profit For You

Metropolitan Radio Corporation
67-71 Goble St. . . . Newark, N. J.

Radio News (Dec. 1922), p. 1241

The Martians: Big 4; Special; Little Gem

The Martians "invaded" the radio world 15 years before H. G. Wells' "War of the Worlds" was convincingly presented by Orson Wells' *Mercury Theatre on the Air* in 1938. The Martian Big 4 crystal set appeared in 1923; and the Martian Special, in 1924. Actually, the Martian crystal sets are not named for the planet Mars but for the founder of the Martian Mfg. Co.: C. L. Marti.

Nonetheless, the Martian Big 4 (Fig. 82), with its unusual metal-tripod support, has an appearance suggestive of a space vehicle. A smaller crystal set on a tripod, the Little Gem (Fig. 83), does not have the word "Martian" in its name, but it has many similarities to the Big 4. The tripods of the Big 4 and the Little Gem are identical in size and appearance as are the set-screw-type crystal cups. The Big 4 has terminals for four (i.e., "Big 4") headphone sets, but the Little Gem accommodates only one phone set. Binding posts on some versions of the Big 4 are the same thumbscrew type as those on the Little Gem. Most Big 4 sets, however, have a different style of metal binding post—a small turn-lever instead of a thumb screw. This lever-style binding post also is present on the Martian Special. In many features, the Martian Special (Fig. 85A) and the Big 4 are alike; but the Special is mounted on a gray painted metal base (instead of having a tripod support), its detector is on the base (not on top of the tuning coil), and it accommodates only one set of headphones (not four).

Two manufacturers (Martian Mfg. Co. and White Mfg. Co.) were involved in the production of the two Martian crystal sets; two other companies (Metro Electrical Co. and United Specialties Co./U. S. Co.) produced the Little Gem. These four companies were all located in Newark NJ; their names appear individually on set labels, and in the early Martian and Little Gem ads.

Mengel Type M.R. 101; Shamrock Radiophone Crystal Model A; United Diamond Crystal Unit

Both the Shamrock Radiophone Crystal Model A (Fig. 135) and the United Diamond Crystal Unit are the same crystal set as the Mengel M.R. 101. On all three, the name of the manufacturer, "The Mengel Company, Inc.," is impressed in the black wood panel above the tuning dial. A lid label identifies the Shamrock set and its agency, Shamrock Radiophone Sales Corporation. The lid label

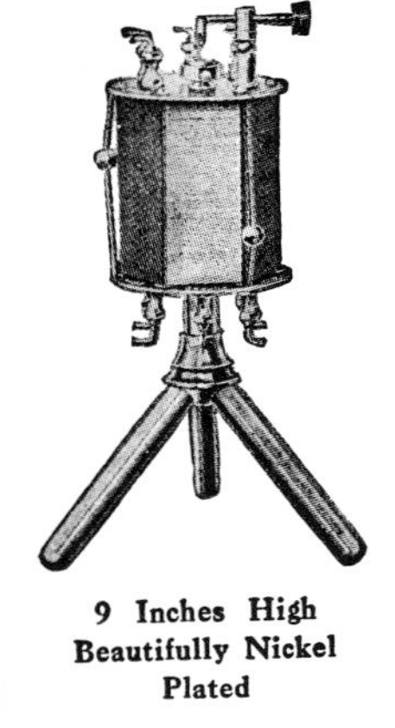

9 Inches High
Beautifully Nickel Plated

MARTIAN BIG FOUR

The Wonderful High Grade Crystal Radio Receiving Set

SIMPLICITY EFFICIENCY QUALITY

The Martian Sliders (2) for sharp tuning produce clearness and volume

One, two, three or four headsets can be used. One to eight persons can "listen in" at one time.

Tuning Coil is non-shrinkable.

Anyone Can Install and Operate the Big Four

Guaranteed to receive concerts from all broadcasting stations within a radius of 30 miles.

If your dealer cannot supply you, we will send you a Martian Big Four set prepaid anywhere in the United States upon the receipt of Price, $7.50

JOBBERS and DEALERS—WRITE OR WIRE FOR DISCOUNT SHEET

WHITE MANUFACTURING COMPANY

93-107 Lafayette Street Newark, New Jersey

Radio News (Dec. 1923), p. 835

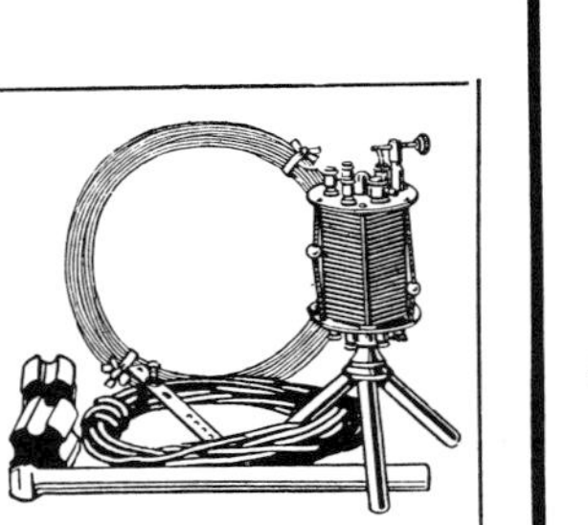

BlairCO "4" Crystal Set

Takes 4 head phones—Enables 4 to listen as well as 1. All nickeled steel 9 in. high. Price only **$7.50.**

Complete with copper clad aerial, waterproof lead-in and ground wire, strain and wall tube insulators, lightning arrester ground wire clamp and set of $9.00 Blairco Headphones, $17.50.

Dealers: An unbeatable profit opportunity is offered you in the Blairco Proven Products. Write quick for our proposition.

No matter what you want in Radio, buy at the "Blairco" Radio store and get dependable value. Every article sold under the Blairco trade name, whether our own make or others, is of proved superiority. Exacting tests have proved it the best of its kind, bar none!

If you have no Blairco dealer, write us now for Folder and Prices

Mitchell Blair Co.

"First With the Best"

1429 So. Michigan Ave., Chicago

Radio Topics (May 1923), p. 3

Martian Big 4 was also sold under "Blairco 4" name.

of the United Diamond set lists the United Metal Stamping & Radio Co.

Metro Jr. (see *Pal*)

Midget Pocket Radio; Midget Pocket Size Radio; Mitey Pocket Radio (see *Pee Wee*)

The Mighty Atom: Crystal Clear

This crystal set (Figs. 93A and B) with its powerful name, "The Mighty Atom," was marketed with a painted wooden base of a single bright hue—either cardinal red, deep yellow, or medium green. The subtitle, "Crystal Clear," on the paper label of this crude set made in the environs of the (then) small city of Phoenix, Arizona, inspired the title of this book. (Incidentally, these two words, with different spellings, were used in 1922 by the *Krystal-Kleer* Co. for the trade name of its crystal.) Production of this crystal set by Furr Radio exemplifies the efforts of many small enterprises in the 1920's and 1930's. Only 250 sets were manufactured by this two-man radio-shop factory, according to information provided to Ed Sharpe by the entrepreneur Vernon Furr shortly before his death in 1988.

PAL RADIO COMPANY, INC.

1204 Summit Ave., Jersey City, N. J.

Standard Crystal Receiver

Hundreds of thousands of these sets are giving unfailing satisfaction. Simple, substantial construction. All working parts in plain sight. Nothing to get out of order or give trouble. Coil wound on core of special material, not affected by heat or moisture. Stamped metal base finished in aluminum with bright nickelplated fittings. Equipped with sensitive Galena crystal and improved type cat's whisker. Receives up to a maximum distance of 35 miles, depending on power of broadcasting station. Receives over a range from 200 to 600 meters. Specially designed coil permits very sharp tuning. Reception wonderfully clear and loud. Each set packed in attractive 3-color box. Standard packing, 12 and 24. Immediate delivery from stock.

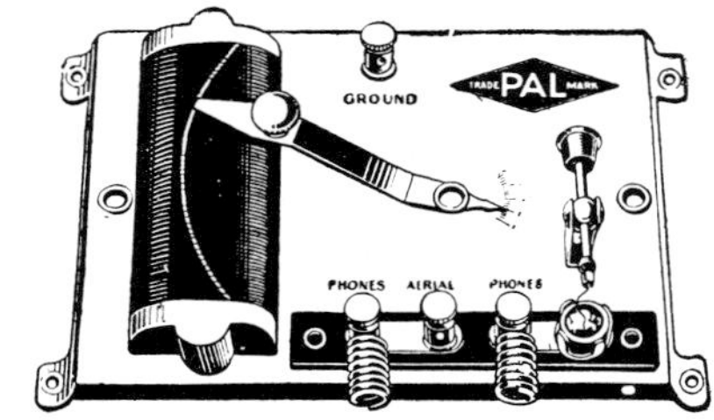

Pal Standard Crystal Receiver
Base .... 7x4½ in.
List price $1.50

McGraw-Hill Radio Trade Directory (Aug. 1926), p. 136

Moore (see *Flivver*)

Pa-Kette; Pakette (see *Pee Wee*)

Pal; Metro Jr.

The inexpensive Pal crystal set (Fig. 89), first listed in *Radio News* (*RN*) during 1926, appears identical to the Metro Jr. (Figs. 90A and B) advertised in *RN* during 1924 (and also listed for 1923 and 1924 in the *Radio Collector's Guide*). Pal Radio Co., which made the Pal crystal set, was located in Jersey City NJ; Metro Electrical Co., manufacturer of the Metro Jr. (and also associated with the Little Gem—see above), was in Newark NJ. The Pal crystal set, unchanged except for label, apparently is the successor to the Metro Jr.

Radio News (Jan. 1925), p. 1335

Pandora

Two closely related crystal sets were called "Pandora." Each has a vertical tuning coil above a round metal base where an open-type crystal detector is mounted. In this discussion, the set with its vertical coil directly attached to the base (Fig. 101) will be called Type 1; the other set, which has a tapered support between the base and the coil, will be referred to as Type 2.

These two types of the Pandora were listed and illustrated in nine available early publications. Each reference, however, has only one version of the Pandora, although both types were marketed during several overlapping years. Possibly Type 1 appeared first. An illustration in *Vintage Radio* (p. 69) shows the Type 1 Pandora and gives 1922 as its date of introduction. *The Radio Trade Directory* of August 1925 lists the Pandora, but does not have an illustration or description. *Radio News* first lists the Pandora (with Type 2 set shown in an accompanying figure) as "approved" in the "Radio News Laboratories" section of the March 1927 issue. The latest year for a retail catalog listing of Type 1 is 1934 and for Type 2, 1933.

The Brooklyn Metal Stamping Corporation made both versions. Indeed, a 1929 mail order

BMS Crystal Set

Very neat in appearance, and selective on all local stations. Tuning is done by the slider arms on the side of the coil. Anyone can afford to have one of these sets by his bedside.

Cat. No.	Price
219—BMS Crystal set	**$1.00**

S. S. Kresge, *Kresge's Radio Catalog and Buyer's Guide* (1929), p. 10

Brooklyn Metal Stamping Corporation
718-728 Atlantic Avenue, Brooklyn, N. Y.

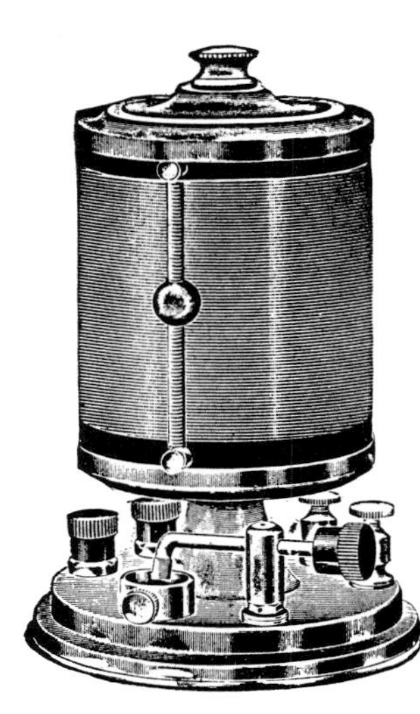

Pandora Crystal Set (Trade Mark Reg.)

Attractive, efficient crystal receiver.

Metal base and top finished in beautiful russet bronze.

Nickelplated trimmings.

List Price, each **$1.75**

McGraw-Hill Radio Trade Catalog (Aug. 1926), p.80

catalog lists the Pandora (Type 1) as the "BMS crystal set."

In addition to the two distinct coil-base styles noted above, other differences of these Pandora sets include: (1) arm of detector on *Type 1* is a slotted, flat-metal strip with a pointed end for crystal contact, and *Type 2* has a rod with a handle and catwhisker; (2) metal terminals on *Type 1* are removable thumb nuts, and *Type 2* has retention-type thumb nuts; and (3) at the top of the coil, *Type 1* has a contoured decorative thumb nut in some versions, but *Type 2* has a less decorative thumb nut similar to that found on the more common variation of Type 1. On both types of the Pandora, the slide-tuner is a suspended spring coil with a ball-type slider. Pandoras were available in at least four solid colors ("russet bronze," medium green, blue, and red) and one speckled finish (black with gold speckles).

Pee Wee Pocket Radio; Civil Defense Pocket Radio; Dick Tracy Two-Way Wrist Radio; Kitcraft Ready Built Radio; Midget Pocket Radio; Midget Pocket Size Radio; Mitey Pocket Radio; Pa-Kette Radio; Pakette Radio; Philmore Pocket Radio; Philmore No. VC-1000; Rada-Phone; Radaradio; Ti-Nee Pocket Radio; Tinymite Radio; Tiny Tone Pocket Radio; Tinytone

The crystal diode, developed during World War II, is a marked improvement over previous fixed crystal detectors. Because it is small, reliable, and inexpensive, the diode began to be used in crystal sets soon after the war. Philmore used a diode detector in its "Selective" Model No. VC-1000 (Figs. 205A and B), and it was used in the Kit Craft Ready Built Radio crystal set (Fig. 198). With the advent of the diode, many crystal radios became quite small in size, and trade names often indicated the diminutive dimensions, e.g., Pee Wee (Fig. 204), Ti-Nee, and Midget Pocket Size Radio. The names of two small sets—the "Rada-Phone" and "Radaradio"—proclaimed the new technology spawned by the recent Second World War. One of the more novel sets was the Dick Tracy Two-Way Wrist Radio that Johnson Smith & Co. marketed by 1950. Several crystal radios using the diode detector had self-contained earphone-speakers.

Small "Pocket (crystal) Radios" were also marketed in the 1930's before the advent of the crystal diode. These early miniature crystal sets have fixed crystal detectors. Included in this category are the Philmore Pocket Radio (Fig. 117), Tiny Tone Pocket Radio (Fig. 118), and Midget Pocket Radio (Figs. 91A, B, and C). The source of the Tiny Tone and Midget was a small manufacturer in Kearney NB that used several company names (e.g., Tinytone Radio Co., Midget Radio Co., Pa-Kette Elec. Co.; Midway Co.) in retailing its crystal sets. This overall enterprise was called Western Manufacturing Company. It first made Pocket Radios in 1932 and rivaled Philmore Mfg. Co. in longevity, and in the numbers of models and sets sold. The sales were almost entirely by mail, fostered by small ads in magazines. This company continued to make and sell crystal sets until at least 1960; later sets using crystal diode detectors included the Civil Defense Pocket Radio, Mitey Pocket Radio (Fig. 202), Pa-Kette/Pakette (Figs. 203A and B), Pee Wee (Fig. 204), Ti-Nee, Tinymite, and Tinytone (Fig. 208). Ads for Tiny Tone Pocket Radio, the first commercial crystal set of this manufacturer, began appearing in January, 1935; but the set was available in late 1932 or early 1933. Three other quite different "Tinytone" versions were marketed in 1937, 1951, and 1959. Both the first and the last of the twelve crystal sets produced by Western Mfg. Co. had the trade name "Tinytone."

Popular Science (Oct. 1949), p. 50

Philmore Crystal Sets (see also *Pee Wee*)

Most of the Philmore crystal sets appeared in different versions (Figs. 105–117, 205A and B, 206) during their long production lives. These variations and other features are described in the section on the Philmore Manufacturing Company.

Quaker Oats Crystal Set

As most collectors know, the typical young radio enthusiast of the 1920's constructed a crystal set with a tuning coil wound on a Quaker Oats box; many examples of these homebrew sets still exist. Less well-known, probably because of their present rarity, is the fact that factory-made Quaker Oats crystal sets (Figs. 123–125) were distributed as promotional items in 1921. They were available with an instruction pamphlet for $1.00 plus two Quaker Oats labels. On these sets, the detector and binding posts are mounted on the top of the box (1 lb. 4 oz. size); the tuning coil is wound beneath the label. On the bottom, this "Notice" is stamped: "This radio absolutely will not be exchanged or replaced if top or bottom has been removed." On the top, beneath the detector, is stamped on three lines: "Pat. Applied For / Marquette Radio Corp. / Chicago, U.S.A." Most surviving examples of this set have an open detector (Figs. 123A and B). Sets with the enclosed detector (Figs. 125A and B) were probably distributed later and much less extensively. The illustration in an ad of the Quaker Oats Co. for its crystal set (Fig. 123) shows only the open-detector set.

In 1977, the Quaker Oats Co. offered a small scale (2½ diam.; 5 ht.) transistor-radio version in plastic "of the famed (1921) Crystal Radio Set" for $9.95.

Rada-Phone; Radaradio (see *Pee Wee*)

Radioceptor (see *Busco Special*)

Radiogem

One of the least expensive of the early crystal sets, the Radiogem, at $1.00, apparently sold very well. Attesting to its continuing commercial success were large ads appearing in almost every issue of *Radio News* from September 1922 through May 1926. Although several thousand of these crystal sets must have been bought, surviving examples are rare. The Radiogem was not durably constructed. It was sold unassembled. By following a booklet of instructions, purchasers wound

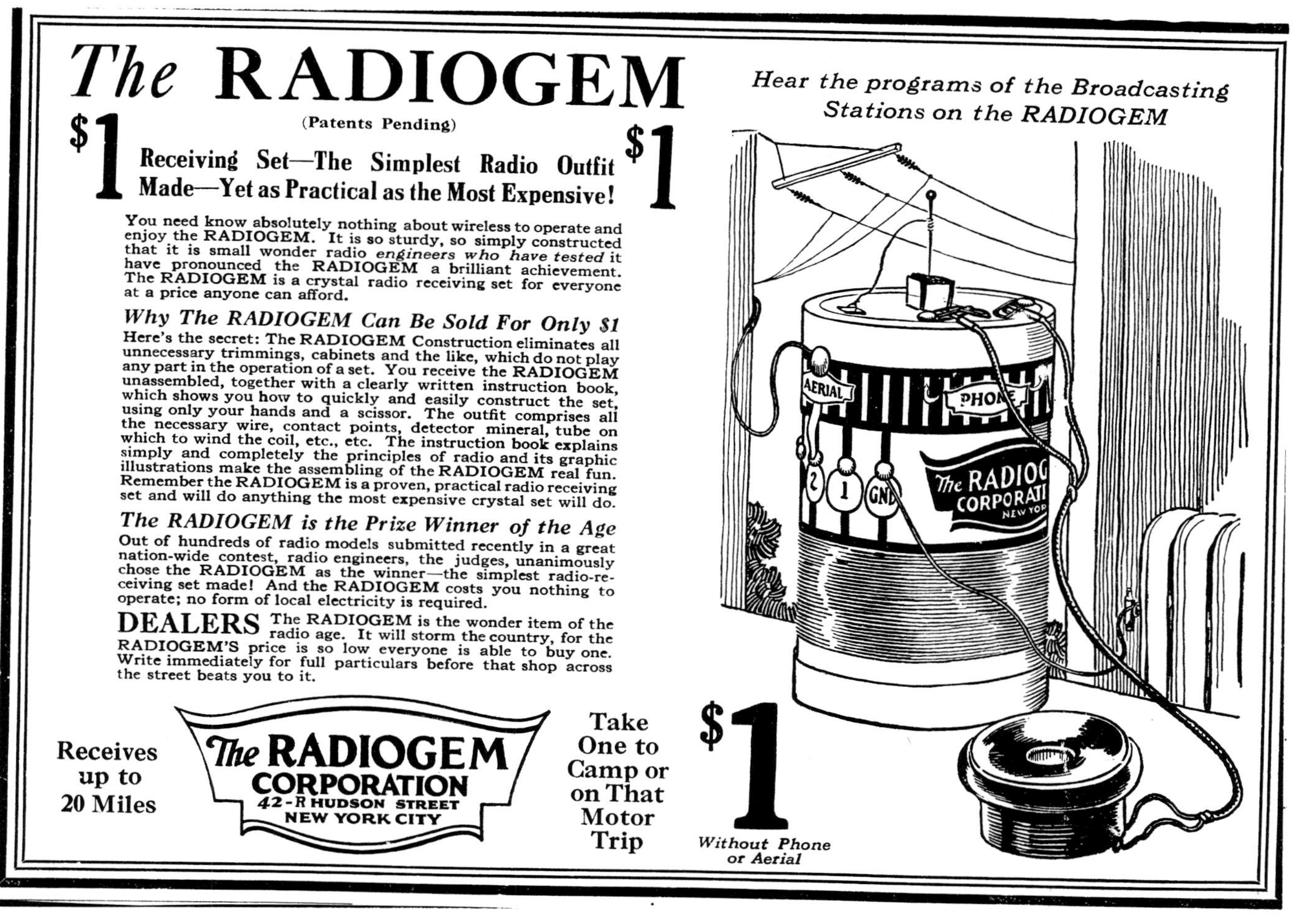

Radio News (Oct. 1922), p. 777

the induction coil on a labelled cardboard tube and made connections with the detector and terminal clips located on the cardboard lid at the top. When assembled, the lidded-cylinder set sits vertically, somewhat resembling a small version of a Quaker Oats box.

Radiola I (see *General Electric ER-753A*)

Radiowunder; Receptor Radionola
(see *California Diamond Jubilee*)

RCA/General Electric Model AR-1300; Radiola V

The General Electric AR-1300 (Fig. 131A) with its panel-mounted Perikon crystal detector is a crystal set if used alone. When used with its companion AA-1400 tube-detector and two-stage amplifier (Fig. 131B), the crystal detector is disconnected, and the AR-1300 serves as a tuner. These coupled sets then form a regenerative radio receiver. The Radiola V, which appeared a few months after the AR1300—AA1400 combination, is an elaborate version of these paired sets. A simulated wood-grain finish and different dials improve the appearance of the Radiola V. Its two metal boxes are held beside each other by a wooden base (with a label) and a fitted lid. (See "RCA," pp. 20–23.)

ROMCO (see *Brownie Airphone*)

Shamrock Radiophone Crystal Model A
(see *Mengel Type M. R. 101*)

Spencer Pocket Radio

This small inexpensive crystal set was marketed in two versions. Both styles have textured fiber cases with "Spencer Pocket Radio, Akron, O."

logos on top of their lids. The style usually seen (Figs. 140A and B) has a case (4⅞ x 3 x 1⅛) that is green stippled on a faint gold background. In the uncommon style (Figs. 141A and B), the case (4 x 3½ x 1⅜) is medium red, and its embossed lid logo is smaller than the one appearing on the more common variation of the set.

An unattached instruction sheet was distributed with both types of the Spencer crystal set. In the green stippled set, the sheet was placed loosely inside the lid; in the red version, it was packed outside the case (along with the set) in a larger orange carton bearing the Spencer label. (See *California Diamond Jubilee* for additional comments on this set and its patent).

Steinite

Each *standard-version* Steinite crystal set (Figs. 145A and B) has an individualized serial number printed on a label attached to the bottom side of its base. Presumably, these numbers were issued sequentially; thus, they indicate the relative order of production. On this basis, the earlier sets (those with lower serial numbers) differ from later ones by having: (1) a somewhat smaller enclosed horizontal detector; (2) two-hole binding posts for the phone terminals to accommodate two sets of headphones (later binding posts have a single-hole); and (3) six solid-end metal taps for the slide tuner (later sets have hollow eyelet taps). Some other nonsystematic variations seem to have occurred in types of tuning dials and heights of the slide-tuner rod above the taps. After the Steinite

CRYSTAL RECEIVER

The Crystal Receiver shown, submitted by the Steinite Laboratories, Radio Bldg., Atchison, Kans., gives unusually good results on local re-

ception; and with a good aerial installation and sensitive headset, distance reception is possible.

AWARDED THE RADIO NEWS LABORATORY CERTIFICATE OF MERIT NO. 1631.

Radio News (Dec. 1926), p. 677

Rectifier with a graphite contact point (Fig. 77B) was marketed in 1925, this detector was used on the Steinite "DeLuxe" crystal set; the standard set also remained available.

The last-marketed version of the Steinite ("New") crystal set (Fig. 144) has a finely finished contoured wood case, a panel window to view the rotating tuner dial indicator within the case, and an instruction label covering the bottom of the base (where the smaller tag with serial numbers appears on the standard set). Its carton is labelled "*NEW* STEINITE CRYSTAL RADIO."

Ti-Nee; Tinymite; Tiny Tone; Tinytone
(see *Pee Wee*)

Tuska No. 4007; Tuska
(see *A. C. Gilbert No. 4016*)

United Diamond Crystal Unit
(see *Mengel Type M.R. 101*)

World; The Crystal-Dyne

The World crystal set (Fig. 190) appeared in two styles of metal cases—nickel-plated or painted (bright yellow, light green, dusky orange). Most sets now in collections are the plated version. The Crystal-Dyne (Fig. 40), made by the same manufacturer, has the same enclosed semi-fixed detector as the World. Bethlehem Radio Corporation also produced a two-stage audio-frequency amplifier for use with its two crystal sets.

Novelty Homebrew Crystal Sets

By the early 1920's, many talented radio enthusiasts were creating small novelty crystal sets. Several of these clever creations were shown and described in radio magazines. A "Finger-Ring Receiver," claimed to be the smallest set made, is shown in *Popular Radio* for November, 1922. The February 1926 issue of this magazine shows small crystal sets mounted on the stem of a corn-cob pipe, on the part below the keyboard of a miniature piano, and within a woman's leather card case. The third set somewhat resembles the California Diamond Jubilee and the Radiowunder (see above text).

The September 1922 *Radio News* (*RN*) pictures a little Kewpie-doll crystal set. Kewpie's body is

An Amateur's Novel Set

Kewpie Has a New Dress All Made of Wire. The New Hat is a Crystal Detector. Kewpie and Her Owner Are Listening In.

A LITTLE Kewpie, tightly wrapped with No. 30 gauge, D. C. C. wire, a crown on its head serving as a crystal detector, a pair of binding posts, a head set and a young inventive genius, have all been used by William Tecklenburg, fourteen years of age, in the making of a complete radio-receiving set He "listens-in" on all of the big Denver stations, which are situated near where he lives, and the little Kewpie gracefully syncopates as the music is pulled from the ether by the novel instrument.

"I guess that, with everything included, it cost about 25 cents, or maybe 30 cents. If I had bought both binding posts it would have cost 30 cents anyhow, but old battery posts are just as good," explained the young enthusiast. It took William almost all of 25 minutes to make the set.

Radio News (Sep. 1922), p. 459

Some STUNT SETS

Peculiar but Practical Receivers Built by Radio Fans

Kadel & Herbert

This miniature set, built in a toy piano by C. W. G. Brown, may play the famous "Prelude" of Rachmaninoff—if the famous "Prelude" of Rachmaninoff is being broadcast from a studio.

Popular Radio (Feb. 1926), p. 124

"A woman is only a woman" the poet Kipling may write, "but a good pipe is a crystal receiver," John Kott of Chicago paraphrases. He built the set for $1.50.

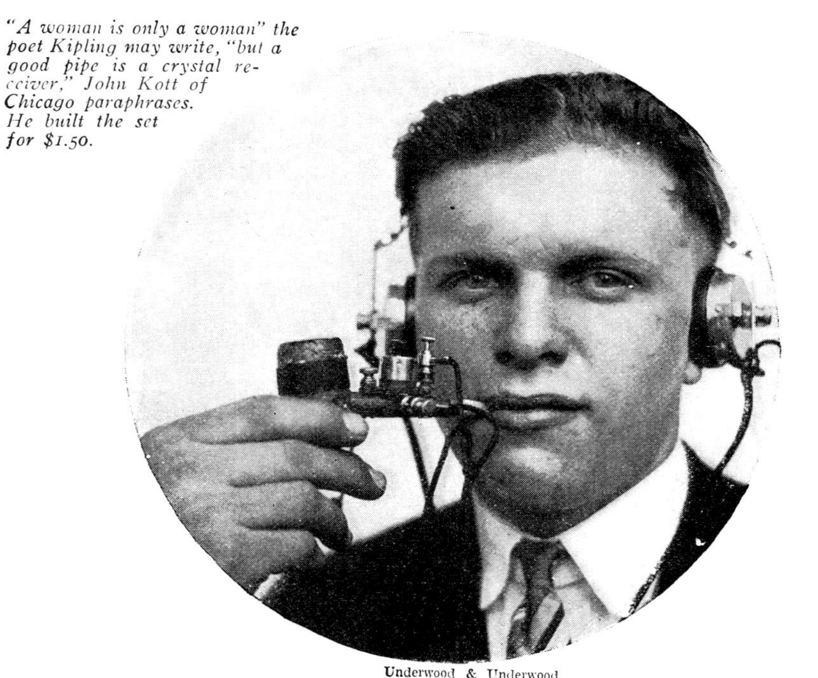

Underwood & Underwood

Popular Radio (Feb. 1926), p. 124

tightly wrapped with No. 30 D. C. C. wire for inductance tuning; crowning its head is the crystal detector. On a tiny platform for Kewpie are the four binding posts. *Radio News* for March, 1923, has an illustration of a flat-top straw hat being used for both the base and case of a crystal set. A photograph in the February 1924 *RN* shows a "nutshell" crystal set; the assembled miniature components are contained within a hinged English walnut shell.

One of the more popular early homebrew novelty items was a lead-pencil crystal set. The detector is mounted on the eraser end. Either D. C. C. or enamelled wire is wound along the length of the pencil for inductance tuning. Many "Old Timers" fondly recall making and operating this clever little crystal receiver.

Crystal Sets as Premiums

During the 1920's, many companies made alluring premium offers to buyers and sellers. In order to attract customers or agents (usually boys) to sell their products, these companies needed popular items for premiums. With the ascendancy of the radio craze, small crystal sets served this purpose well.

In 1921, as noted above, the Quaker Oats crystal set (Figs. 123–125), made by Marquette Radio Corp., was offered as a premium for "$1.00 with 2 Quaker Oats trade-marks cut from packages" (Fig. 123).

An October 1922 ad in *Radio News*, by the Home Supply Co. of New York City, proclaimed, "BOYS! A REAL RADIO SET (i.e., a crystal set:

Radio News (Oct. 1922), p. 766

single-slide tuner, base-mounted detector, unstated mfr.) ABSOLUTELY FREE. RUSH your name and address and we will tell you HOW you can get this RADIO SET ABSOLUTELY FREE. . . . Write today for free Radio Plan."

The December 1922 *Radio Broadcast* advertised, "A complete receiving set for 4 subscriptions to RADIO BROADCAST ($3.00 each). The Audiola Crystal Receiving Set is not to be confused with the ordinary crystal sets on the market. . . ." In the same issue of that magazine, an ad of the manufacturer of this set appeared, showing the sales price of the Audiola as $10. But for only the $12 from four subscriptions, an enterprising boy could get the crystal set free as a premium.

Similarly, *Radio Topics* of May, 1923, offered "States Radio Corporation Crystal Set free for three one-year subscriptions at $2.00 each ($6.00 total)."

THE MOST LIBERAL OFFER OF THEM ALL

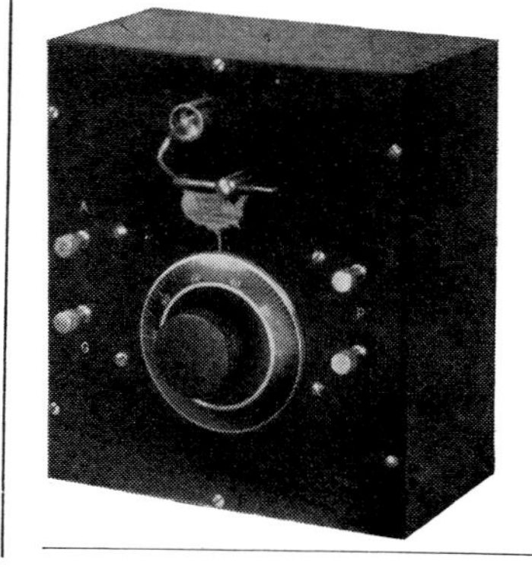

A STATES RADIO CORPORATION CRYSTAL SET FREE

Your opportunity to get this Super Crystal Set

With But Little Effort

THREE one-year subscriptions or one three-year subscription $6.00.

FOUR one-year subscriptions or one four-year subscription includes head phone $8.00.

Radio Topics (May 1923), p. 38

An ad of the Guarantee Sales Co., N. Y. C., in the November 1923 *Radio News*, showed a crystal set (somewhat similar to The Crystal-Dyne in Fig. 40) and exclaimed, "Quick! Send only your name to-day for wonderful FREE RADIO PLAN." Most such "plans" in the 1920's were offers of premiums for selling magazine subscriptions, seed packets, or other merchandise.

Radio News (Nov. 1923), p. 632

THREE UNIQUE CRYSTAL RECEIVERS

The three illustrations show novel ideas in simple receivers. These are primarily for the person who does not wish to be

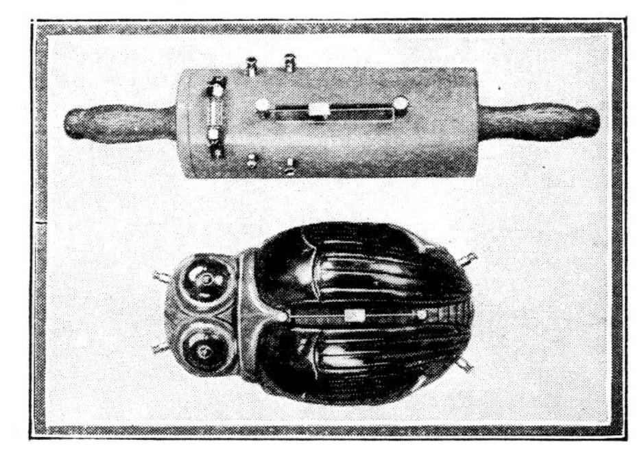

Two unique crystal receivers, the upper one being constructed inside a wooden rolling-pin and the other in a "bug" of burned stone.

bothered with batteries and the care that must be exercised upon them. It is also a well known fact that reception with a crystal receiver gives as clear reproduction of signals as can be obtained; so that, all in all, these sets are good from the viewpoint of novelty of make-up as well as reception.

The body of the "bug" and the wall receiver are made of hard, burned stone, which has excellent insulating properties. In the case of the bug the crystal detector is of the fixed type and is mounted in the head, the mountings being the pupils of the eyes. The head-phones are connected in the place where the antennae of the bug would be, and in place of hind legs the antenna and ground connections are made.

The rolling-pin receiver has the set mounted within the cylinder with only the slide tuner, the detector, and the binding posts showing. This receiver is made of wood, following the model so skillfully used by housewives.

The third set is to be hung on the wall and, as may be seen, it is extremely simple to install. It is claimed that the insulating properties of this material are so high that exceptional reception results.

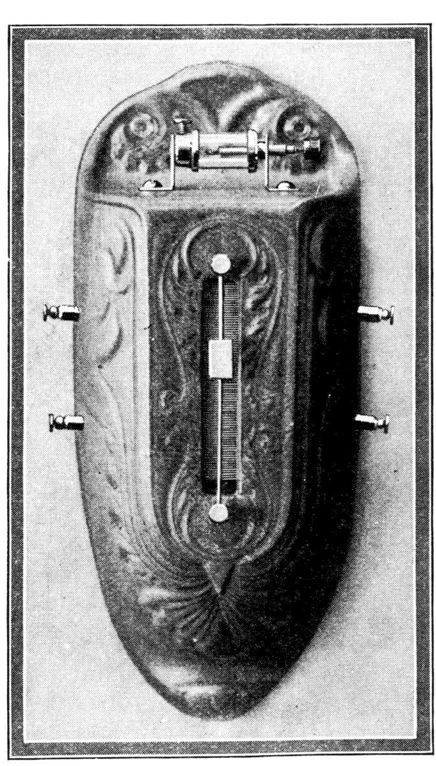

A wall-type crystal receiver, of the same insulating material as the "bug" shown above.

Photos by courtesy of Brush Pottery Co.

Radio News (Dec. 1926), pp. 647-648

Crystal Sets

Dimensions for case or base = length x width x height (length is the larger and width, the smaller of the two horizontal measurements). Consult crystal sets table for additional details including date of manufacture. See text for discussion of some special features.

Fig. 1. ABC (Jewett Mfg. Co.; Wireless Equipment Co.); 9″ x 7 1/2″ x 4 1/2″. Al Reymann collection and photo.

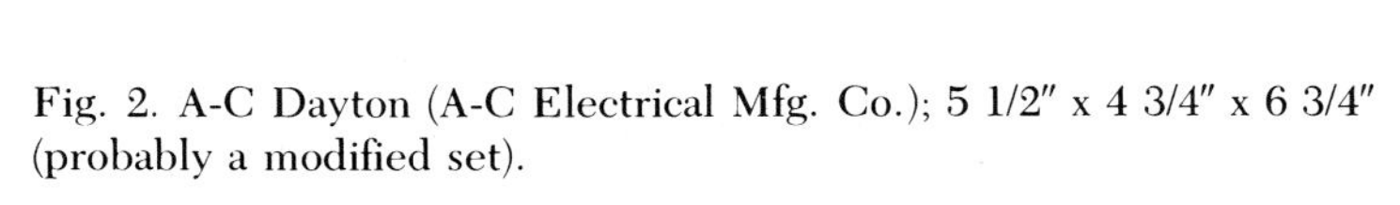

Fig. 2. A-C Dayton (A-C Electrical Mfg. Co.); 5 1/2″ x 4 3/4″ x 6 3/4″ (probably a modified set).

Fig. 3. Adams-Morgan Type RC (Adams-Morgan Co.); base = 15″ x 8 1/2″ x 3/4″; coil = 3″ diam., 3 5/8″ length.

Fig. 4. AMCO Double Crystal, The Arlington (Adams-Morgan Co.); 9 1/2″ x 7 1/2″ x 9 1/2″. Al Reymann collection and photo.

Fig. 5. Aeriola Jr., Model RE (Westinghouse Elec. & Mfg. Co.; RCA); 8 1/2″ x 7 1/4″ x 6 3/4″.

Fig. 6. Aeriola X (Crain Bros. Radio Shoppe); 6 7/8″ x 4 7/8″ x 5 1/2″. The trade name of this set was changed soon after it was marketed (see text and crystal set table).

Fig. 7. Air Bug (Custer); 3″ diam., 3/4″ ht. No provisions for internal tuning. Ed Sharpe collection.

Fig. 8. Air-Scout Senior (unknown mfr.); 5″ x 2 3/4″ x 4″.

Fig. 9. Ajax (Ajax Elec. Specialty Co.); 5″ x 3 3/4″ x 2 7/8″.

Fig. 10. American Leader Pocket Radio (American Leader Products Co.); 5 1/2″ x 2″ x 5 1/4″. Compartment at right end of case (with knob visible) holds a single headphone.

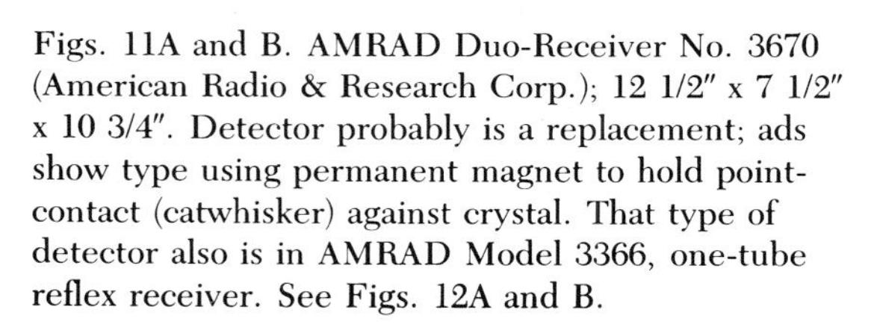

Figs. 11A and B. AMRAD Duo-Receiver No. 3670 (American Radio & Research Corp.); 12 1/2″ x 7 1/2″ x 10 3/4″. Detector probably is a replacement; ads show type using permanent magnet to hold point-contact (catwhisker) against crystal. That type of detector also is in AMRAD Model 3366, one-tube reflex receiver. See Figs. 12A and B.

Figs. 12A and B. AMRAD Reflex Receiver 3366 (American Radio & Research Corp.); 10″ x 5 7/8″ x 5″. Magnetic detector shown on this receiver also was used on AMRAD Duo-Receiver No. 3670. See Figs. 11A and B.

Fig. 13. AMRAD #2575 (American Radio & Research Corp.); 6″ x 5″ x 5 1/4″. Al Reymann collection and photo.

Fig. 14. Atwater Kent (Atwater Kent Mfg. Co.); 16″ x 10″ x 1″. Model 2A Twin Unit (Part No. 4200) detector; Type 11 tuner. The manufacturer apparently did not assemble the Atwater Kent crystal sets. Paul Thompson collection.

Figs. 15A and B. Audiola Model A No. 1 (Audiola Radio Co.); 7 1/2″ x 5 3/4″ x 3 1/4″. See Fig. 16 and text.

Fig. 16. Splendid (Splendid Radio Co.); 7 1/2″ x 5 3/4″ x 3 1/4″. See Figs. 15A and B and text. Al Reymann collection and photo.

Figs. 17A and B. BASUB/Simplex (Simplex Electrical Lab.); 5 3/4″ x 5 1/2″ x 5 5/8″. The set is double labelled. Engraved on its panel (Fig. 17A) is "BASUB / SIMPLEX ELECTRICAL LAB. INC. / BROOKLYN N.Y.," but a metal tag attached to the back side of the case (Fig. 17B) states, "TRADE MARK / SIMPLEX RADIO COMPANY, PHILA. PA."

Fig. 18. B and C (B & C Radio Comm. Co.); 12″ x 8″ x 8″. Painted wood panel.

Fig. 19. BC-14A Signal Corps U. S. Army Radio Receiving Box (General Radio Co.; also made by DeForest Radio Tel. & Tel. Co., and by Liberty Elec. Corp.); 12 1/2″ x 8 1/2″ x 8 1/4″. Available to general public as an Army surplus item as late as 1931.

Figs. 20A and B. Beaver Baby Grand Vest Pocket (Beaver Machine & Tool Co.); 3 3/4″ x 2″ x 3/4″. Case of set in 20A is oak; in 20B, it is molded hard rubber—a less common version. Fig. 20B, Ed Sharpe collection.

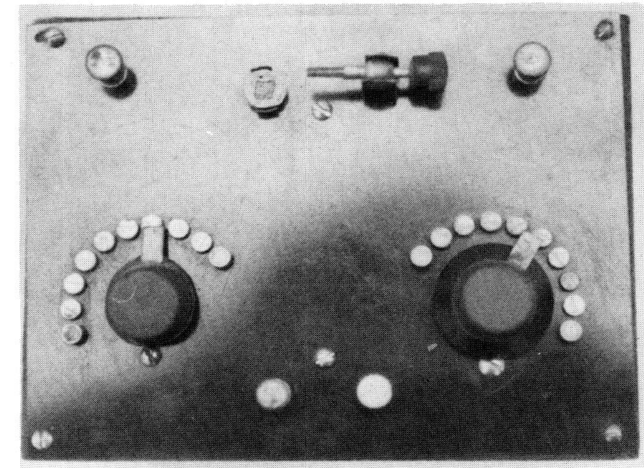

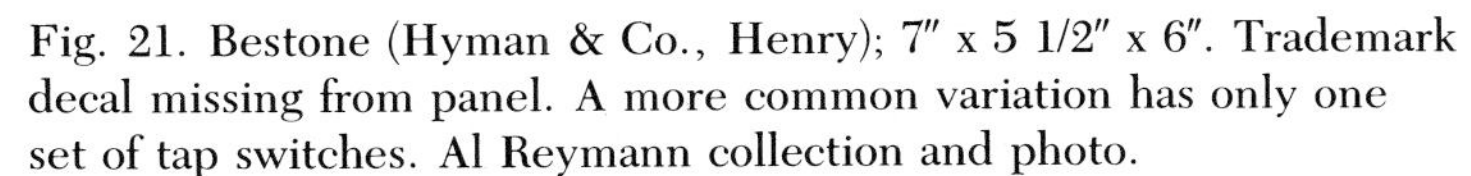

Fig. 21. Bestone (Hyman & Co., Henry); 7″ x 5 1/2″ x 6″. Trademark decal missing from panel. A more common variation has only one set of tap switches. Al Reymann collection and photo.

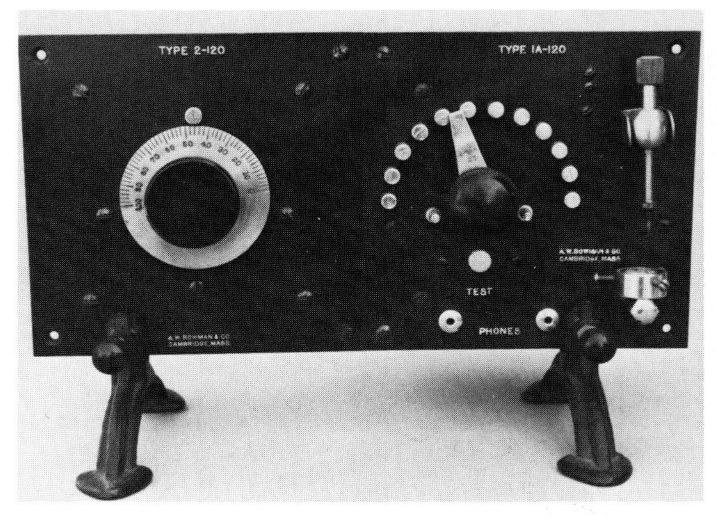

Figs. 22A and B. Bowman Type 1A-120, Unit Panel Set with Type 2-120 Tuner Unit attached (Bowman & Co., A. W.; Sears, Roebuck & Co.); each panel 5″ x 5″. Ed Sharpe collection.

Figs. 23A and B. Brownie Airphone, in 23A (Brownie Mfg. Co.; Rivero & Co.) ROMCO, in 23B (R & O Mfg. Co.); both = 7″ x 5″ x 1 5/8″. See text.

Fig. 24. Bronswick (unknown mfr.); breadboard 5″ x 4″ x 1″, coil ht. = 4″. Al Reymann collection and photo.

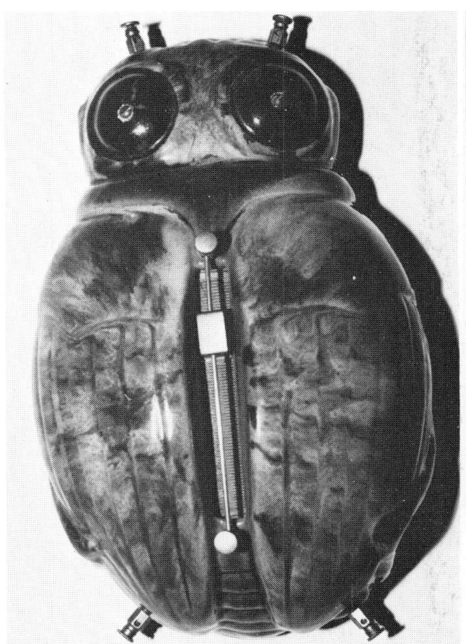

Fig. 25. The Bug (Brush Pottery Co.); 9 1/4″ x 3 1/2″ x 5 3/4″. Greg Farmer collection and photo. The body of the "bug" is pottery. A fixed detector is mounted behind the eyes; some "Bug" crystal sets have an enclosed, horizontal detector.

Fig. 26. BUSCO Special (Buscher Co., C. A.); 7 1/2″ x 5 1/4″ x 6 1/8″. See Figs. 27, 30, and text.

Fig. 27. BUSCO Special, variant with semi-fixed detector (Buscher Co., C. A.); 7 1/2″ x 5 1/4″ x 6 1/8″. See Fig. 26.

Fig. 28. BUSCO Jr. (Buscher Co., C. A.; Kansas City Radio Mfg. Co.); 7 1/2″ x 5 1/4″ x 6 1/8″. See Figs. 29, 30.

Fig. 29. Unknown mfr.; similar to BUSCO Jr.; 7 1/2″ x 5 1/4″ x 6 1/8″. See Fig. 28.

Fig. 30. Radioceptor (Kansas City Radio Mfg. Co.); 7 1/2″ x 5 1/4″ x 6 1/8″. Same or very similar features to BUSCO Special and BUSCO Jr. (Figs. 26-29); apparently, all of these sets were made by K. C. Radio Mfg. Co.

Fig. 31. Radiowunder (Radiowunder Specialty Co.); 4 1/2″ x 3 3/4″ x 1/2″. See Figs. 32A and B, 34A and B.

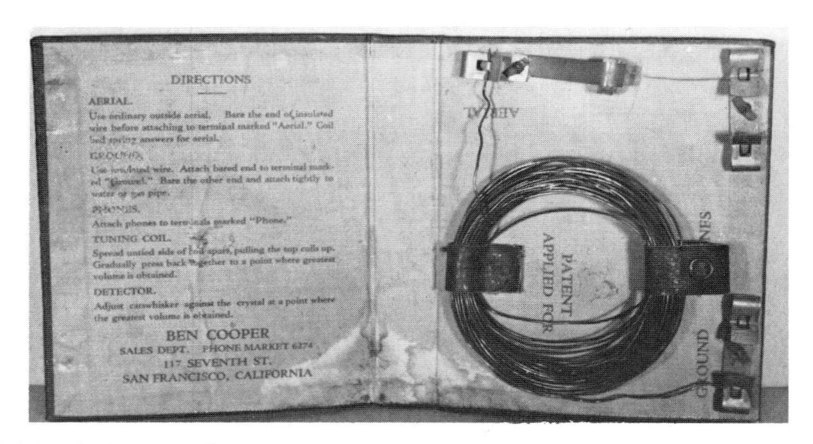

Figs. 32A and B. California Diamond Jubilee (unknown mfr.; Ben Cooper Sales Dept.); 4 1/2″ x 3 3/4″ x 1/2″. See Figs. 31, 34A and B, and text.

Fig. 33. Cannon Ball Baby Grand (Cannon Co., C. F.); base = 5″ x 4 1/4″ x 1/2″, coil = 2 3/4″ diam., 3″ ht.

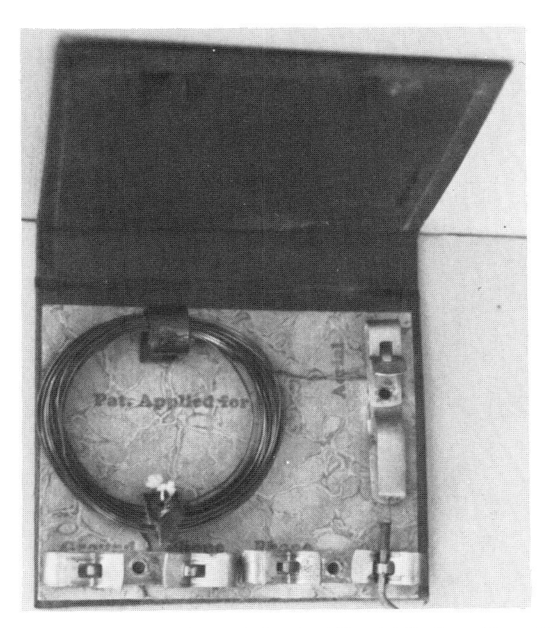

Figs. 34A and B. Receptor Radionola, The Pocket Radio (Radionola International Corp.); 4 1/2" x 3 3/4" x 3/4". See Figs. 31, 32A and B.

Figs. 35A and B. CARCO (Carter Mfg. Co.); 7 1/4" x 4 1/4" x 7/8". CARCO 1-tube receiver shown in Fig. 35B beside the crystal set. Ed Sharpe collection.

Fig. 36. Carter (Carter Radio Co.); 9 3/4″ x 9 3/4″ x 7″. Impressed in the round, black fiber disc attached to left front of panel is the label "Carter Radio Co.—Chicago." See Cunningham (Figs. 41A and B)—same crystal set as Carter.

Fig. 37. Champion (Champion Radio Products Co.); 6″ x 4 3/4″ x 5″. Les Rayner collection.

Fig. 38. Cherington (Cherington Radio Labs).; 3 3/4" diam., 1 1/4" ht.

Fig. 39. Commerce Radiophone (Commerce Radiophone Co.); 8″ x 7 5/8″ x 10 1/2″. Detachable lid of case (positioned).

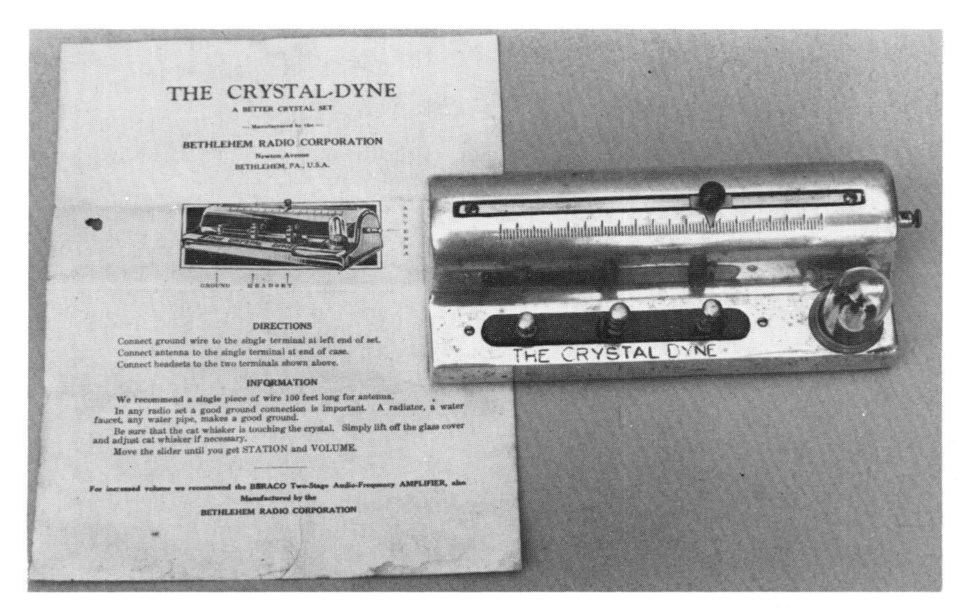

THE CRYSTAL-DYNE

A BETTER CRYSTAL SET

—Manufactured by the—

BETHLEHEM RADIO CORPORATION

Newton Avenue

BETHLEHEM, PA., U.S.A.

DIRECTIONS

Connect ground wire to the single terminal at left end of set.
Connect antenna to the single terminal at end of case.
Connect headsets to the two terminals shown above.

INFORMATION

We recommend a single piece of wire 100 feet long for antenna.

In any radio set a good ground connection is important. A radiator, a water faucet, any water pipe, makes a good ground.

Be sure that the cat whisker is touching the crystal. Simply lift off the glass cover and adjust cat whisker if necessary.

Move the slider until you get STATION and VOLUME.

For increased volume we recommend the BERACO Two-Stage Audio-Frequency AMPLIFIER, also Manufactured by the BETHLEHEM RADIO CORPORATION

Fig. 40. The Crystal-Dyne (Bethlehem Radio Corp.); 8 1/2″ x 3 7/8″ x 2 1/8″. Instruction sheet at left of set. See Fig. 190.

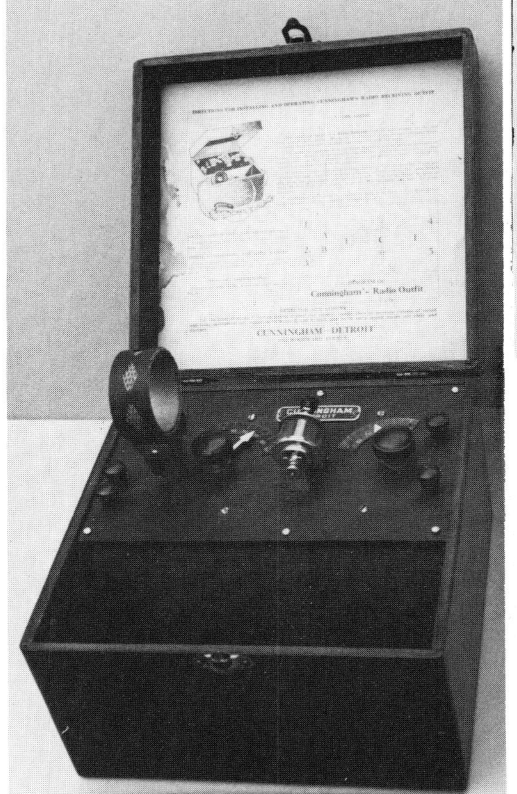

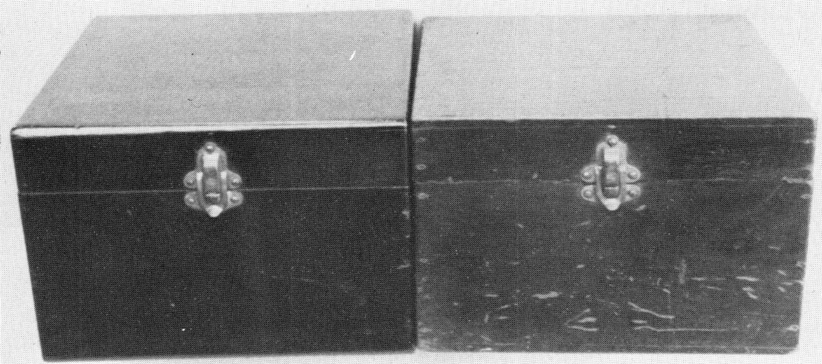

Figs. 41A and B. Cunningham's Radio Receiving Outfit (Cunningham); 9 3/4″ x 9 3/4″ x 7″. Lid not detachable. See Carter (Fig. 36)—same crystal set; cases of both sets have Corbin lid latches mounted *upside down* (Fig. 41B—Cunningham on left, Carter on right). See also Fig. 42 for a set with several similarities.

Fig. 42. DeForest Everyman DT-600 (DeForest Radio Tel. & Tel. Co.); 9 3/4″ x 8 1/2″ x 7″. Detachable lid (positioned).

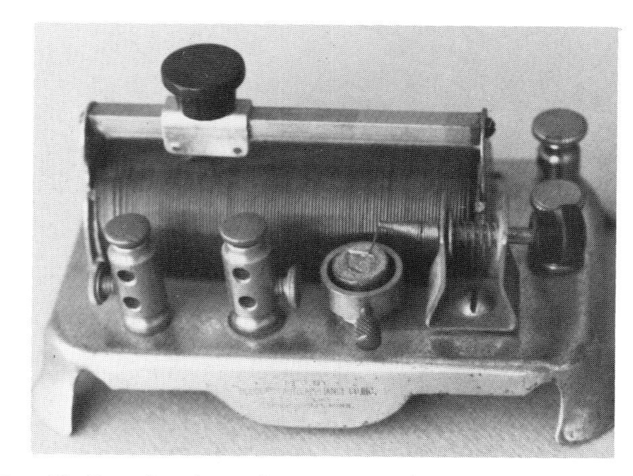

Fig. 43. Douglass (Douglass Auto Appliance Co.); base = 4 1/2″ x 3″ x 3/4″, coil = 1″ diam., 3″ length.

Fig. 44. Duncan Simple (Duncan, Donald F.); 7″ x 3 3/4″ x 6″. Has same type detector as Magnus crystal detector (Fig. 255). Al Reymann collection and photo.

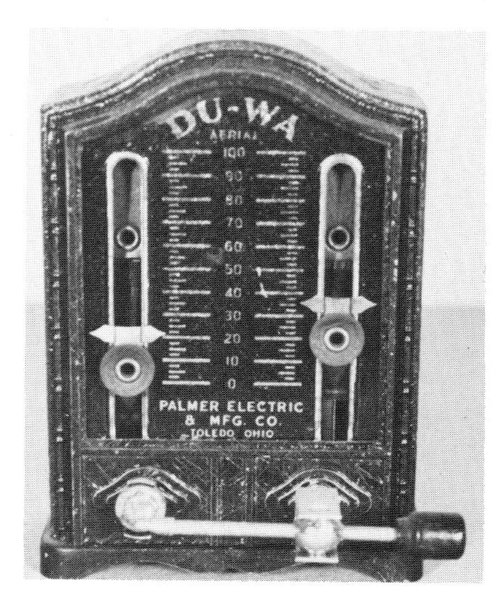

Fig. 45. Du-Wa, "Dual Wave" (Palmer Elec. & Mfg. Co.); 3 1/2″ x 2″ x 5 1/4″. Metal case; detector of the Philmore type but with a longer detector arm (2 7/8″) than that (1 1/2″) on the usual Philmore detectors (see Figs. 268, 300).

Fig. 46. The Eagle (C & R Radio Shop); 5 1/4″ x 5 1/8″ x 6″.

Fig. 47. Entertain-A-Phone Model 1 (Pioneer Wireless Mfrs.; New York Coil Co.); base = 10 1/4″ x 6 1/2″ x 1/2″, coil = 4″ diam., 4″ ht. Round metal plate on breadboard is the phone condenser. Ed Sharpe collection.

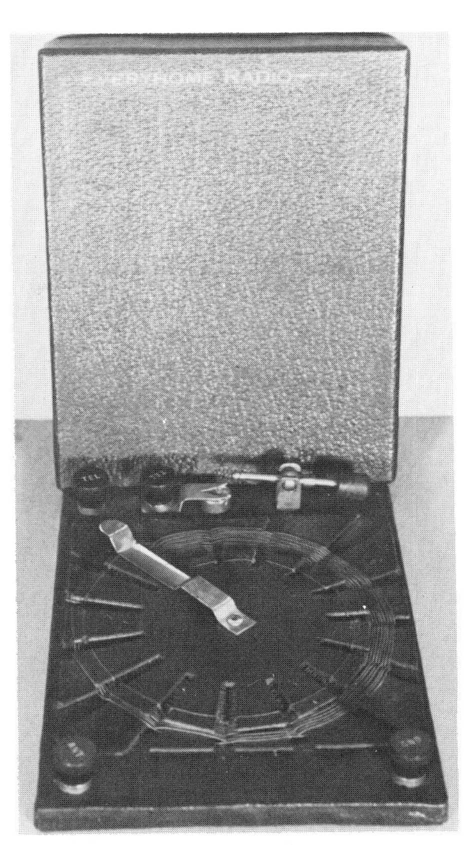

Fig. 48. Everyhome Radio (Everyhome Radio Co.); 7″ x 5 1/4″ x 1 1/4″. Removable lid (standing at end of set).

Fig. 49. Excello Galena (Sterling Radio Mfg. Co.); 6 3/8″ x 4 1/8″ x 5 5/8″. "LEMCO type" catwhisker (see Figs. 71B, 72B).

Fig. 50. Ferro (Ferro Mfg. Co.); 9 1/2″ x 4 5/8″ x 5 3/8″. Instruction sheet in foreground.

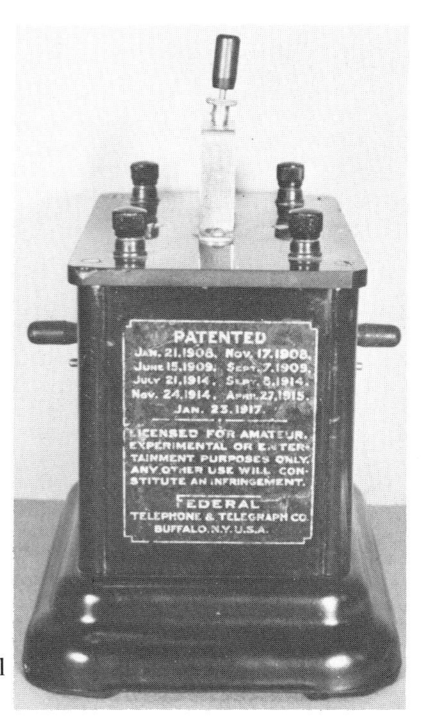

Figs. 51A and B. Federal Jr. (Federal Tel. & Tel. Co.); 8 1/2" x 5 1/2" x 6 1/2". As shown in 51B, patent dates are listed on the end of *some* Federal Jr. crystal sets.

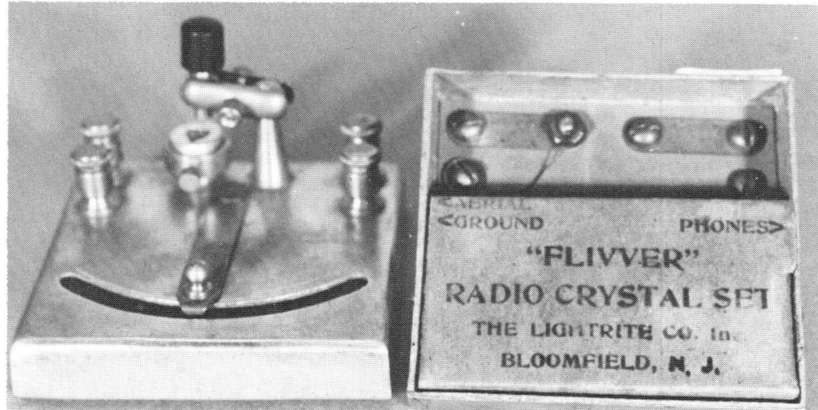

Fig. 52. Flivver (Lighthouse Co.); 3" x 3" x 1/2". Set on right, view of underside with label on coil cover.

Fig. 53. Gardner Broadcast Transformer (Gardner Labs.); 7 3/8" diam., 7 1/2" ht. Loose coupler and variable capacitor tuning.

Fig. 54. General Electric /RCA Model ER-753 (General Elec. Co.; RCA); 5 1/2" x 4 5/8" x 8".

Fig. 55. General Electric /RCA Model ER-753A, Radiola I (General Elec. Co.; RCA); 6" x 6" x 11". Paul Thompson collection.

Figs. 56A and B. Giblin Radiocar (Standard Radio & Elec. Co.); 10 3/4" x 7 1/2" x 5 5/8".

Figs. 57A, B and C. A. C. Gilbert No. 4016 (Gilbert Co., A. C.); 7 3/4" x 6 3/4" x 4 1/2". See text for discussion of variations.

Fig. 58. Glen Radio Model K-12 (Lyons Co., G. E.); 10" x 5 3/4" x 6 1/2". A model variation with tap switches also exists.

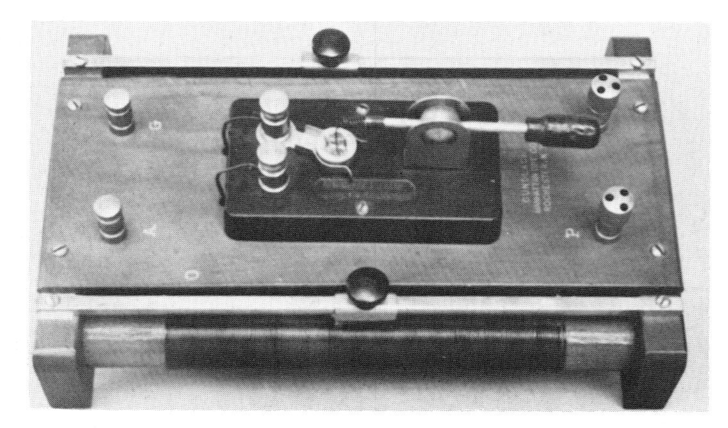

Fig. 59. Gundlach (Gundlach Manhattan Optical Co.); 7 3/4" x 5" x 1 3/4".

Fig. 60. G. W. (Gehman & Weinert); base = 5″ x 5″ x 3/4″, coil = 1 1/2″ diam., 2 5/8″ ht.

Fig. 61. Heard (Heard Co.); 11 1/8″ x 7 7/8″ x 8 1/2″.

Fig. 62. Heliphone (Gardner-Rodman Corp.; RADISCO); 6 3/4″ x 4 1/2″ x 1 1/8″.

Figs. 63A and B. Howe (Howe Auto Products Co.); 3″ x 3″ x 2 1/2″. Label and handle-knob variations; set at left is earlier. In Fig. 63B, Howe store-display box that holds six sets in cartons—one carton shown. Fig. 63B, Al Reymann collection and photo.

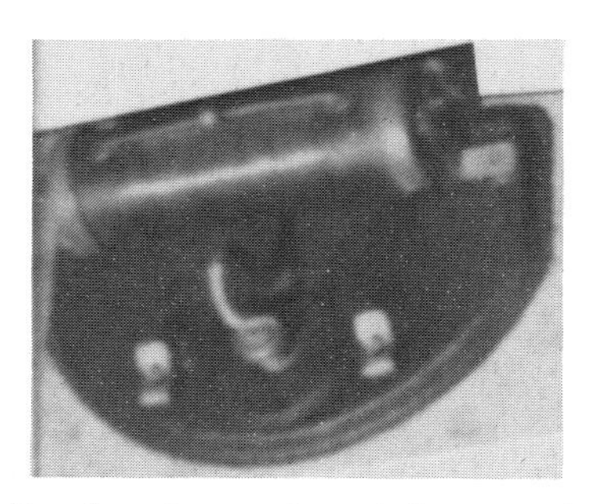

Fig. 64. ICA (Insuline Corp. of Am.); base = 6″ x 3 3/4″ x 1/2″. Al Reymann collection and photo.

Fig. 65. Hunt & McCree No. 797 (Hunt & McCree); base = 7 1/2″ x 4 1/2″ x 3/4″ (with legs), coil = 2 1/4″ diam., 4 3/4″ length. Electrolytic detector. Les Rayner collection.

Fig. 66. Jubilee (Jubilee Mfg. Co.); base = 5 1/4″ x 4 1/4″ x 7/8″, coil = 2 1/2″ diam., 3 1/8″ length.

Fig. 67. Kleer-Tone Model B (Buscher Co., C. A.); base = 5 1/8″ x 5 1/8″ x 3/4″, coil = 4″ diam., 7 1/4″ ht.

Figs. 68A and B. Kismet (Radio Tel. & Tel. Co.); 5″ x 3″ x 7″. Set came equipped with Kilbourne & Clark detector as shown.

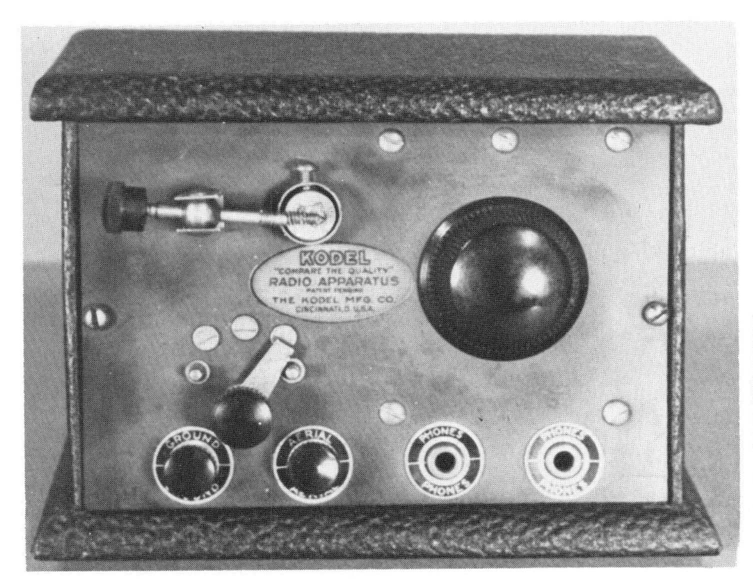

Fig. 69. Kodel Model S-1 (Kodel Mfg. Co.); 6 1/2″ x 4″ x 4 1/2″.

Fig. 70. Last Word Model B-12 (Kenrad Radio Corp.); 8 3/4″ x 5 1/4″ x 5 1/4″.

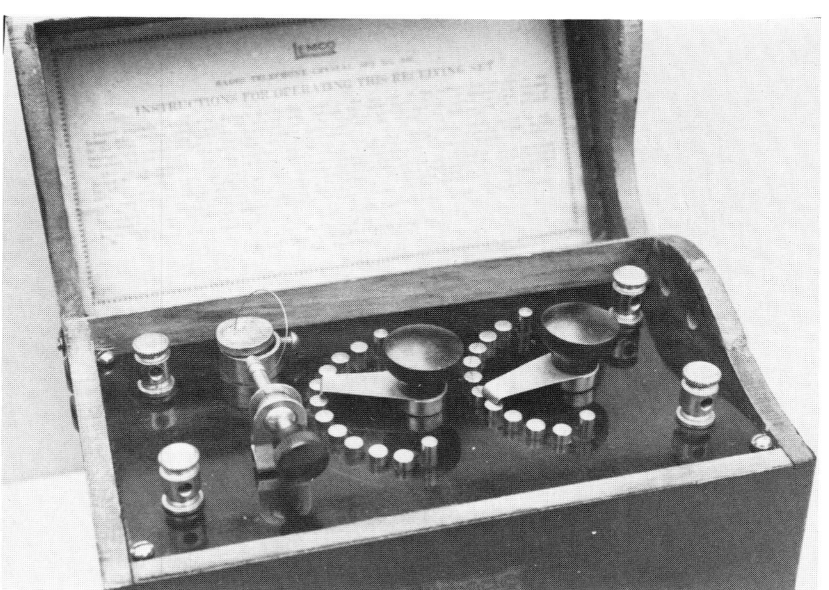

Figs. 71A and B. LEMCO No. 340, tap-switch version (Lee Elec. & Mfg. Co.); 6 5/8" x 3 3/4" x 5 1/4". Apparently, also called No. 340-B (see text). Fig. 71B, close-up of catwhisker and lid label.

Figs. 72A and B. LEMCO No. 340, tuning dial version (Lee Elec. & Mfg. Co.); 6 5/8" x 3 3/4" x 5 1/4". Apparently, also called No. 340-A. Fig. 72B, close-up of catwhisker and lid label.

Figs. 73A and B. LEMCO Portable (Lee Elec. & Mfg. Co.); 11″ x 7″ x 7″. Set came equipped with an R-W detector as shown. In Fig. 73A, detachable lid is removed (appears at top); in Fig. 73B, lid is closed. See Figs. 134, 281 (R-W detectors).

Fig. 74. Lawsam Baby (Lawsam Elec. Co.); 7″ x 4 7/8″ x 4 1/2″.

Fig. 75. (Genuine) Leon Lambert Long Distance (Lambert Radio Co., Leon); 6 1/4″ x 4 1/2″ x 4 1/2″. A later version has Fahnestock-clip terminals.

Figs. 76A and B. Lambert Long Distance (Lambert Radio Co., Leon); base = 4 1/2″ x 3 3/4″ x 1/2″, coil = 3 1/2″ diam., 3″ ht. Both Lambert crystal sets (Fig. 76B) have similar basic components as discussed in text.

Figs. 77A and B. Le Roy (Le Roy Elec. Co.); 6″ x 5 3/8″ x 5 3/8″. Set came equipped as shown with a "Steinite Rectifier" (Fig. 77B), a detector marketed in 1925 (the 1922 date shown on tag attached to panel of set is incorrect).

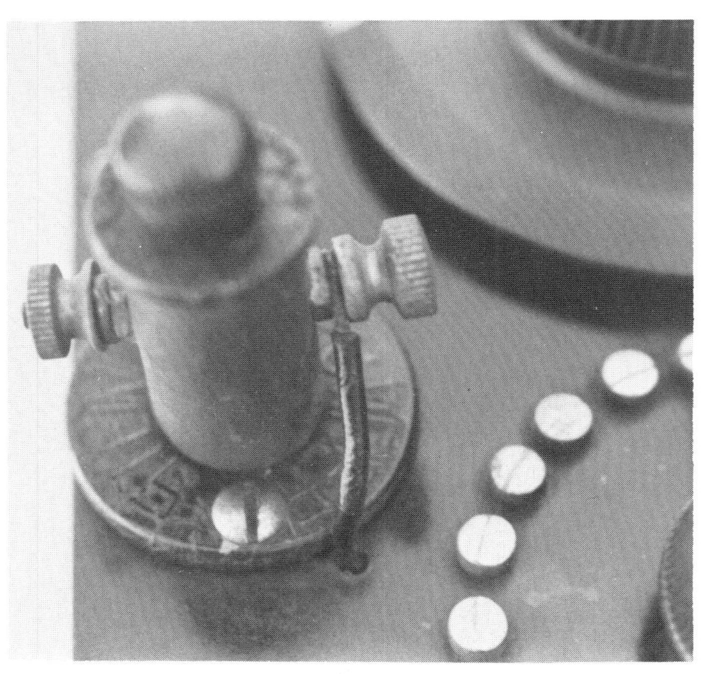

Fig. 78. Liberty Jr., Freedom of the Air (Parker Radio Co.); base = 5 1/2″ diam., set ht. = 7 1/2″.

Fig. 79. The Little Giant Pocket Radio (Little Giant Radio Co.); diam. = 3 1/2″, ht. = 2″. Rear view (right) shows built-in phone.

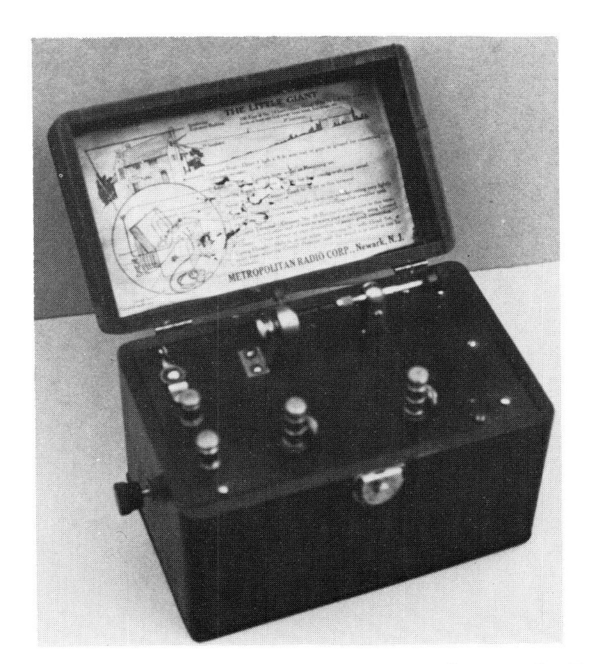

Figs. 80A and B. The Little Giant (Metropolitan Radio Corp.); 6 1/2″ x 4″ x 5 1/4″. Fig 80B, matching Metropolitan Antenna Tuner for use with The Little Giant crystal set.

Fig. 81. Louisville Electric (Louisville Elec. Mfg. Co.); 7 1/4″ x 4″ x 6″. Al Reymann collection and photo.

Fig. 82. Martian Big 4 (Martian Mfg. Co.; White Mfg. Co.); tripod base = 5″ x 5″ x 5″, tripod ht. = 4 3/8″, coil = 2 7/8″ diam., 3 1/4″ ht.

Fig. 83. Little Gem (United Specialties Co./U. S. Co.; Metro Electrical Co.); tripod base = 5″ x 5″ x 5″, tripod ht. = 4 1/8″, coil = 2 1/2″ diam., 2 7/8″ ht.

Fig. 84A and B. MacKay Marine Radio Receiver Type 123 BX (Federal Telephone & Radio Corp.); metal panel = 9 1/2″ x 6 1/2″ (no case). For shipboard use; also made for U. S. Maritime Commission. Detector on this set (Fig 84B) is very similar to the Puretone (Fig. 273). Ed Sharpe collection.

Figs. 85A and B. Martian Special (Martian Mfg. Co.); base = 3 7/8″ x 3 7/8″ x 5/8″, coil = 2 7/8″ diam., 3 1/4″ ht. In 85B, Martian Beauty one-tube set (base = 5 1/2″ x 5 1/2″ x 1 1/4″; coil = 3 3/8″ diam., 3 3/4″ ht.). It appears similar to the Martian Special crystal set except that the larger base is marbleized Bakelite instead of painted metal. (Fig. 85B, Les Rayner collection.)

Fig. 86. Marvel Model 101 (Freed-Eisemann Radio Co.; also called Marvel Radio Mfg. Co., Radio Mfg. Co., Radio Receptor Co.); 4 1/4″ x 4 1/4″ x 4 1/2″. Les Rayner collection.

Fig. 87. Maxitone No. 2 (Detroit Radio Co.); 9″ x 5″ x 5″. Al Reymann collection and photo.

Fig. 88. Maxitone No. 1 (Detroit Radio Co.); 5 1/4″ x 5 1/4″ x 5″. Metal tag shown on panel was originally on side of case (top side in photo). Al Reymann collection and photo.

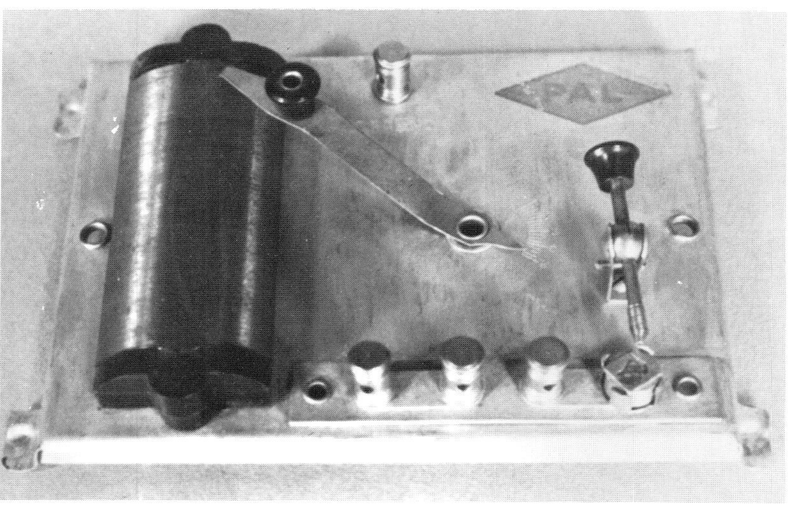

Fig. 89. Pal (Pal Radio Co.); base = 6″ x 5″ x 1/2″, coil = 3/4″ x 1/2″ x 5/8″. See Figs. 90A and B, and discussion in text.

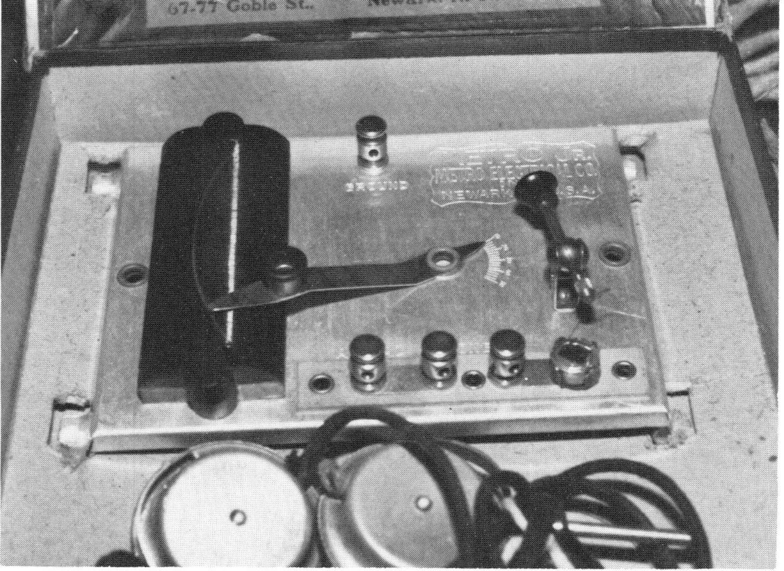

Figs. 90A and B. Metro Jr. (Metro Electrical Co.); base = 6″ x 5″ x 1/2″, coil = 3/4″ x 1/2″ x 5/8″. New in original box. See Fig. 89. Fig. 90B, Joe Knight collection.

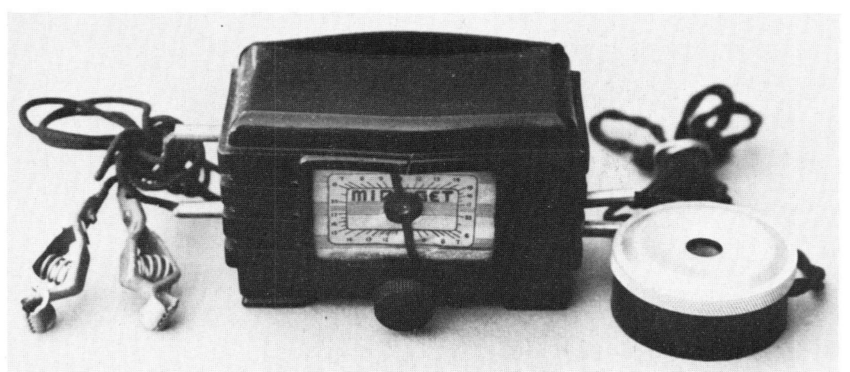

Figs. 91A, B, and C. Midget Pocket Radio, three versions (Midget Radio Co.); 3″ x 3″ x 1/2″. Other versions also exist. These crystal sets were first marketed in 1939. Fig. 91B, Ed Sharpe collection; Fig. 91C, Don Iverson collection.

Fig. 92. Miracle (Uncle Al's Radio Shop); 15 1/2″ x 7 3/4″ x 8″.

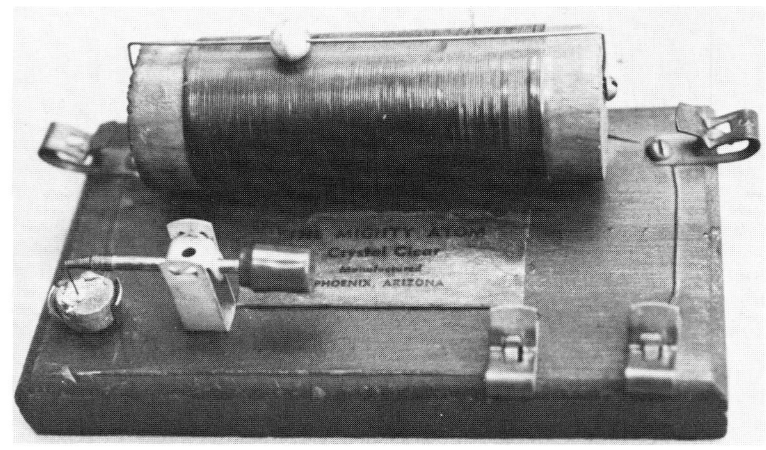

Figs. 93A and B. The Mighty Atom Crystal Clear (Furr Radio); base = 6″ x 3 1/2″ x 3/4″, coil = 1 3/8″ diam., 4″ length. Fig 93B shows later version label and detector. Fig. 93B, Ed Sharpe collection.

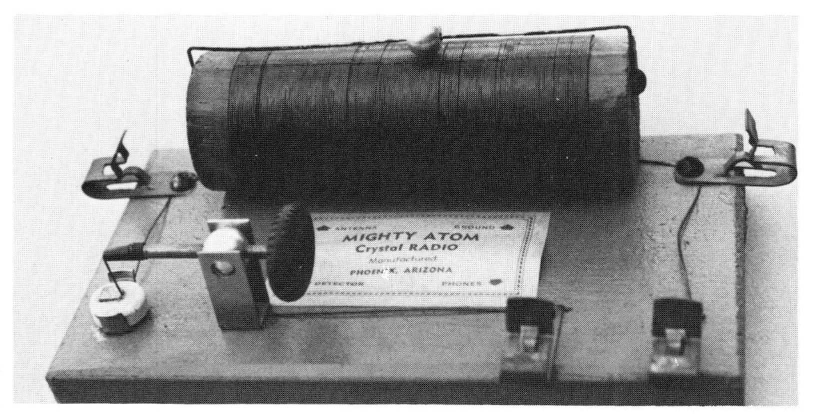

Fig. 94. Mohawk (Mohawk Battery & Radio Co.); 6″ x 5″ x 6″. Another version of the Mohawk has a slightly larger case and tap switches on panel. Al Reymann collection and photo.

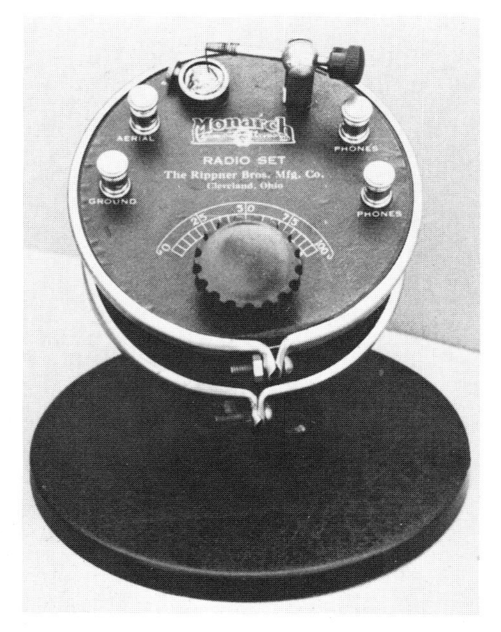

Fig. 95. Monarch (Rippner Bros. Mfg. Co.); base = 5 3/4″ diam., 3/8″ ht., support = 1″, case = 4 1/4″ diam., 1 7/8″ ht. Les Rayner collection.

Figs. 96A and B. Multiphone (Multiphone Co.); 6 1/4″ x 5″ x 1/2″; case = 8 5/8″ x 5 5/8″ x 3 5/8″. Ed Sharpe collection. See Fig. 200, "Lombard Multiphone." Both the Multiphone Co. and E. T. Lombard Co. were in Oakland CA.

Fig. 97. Musio (Musio Radio Co.); 8 1/4″ x 6 1/4″ x 6″. The knob at the right of the tuning dial controls the catwhisker for the galena detector located behind the panel.

Fig. 98. NEWCO (New Era Wireless Corp.); 9 1/2″ x 6 3/4″ x 8″. Al Reymann collection and photo.

Fig. 99. O-SO-EZ (unknown mfr.); base = 6″ x 3 1/4″ x 1/24″, coil = 1″ diam., 4 3/8″ length. Catwhisker on set is a replacement. Has unusual tuning device—a twisted, slightly spiralled rod ("eccentric") that permits variation of its point of contact with the coil by turning handles at ends.

Fig. 100. Palset (unknown mfr.); 6″ x 6″ x 6″.

Fig. 101. Pandora, also called "BMS" (Brooklyn Metal Stamping Corp.); base = 5″ diam., 7/8″ ht., coil = 2 3/4″ diam., 3″ ht. A somewhat different version also was available; see discussion in text.

Fig. 102. Parkin (Parkin Mfg. Co.); 4 1/2″ x 4 1/2″ x 2 3/4″. The catwhisker-galena detector, which is below the panel, is controlled by the knob that is above the tuning dial. Ed Sharpe collection.

Fig. 103. Peerless (unknown mfr.); 9 3/4″ x 5″ x 5 1/2″. Location of terminals (phone on left; antenna, ground on right) is reversed from the usual. Al Reymann collection and photo.

Fig. 104. Penberthy Model 4R (Penberthy Injector Co.; Penberthy Products); 5 3/4" x 3/4" x 5". Metal case. Terminals for two phone sets.

Fig. 105. Philmore, The Blackbird (Philmore Mfg. Co.); 6 1/8" x 6 1/8" x 6 1/8"; front ht. = 2 1/4". Earlier version on left.

Fig. 106. Philmore Little Wonder, early, metal case, two versions (Philmore Mfg. Co.); 3 1/2" x 3" x 5/8". Left set is earlier.

Figs. 107A and B. Philmore Little Wonder, later, Bakelite case (Philmore Mfg. Co.); 3 3/4″ x 3 1/4″ x 5/8″. In 107A, left set is earlier. One of the latest versions (Fig. 107B) has variegated coloring with white streaking ("marbling") of the light green case.

Fig. 108. Philmore Peerless (Philmore Mfg. Co.); 8″ x 6 7/8″ x 6 7/8″; slant front = 7 1/4″ length.

Fig. 109. Philmore Selective, early (Philmore Mfg. Co.); 8″ x 6 7/8″ x 6 7/8″; slant front = 7 1/4″ length.

Fig. 110. Philmore Selective, later (Philmore Mfg. Co.); 6″ x 5″ x 7″.

Fig. 111. By 1931, Philmore Mfg. Co. had marketed a "cathedral" radio in 1-, 2-, and 3-tube versions, all with the same metal cabinet (Philmore 2-tube radio, left in photo). By 1932, the case style of the Philmore Selective crystal set had changed from slant-front (Fig. 109) to "cathedral" (right, Fig. 111), similar to that of the Philmore tube radios (7 1/2″ x 5 1/2″ x 9″) but smaller (6″ x 5″ x 7″).

Fig. 112. Philmore Super Tone / Supertone, early, metal case (Philmore Mfg. Co.); 5 5/8″ x 4 1/8″ x 7/8″.

Fig. 113. Philmore Super, early, metal case (Philmore Mfg. Co.); 5 5/8″ x 4 1/8″ x 7/8″. Marketed later than set in Fig. 112.

Fig. 114. In 1934 or earlier, Philmore Mfg. Co. marketed a "Double Tuned Aerial Eliminator"; 5 5/8" x 4 1/8" x 5/8". Its metal case is the same in size and finish as the early Philmore Supertone crystal set (right), but the Aerial Eliminator functioned much better for vacuum-tube receivers than for crystal sets.

Fig. 115. Philmore Super / Little Giant, later, Bakelite case (Philmore Mfg. Co.); 5 7/8" x 4 1/4" x 7/8".

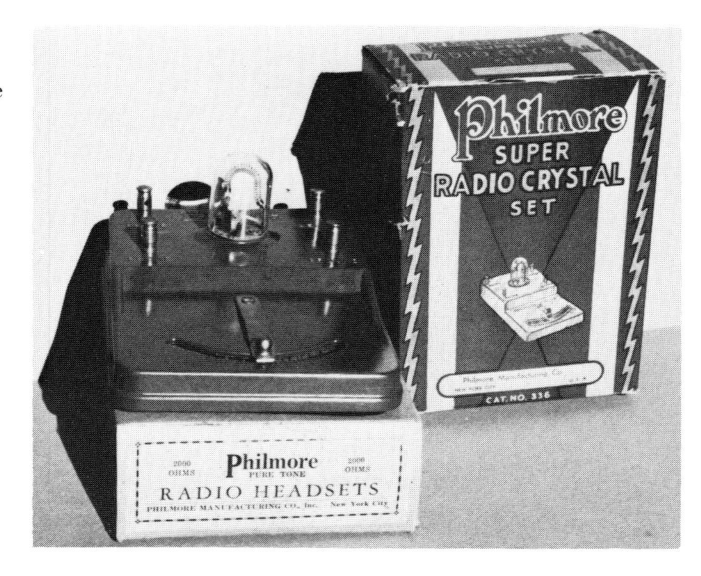

Figs. 116A and B. Philmore Super / Little Giant, later, Bakelite case (Philmore Mfg. Co.); 5 7/8" x 4 1/4" x 7/8". Marketed later than set in Fig. 115. An interesting color variation (Fig. 116B) has white streaking ("marbling") of a salmon-colored case.

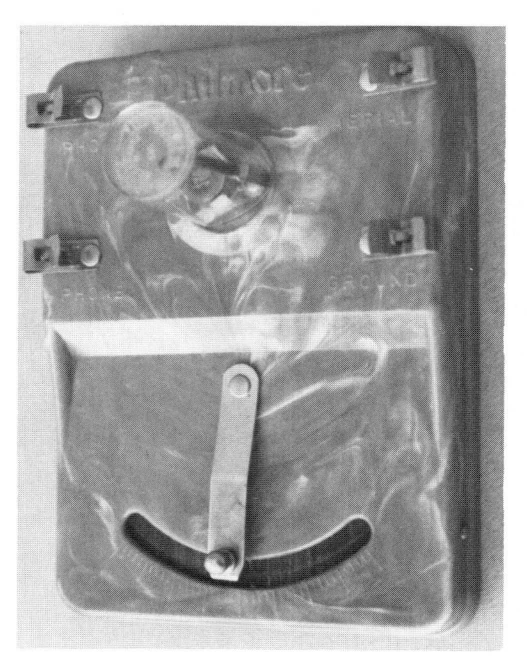

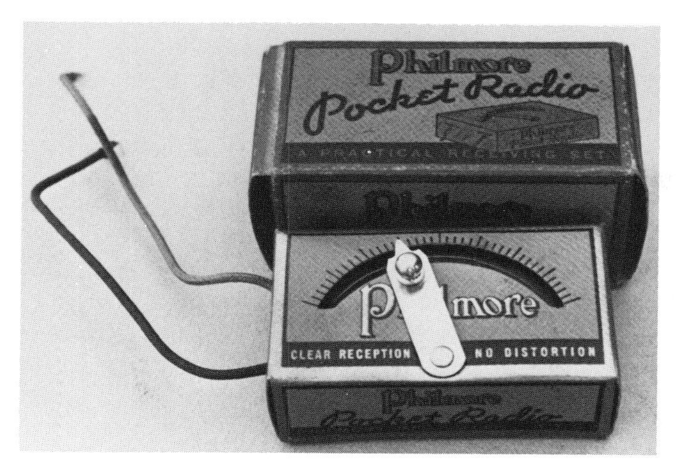

Fig. 117. Philmore Pocket Radio (Philmore Mfg. Co.); 3″ x 1 3/4″ x 1″. (Tiny Tone Radio [see Fig. 118] is a similar crystal set.) Ed Sharpe collection.

Fig. 118. Tiny Tone Pocket Radio (Tinytone Radio Co.); 3″ x 1 3/4″ x 1″. Dimensions of this set and the Philmore Pocket Radio (Fig. 115) are the same. Don Iverson collection. See Fig. 208 for a different, later "Tinytone" crystal receiver. The set in Fig. 118 is the first model marketed by this company (in 1932), and the set in Fig. 208 is the last (in 1959). Two other models (1937, 1951) of this manufacturer were also called "Tinytone."

Fig. 119. Pink-a-tone (Pinkerton Elec. Equipment Co.); 7 5/8″ x 7 1/2″ x 8″. Instruction sheet in foreground. Al Reymann collection and photo.

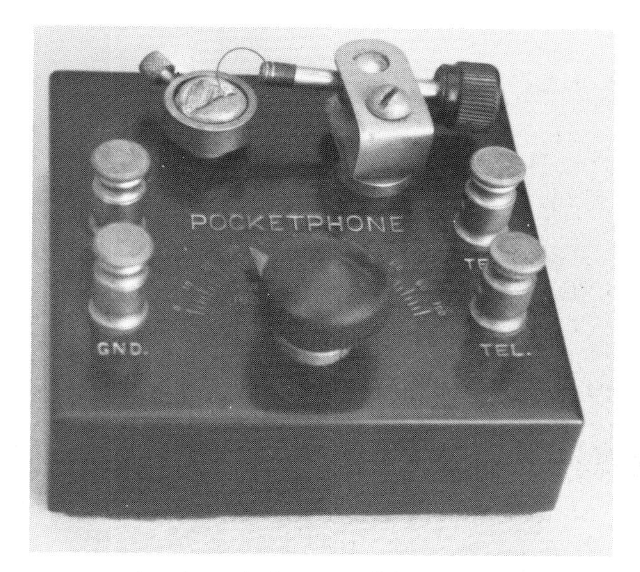

Fig. 120. Pocketphone (unknown mfr.); 3 1/4″ x 3 1/4″ x 1″.

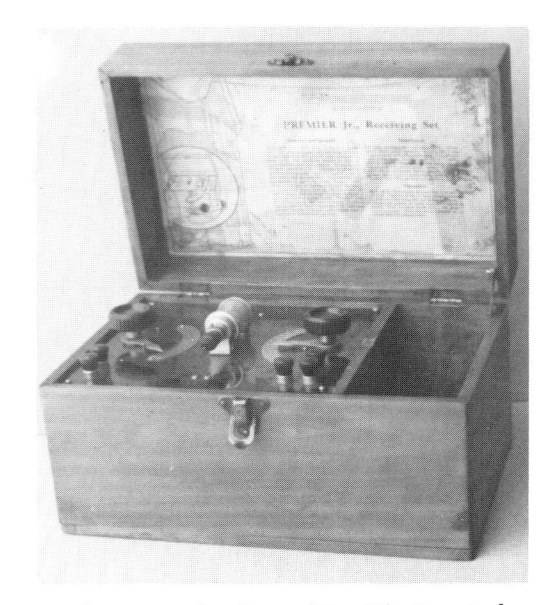

Figs. 121A and B. Premier Jr. (unknown mfr.); 12 1/4″ x 8″ x 7 3/8″. Many similarities to the Carter (Fig. 36), Cunningham (Figs. 41A and B), and DeForest Everyman DT-600 (Fig. 42). The Corbin lid latch of the Premier Jr. (Fig 121B), unlike that of the Carter or Cunningham, is not mounted upside down.

Fig. 122. Power-Tone (Power-Tone); breadboard = 17″ x 11 3/4″ x 1″. Al Reymann collection and photo.

Fig. 123. Poster showing the 1921 ad for Quaker Oats crystal set. Courtesy of Floyd Engels and Les Rayner.

Figs. 124A and B. Quaker Oats (Marquette Radio Corp.); 4 1/8″ diam., 7 1/4″ ht. Open detector-type set. Fig. 124B, close-up view of detector. Les Rayner collection.

Figs. 125A and B. Quaker Oats (Marquette Radio Corp.); 4 1/8″ diam., 7 1/4″ ht. Enclosed detector-type set. Fig. 125B, close-up view of detector.

Fig. 126. Radio Eng. Labs (Radio Engineering Labs); panel = 6 3/4″ x 5″ (no case). Al Reymann collection and photo.

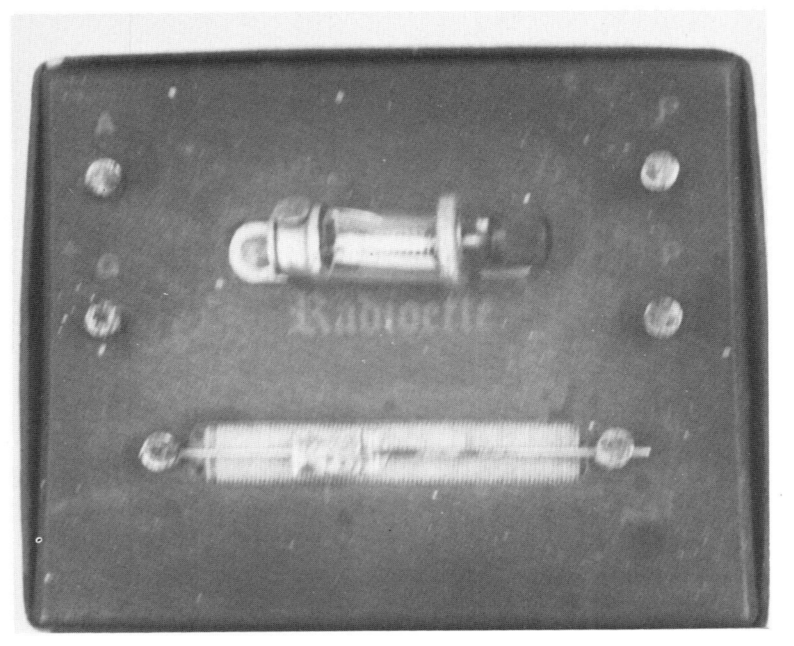
Fig. 127. Radioette (Radioette Mfg. Co.); 6 1/4″ x 5 1/4″ x 4 1/2″. Al Reymann collection and photo.

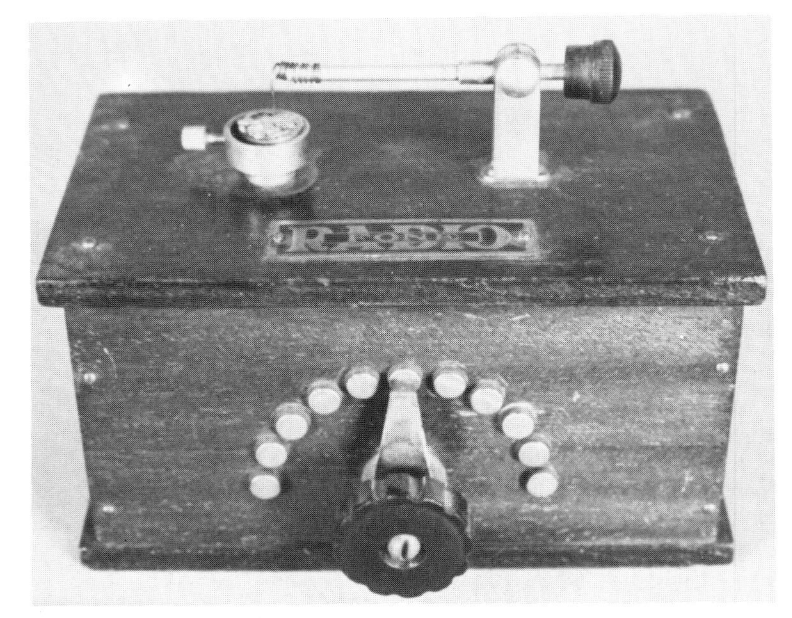

Fig. 128. Radiofone Type No. 1-B (Radiofone Corp.); 7 1/2″ x 2 7/8″ x 2 3/4″.

Fig. 129. Radio Service Type CR (Radio Service Co.); 5 3/4″ x 6 1/2″ x 7 1/4″. The enclosed vertical detector is missing from set except for upright support and enclosure base. Ed Sharpe collection.

Figs. 130A and B. RADISCO CRYS-TON Type R-300A (Radio Distributing Co.); 7 1/2" x 6 3/4" x 6 1/4".

Figs. 131A and B. RCA / General Electric Model AR-1300 (General Electric Co.; RCA); 11" x 5 1/2" x 9". A matching Detector-Amplifier, Model AA-1400 (Fig. 131B, right), was available. See discussion in text.

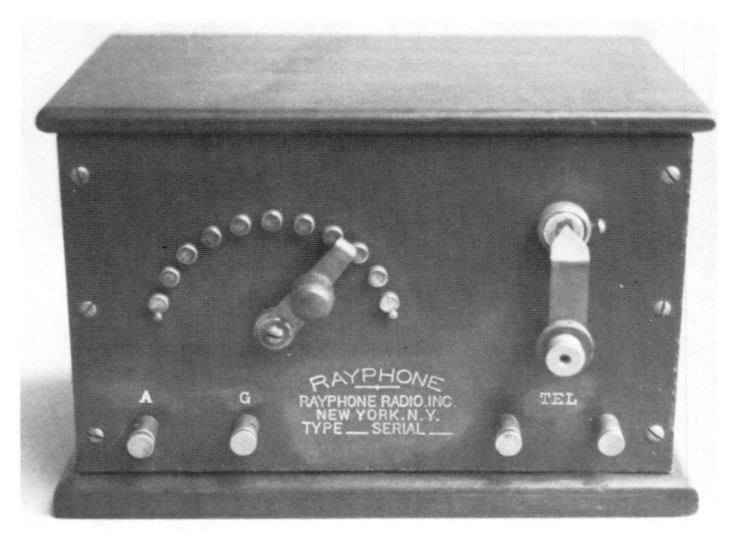

Fig. 132. Rayphone (Rayphone Radio); 8″ x 5 1/2″ x 5 3/8″.

Fig. 133. Roll-O Super Set, 5 in 1 Detector (Roll-O Radio Co.; Roll-O Radio Corp.); 6″ diam., 1″ ht. Ed Sharpe collection.

Fig. 134. R-W (R-W Mfg. Co.); 7″ x 3 3/4″ x 7 1/2″. Al Reymann collection and photo. See Figs 73A, 281.

Fig. 135. Shamrock Radiophone "Crystal" Model A (Shamrock Radiophone Sales Corp.; Mengel Co., Mfr.); 5 7/8″ x 3 7/8″ x 1 5/8″. Replacement catwhisker.

Fig. 136. Signal (Signal Elec. Mfg. Co.); base = 18″ x 4 3/4″ x 3/4″, outer coil = 4″ diam., 8″ length.

Fig. 137. Signal Jr. with Grewol detector (Signal Elec. Mfg. Co.); base = 12'' x 3 1/2'' x 1/2'', outer coil = 2 3/4'' diam., 3 1/4'' length.

Fig. 138. Simple-X (ARJO Radio Products Co.); 7 3/4″ x 5 3/8″ x 1 1/2″. Ed Sharpe collection.

Fig. 139. SPCO (Steel Products Corp. of California); 5″ x 4″ x 2″. Note compartment for single phone.

Figs. 140A and B. Spencer Pocket Radio (Spencer Radio Labs.); 4 7/8" x 3" x 1 1/8". See discussion in text.

Figs. 141A and B. Spencer Pocket Radio (Spencer Radio Labs.); 4" x 3 1/2" x 1 3/8".

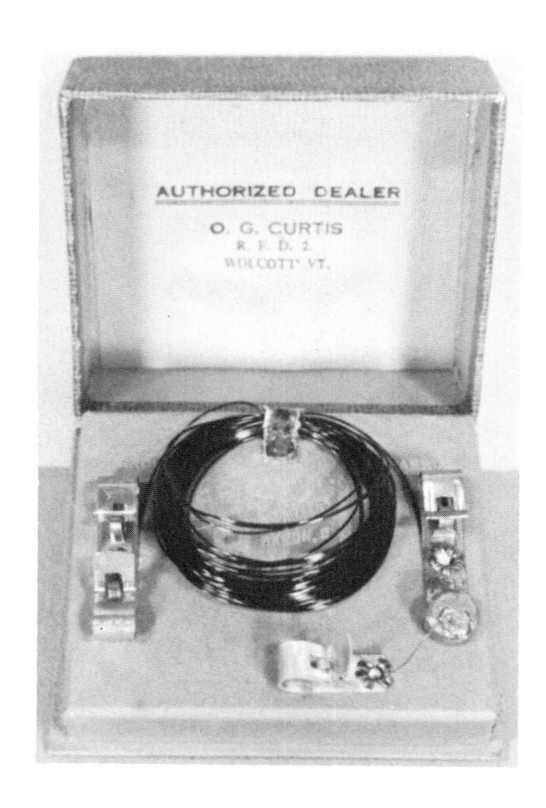

Fig. 142. Stanford (Stanford Elec. Co.); base = 4 1/4" diam., 1 1/8" ht. A variation from set shown has an open, horizontal detector. Al Reymann collection and photo.

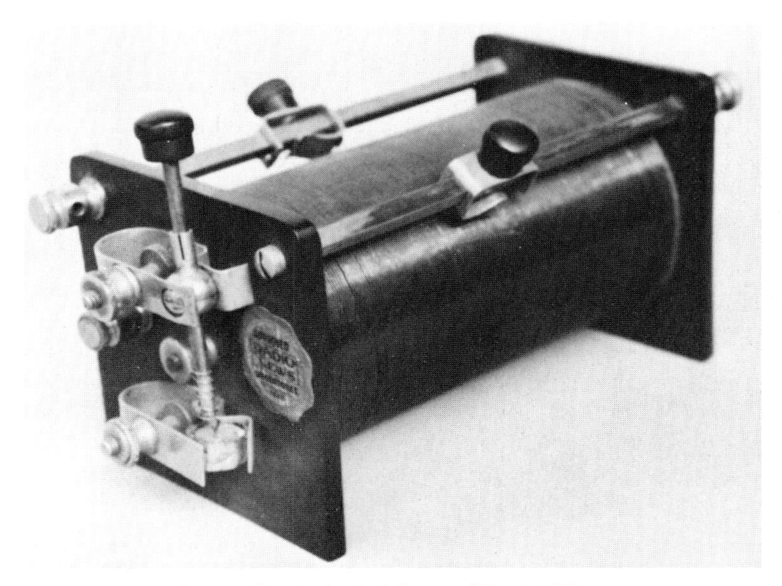

Fig. 143. Star (Star Mfg. Co.); 5 1/4″ x 3 1/8″ x 3 1/8″. Some sets were sold with Radio News "approved" seals as shown.

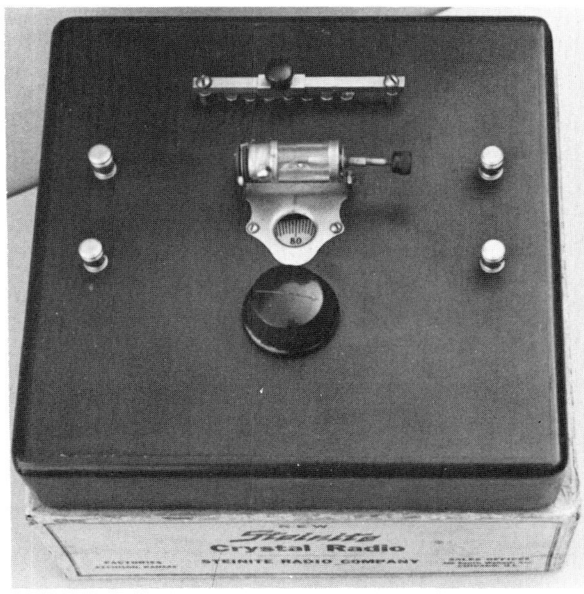

Fig. 144. Steinite, "New" (Steinite Labs.); 8″ x 8″ x 2 1/4″. Later than sets in Figs. 145A and B. Don Iverson collection.

Figs. 145A and B. Steinite, standard version (Steinite Labs.); both cases = 8 3/8″ x 7 1/2″ x 2 1/8″. Set in Fig. 145A is earlier than the one in Fig. 145B.

Fig. 146. Superior (Superior Mfg.); 6″ x 3 3/4″ x 1″. Al Reymann collection and photo.

Fig. 147. Sure-set Radio-de-tectaphone (Ward & Bonehill); 7″ x 4 5/8″ x 5″.

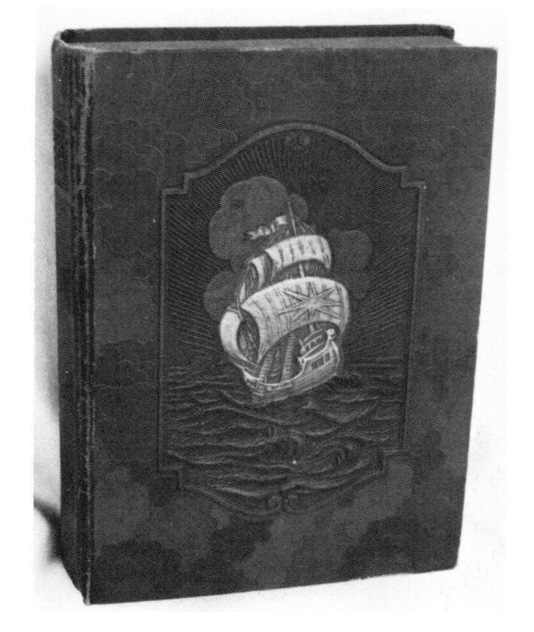

Figs. 148A and B. Teletone, Model C (Teletone Corp. of Am.); book type, 7 1/2″ x 5 1/2″ x 1 3/4″. Al Reymann collection and photo.

Fig. 149. Teleradio No. 1 (Teleradio Co.); 9 3/4″ x 7 1/8″ x 7 7/8″.

Fig. 150. TECLA "Thirty" (Clark Co., Thos. E.); 12″ x 7 1/4″ x 9 1/2″. Greg Farmer collection and photo.

Fig. 151. TESCO Type B (Eastern Specialty Co., The); 7″ x 5″ x 3 5/8″. Detector mounted in position for left-handed operator. Al Reymann collection and photo.

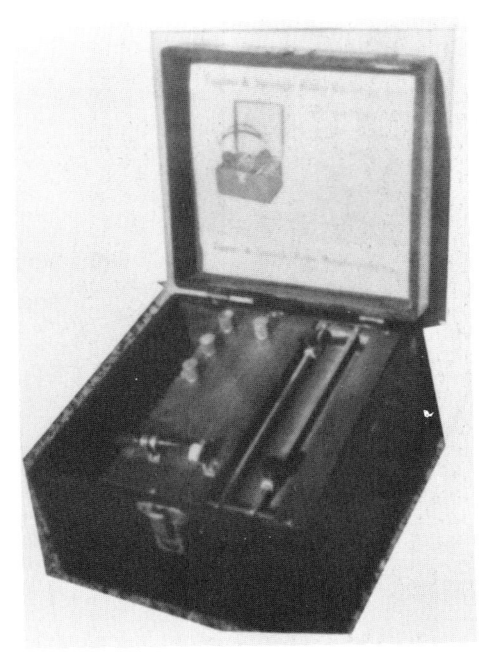

Fig. 152. Tippins & Sprengle (Tippins & Sprengle Radio Mfg. Co.); 9″ x 8 1/2″ x 6″. Al Reymann collection and photo.

Fig. 153. Tuska (Tuska Co., C. D.); 7 1/2″ x 5″ x 4 3/4″.

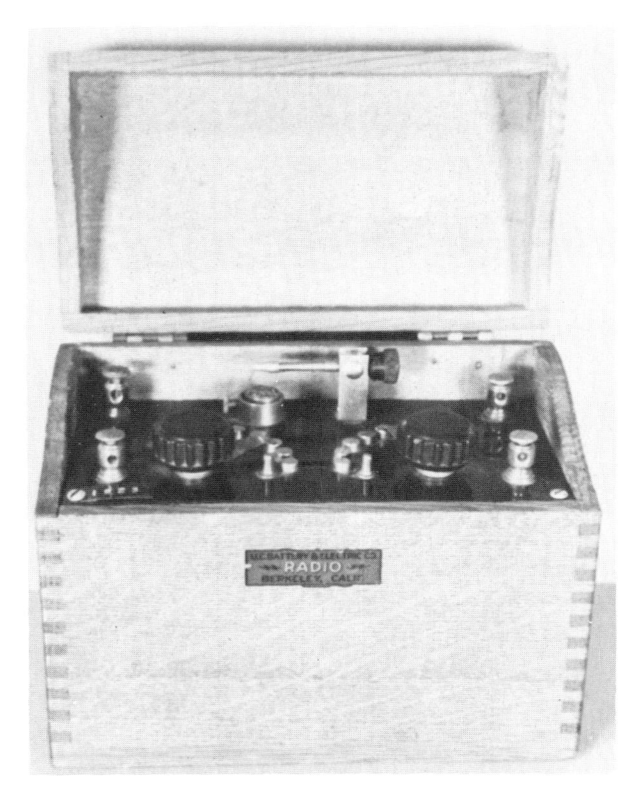

Fig. 154. U. C. (U. C. Battery & Elec. Co.); 6 5/8″ x 3 3/4″ x 5 3/8″.

Fig. 155. United Cabinet Crystal Unit (United Metal Stamping & Radio Co.); 5 1/2″ x 3″ x 4″. Al Reymann collection and photo.

Fig. 156. Unknown mfr., leatherette case; 5″ x 5″ x 2 1/2″.

Fig. 157. Unknown mfr., leatherette case; 7 3/4" x 5" x 5 1/2". Al Reymann collection and photo.

Fig. 158. Unknown mfr.; 8 1/8" x 5 1/2" x 5 1/2".

Fig. 159. Unknown mfr.; 7 1/4" x 4 1/2" x 4 7/8".

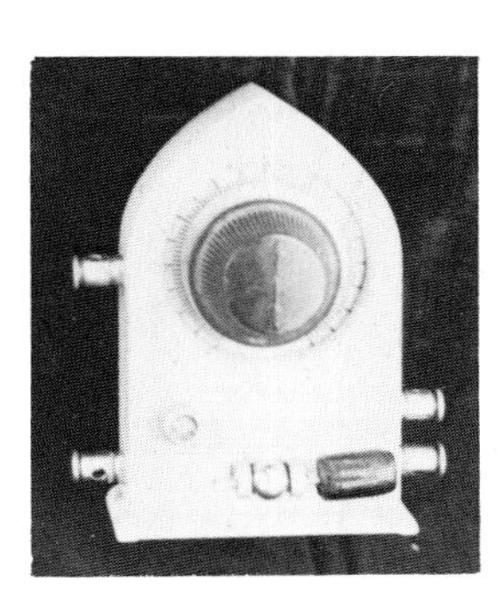

Fig. 160. Unknown mfr., white bakelite case; 2 1/4" x 1 1/2" x 3 1/2". Al Reymann collection and photo.

Fig. 161. Unknown mfr.; 6 3/8″ x 4 1/2″ x 3 3/8″.

Fig. 162. Unknown mfr.; 4 1/2″ diam., 1 1/4″ ht.

Fig. 163. Unknown mfr.; 10 1/2″ x 8″ x 5″.

Fig. 164. Unknown mfr.; 6″ x 6″ x 8″.

Fig. 165. Unknown mfr.; base = 3 7/8″ x 3 3/8″ x 1/2″, coil = 1 1/4″ diam., 2 5/8″ length.

Fig. 166. Unknown mfr., metal case (Howe Auto Products Co.?); base = 3 1/2″ x 3″ x 3/8″, mounted box = 1 15/16″ x 1 11/16″ x 1 1/2″.

Fig. 167. Unknown mfr.; 2 3/4″ diam., 4″ ht.

Fig. 168. Unknown mfr.; wood base = 6 1/2″ x 3 3/4″ x 1 1/4″, coil = 1 1/4″ diam., 4″ length. Probably only a few sets were made (in 1930's?) by a small enterprise; at least one other identical set is known to exist now.

Fig. 169. Unknown mfr. (Bowman Co., A. W. or homebrew with Bowman detector?); base = 11 3/4″ x 9 1/2″ x 3/4″, coil = 2 1/2″ diam., 4 1/4″ length.

Fig. 170. Homebrew with early DeForest D-101 detector; 10″ x 8″ x 7″.

Figs. 171A and B. Unknown mfr. with Ghegan detector (Bunnell & Co., J. H.?); 6 1/4″ x 3 1/4″ x 9 1/8″.

Fig. 172. Homebrew with Grewol detector; 10″ x 5 1/2″ x 5 1/2″.

Fig. 173. Unknown mfr. or homebrew with Merit detector; 6 1/2″ x 6 1/2″ x 3 1/2″.

Fig. 174. Unknown mfr. or homebrew with Schwartz Perikon detector; 7 1/8″ x 4 5/8″ x 4 5/8″; front ht. = 3 1/4″.

Fig. 175. Homebrew; 7″ x 4 1/2″ x 5″.

Fig. 176. Van Fixed (Van Valkenburg Co., J. D.); 6 1/2″ x 5″ x 1 1/8″.

Figs. 177A and B. Homebrew miniature treasure chest; 4 3/8″ x 2 3/4″ x 2 3/8″.

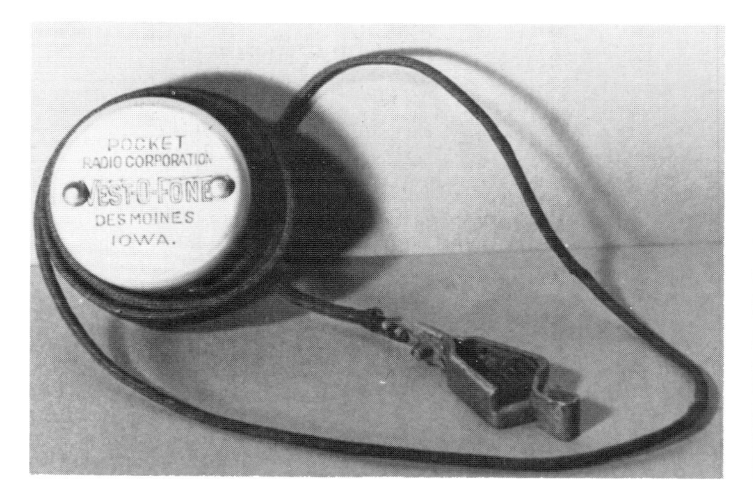

Figs. 178A and B. Vest-O-Fone (Pocket Radio Corp.); 2″ diam., 1 1/8″ ht. Fixed detector and driver inside headphone case.

Fig. 179. Victor (J. K. Corp.); 7 1/2″ x 6 1/2″ x 5 1/2″. The detector of this set is the same type as the Magnus crystal detector (Fig. 256). Al Reymann collection and photo.

Fig. 180. Volta (Volta Engineering Co.); 7 1/2″ x 6 1/2″ x 3 1/4″; front ht. = 1″.

Fig. 181. Wade (Wade-Twichell Co.); 7 1/2″ x 6 1/2″ x 3 1/2″. Al Reymann collection and photo.

Fig. 182. WEB (unknown mfr.); 3″ x 3″ x 2 1/2″.

Figs. 183A and B. Wecco Gem (Cheever Co., Wm. E.); base = 5″ x 3″ x 3/8″, coil = 1 3/4″ diam., 3 3/4″ length; sold with either binding posts (Fig. 183A, probably earlier) or Fahnestock-clip terminals (Fig. 183B).

Fig. 184. Western Coil Type A (Western Coil Co.); 8 1/2" x 7 1/4" x 7". Detector on set is Signal Standard Galena No. 41 (Fig. 284).

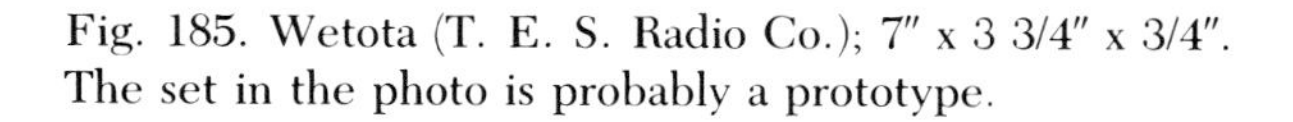
Fig. 185. Wetota (T. E. S. Radio Co.); 7″ x 3 3/4″ x 3/4″. The set in the photo is probably a prototype.

Fig. 186. Willemin 1100 (Willemin, E. A.); 7″ x 4 3/4″ x 3/4″, coil ht. = 1 1/4″. Al Reymann collection and photo.

Fig. 187. Wilson (Wilson, K. K.); 5″ x 2 3/4″ x 6″. Note that, unlike most sets, the detector handle is at the operator's left. Al Reymann collection and photo.

Fig. 188. Winchester (Winchester Repeating Arms Co.); 8″ x 4 1/4″ x 3/8″, coil housing = 3 5/16″ ht.

Fig. 189. Wireless Shop (Wireless Shop, The / Edgecomb, A. J.); 7″ x 7″ x 7″. Al Reymann collection and photo.

Fig. 190. World (Bethlehem Radio Corp.); base = 5″ x 4″ x 5/8″, coil housing = 2 1/4″ ht. See Fig 40.

More Recent (and Kit) Crystal Sets

Fig. 191. Air-Champ (Electronic Age Mfrs.); 7 7/8" x 7 7/8" x 1/8", coil = 1 1/8" diam., 3 7/8" length. In foreground, assembled set beside the instruction pamphlet.

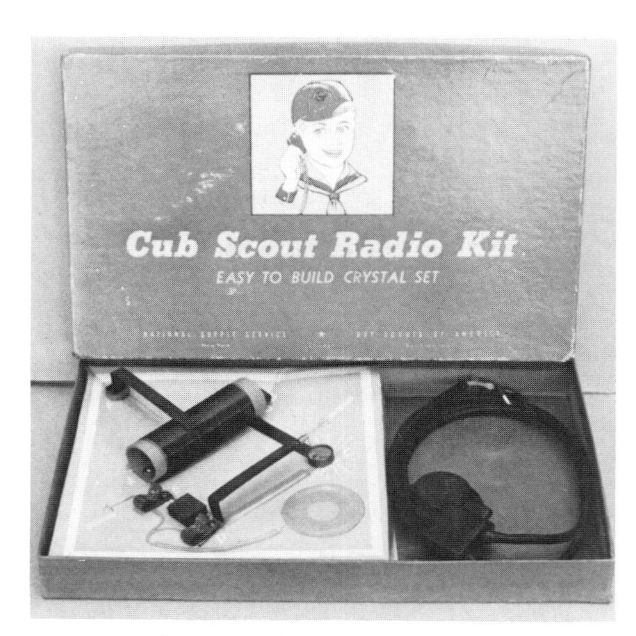

Fig. 192. Cub Scout Radio Kit (National Supply Service); base = 7 7/8" x 7 7/8" x 1/8", coil = 1 1/8" diam., 3 7/8" length.

Fig. 193. Comet (REMCO Sales); base = 5 1/2" x 4 1/4" x 3/4". Coil is wound around sides of base.

Fig. 194. Dank (Dank & Co., M. Carlton); breadboard = 10″ x 5″ x 1″, ht. = 5″. Al Reymann collection and photo.

Fig. 195. Distite No. 1 (Allen, Alva F.); 4 1/2″ x 3″ x 5/8″. Al Reymann collection and photo.

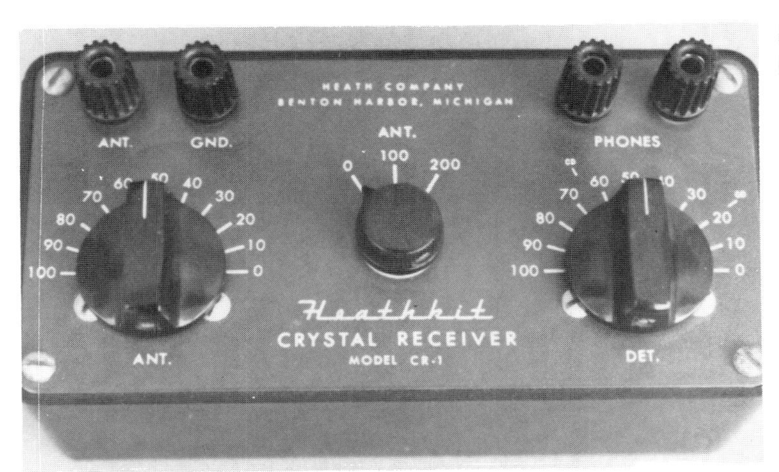

Fig. 196. Heathkit Model CR-1 (Heath Co.) 5 3/4″ x 3″ x 2″. A later version listed on the panel in a line between "HEATH COMPANY" and "BENTON HARBOR, MICHIGAN" the words "A SUBSIDIARY OF DAYSTROM, INC."

Fig. 197. Kitcraft Radio Crystal Set No. 102 (Kitcraft Products Co.); 3 3/16″ x 2 15/16″ x 5/8″.

Fig. 198. Kitcraft Ready Built No. 130 (Kitcraft Products Co.) 3 3/16″ x 2 15/16″ x 5/8″.

Fig. 199. Kitcraft Minature Tube Radio Model 200 (Kitcraft Products Co.); 3 3/16″ x 2 15/16″ x 5/8″. The case and many components of this one-tube radio are the same as those of the two Kitcraft crystal sets (Figs. 197, 198).

Fig. 200. Lombard Multiphone (Lombard Co., E. T.); 6 1/8″ x 4″ x 5 3/4″. See Figs. 96A and B for an earlier Multiphone set.

Fig. 201. Magna Crystal Radio (Magna Kit Co.); 7 3/8″ x 6″ x 2″.

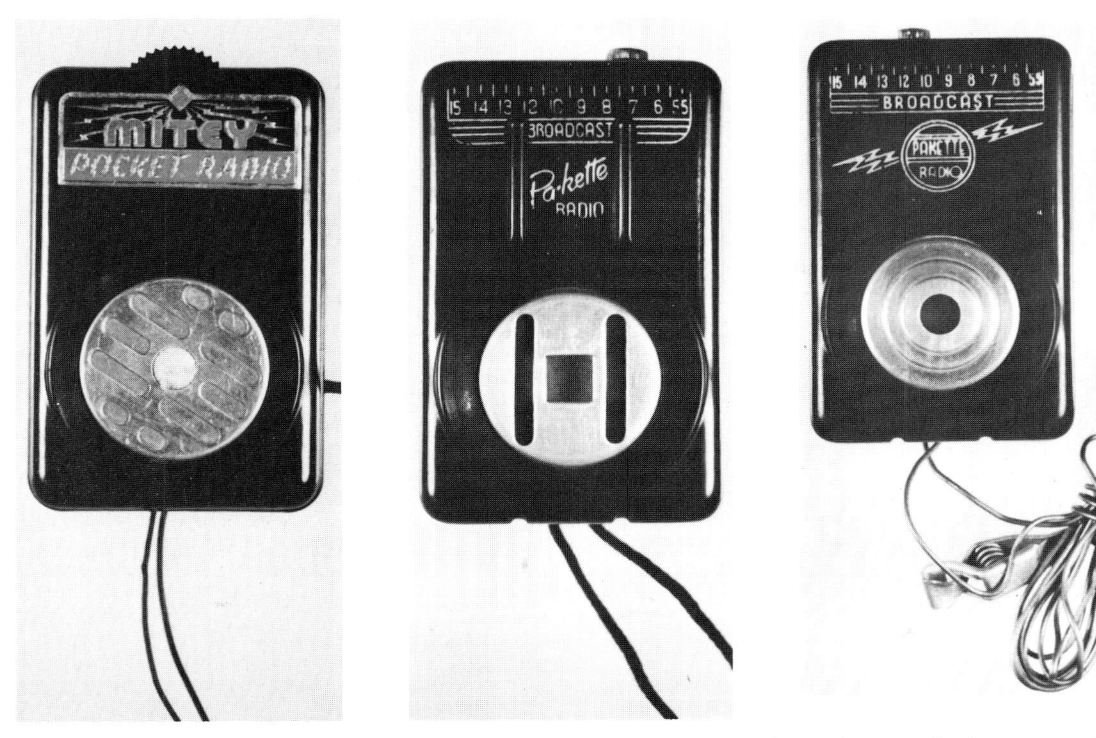

Fig. 202. Mitey Pocket Radio (Midway Co.); 3 1/8″ x 2″ x 1″. Mike and Jeremy Schiffer collection and photo.

Figs. 203A and B. Pa-Kette Radio, Pakette Radio (Pa-Kette Elec. Co.); 3 1/8″ x 2″ x 1″. Marketed earlier, the Pa- kette varies only slightly from the Pakette. Mike and Jeremy Schiffer collection and photos.

Fig. 204. Pee Wee Pocket Radio (Midway Co.); 3 1/8″ x 2″ x 1″. Ads for this midget crystal radio stated that it uses the "new" crystal diode. Only the horizontal version of the Pee Wee appears in all the old ads; surviving examples of the vertical version, which was marketed later, are rare.

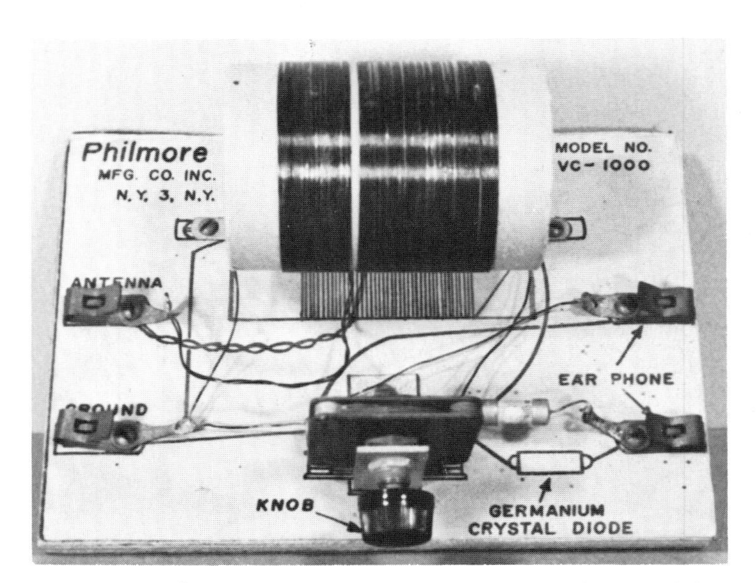

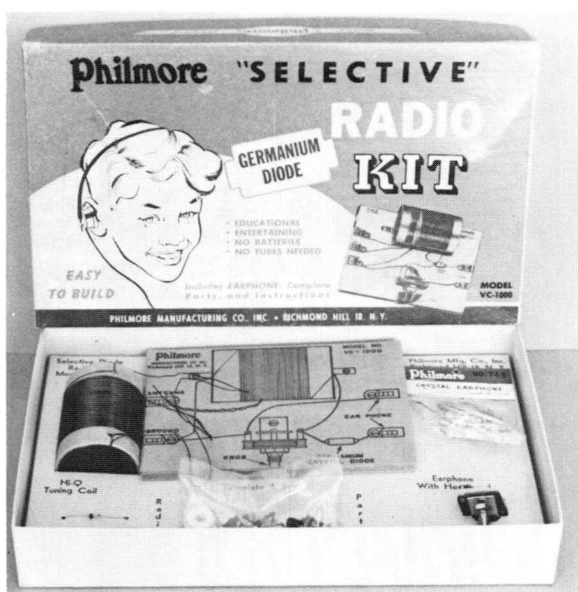

Figs. 205A and B. Philmore Selective No. VC-1000, kit set (Philmore Mfg. Co.); base = 6 3/8″ x 4 1/2″ x 3/8″, coil = 2″ diam., 2 3/4″ length. Fig. 205B, unassembled, is a later kit than one shown in Fig. 205A.

Fig. 206. Philmore Panel Crystal Set (Philmore Mfg. Co.); base = 3 3/4″ x 2 3/8″ x 3/8″, panel ht. = 4 1/2″; no case.

Fig. 207. Tee-Nie Ra-Di-O Model 250 (Carron Mfg. Co.); 5 3/4″ x 3″ x 4 3/4″.

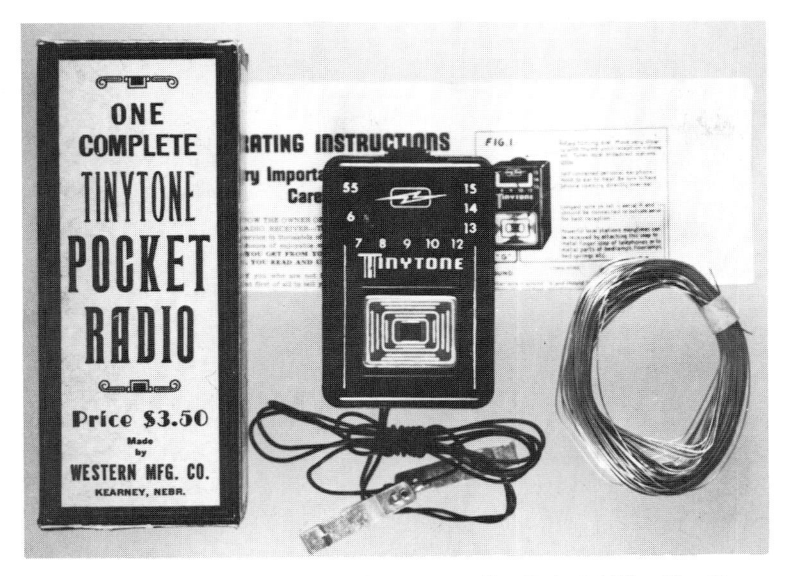

Fig. 208. Tinytone (Midway Co./Western Mfg. Co.); 3 1/8" x 2" x 1". This is the last of at least 12 models of crystal receivers marketed by this manufacturer between 1932 and 1960 (see Fig. 117 for the first marketed crystal set also called "Tiny Tone"). Mike and Jeremy Schiffer collection and photo.

Novelty: Porcelain "Receiving" Set

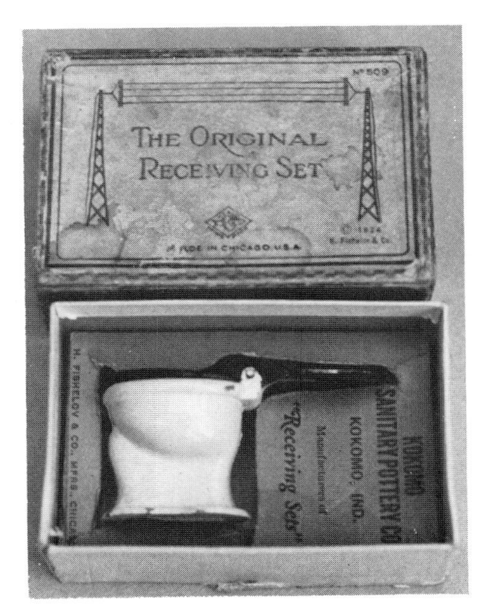

Fig. 209. "The Original Receiving Set" (Fishelov & Co., M., Chicago IL); box = 3" x 2" x 1". Promotional item for the Sanitary Pottery Co., 1924. The box of the 1925 version of this novelty item has the appearance of a 3-dial radio and is labelled "The World's Smallest Receiver."

Crystal Detector Differences and Details

Noteworthy differences or special features of some models of crystal detectors are described below.

AMRAD Duplex Type C; AMRAD Single Type C-1

As their names indicate, the AMRAD Duplex Type C has two detectors, and the AMRAD Single Type C-1 has but one. Only a few early American crystal detectors have two detector units; these include the Grebe RPDA (see below), MESCO Duplex, Wireless Specialty Apparatus (WSA) Type IP-202, and the Westinghouse DB (Figs. 321A and B). The WSA *Triple* Detector (Fig. 328), of course, has three detector units.

Carborundum: Fixed Detector; Adjustable Detector; Stabilizing Detector Unit

The cylindrical, fuse-type Carborundum detector is generally seen in its fixed version (Fig. 215), but an adjustable type (Figs. 216A and B) was also marketed. Both types were listed as "No. 30" by the Carborundum Co. The adjustable version, which appeared later, was called the "improved" type; surviving examples of it are rare. The Carborundum Fixed Detector is used in the Carborundum Stabilizing Detector Unit (Figs. 218A and B).

Clapp-Eastham Universal Ferron; J. J. Duck Ferron; Wm. B. Duck Type E Universal Ferron

The popular, marble-based Clapp-Eastham Universal Ferron detector was marketed in two styles (Figs. 240, 244). Appearing earlier (1913)

"AMRAD"
DETECTOR STANDS

Duplex stand, on Bakelite base, with two selecting switches and two binding posts—$4.50—shipping weight 1 lb.

Single stand, on Bakelite base, with two binding posts—$2.50—shipping weight ¾ lb. Your dealer should be able to supply "Amrad" Detector Stands—if not, send us his name and order direct.

Folders describing any pieces of "Amrad" apparatus will be sent on request—write for them.

AMERICAN RADIO AND RESEARCH CORPORATION

25 PARK ROW **NEW YORK**

Factory at Medford Hillside, Mass.

QST (Mar. 1920), outside back cover

Radio News (Mar. 1925), p. 1783

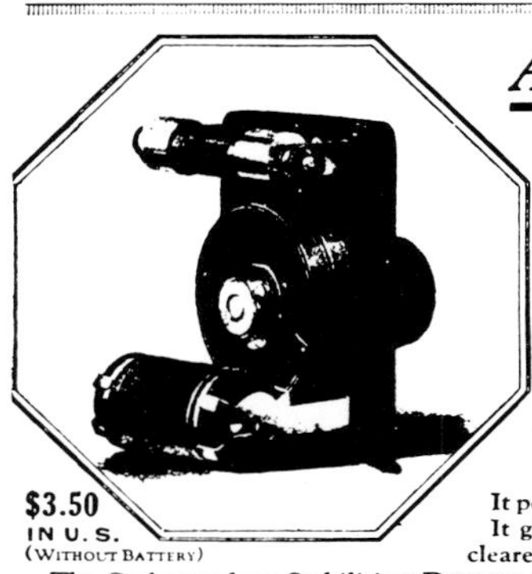

And Now—

The Carborundum Stabilizing Detector Unit

BUILT around the Carborundum Fixed Detector is this simple, highly efficient stabilizing device. By adjusting the detector resistance to match the circuit it absolutely controls self oscillation in the radio frequency tubes.
It permits operation at peak of regeneration.
It gives greater sensitivity—increased selectivity—clearer tones.

$3.50 IN U. S. (WITHOUT BATTERY)

The Carborundum Stabilizing Detector Unit gives a potentiometer controlled booster voltage to the Carborundum Fixed Detector.
A small sized flash-light battery is all it needs and of course it comes to you equipped with the genuine Carborundum Detector.

Send for Descriptive Circular Showing Hook-Ups. From your Dealer or Direct

MADE BY THE CARBORUNDUM COMPANY, NIAGARA FALLS, N.Y.
New York : Chicago : Boston : Philadelphia : Cleveland : Detroit : Cincinnati : Pittsburgh : Milwaukee : Grand Rapids

BANISHES DETECTOR TROUBLES

Radio News (Jan. 1926), p. 1056

was the vertical type (Figs. 244A, B, and C) with its double adjustment for point contact with the crystal. The later (1914), horizontal version (Figs. 240, 241) provided interchangeable metal-point or crystal-to-crystal (Perikon) contact. On both the horizontal and vertical versions, an identical metal name plate with the words "Ferron Detector" (plus the name and address of the manufacturer, patent dates) was mounted on the base. Both styles of this detector were sold by other companies, usually without the manufacturer being identified either in ads or on the detectors; the base name plate was often absent. In 1913, J. J. Duck Co. marketed the vertical version (Figs. 244B and C) bearing the "Ferron Detector" metal nameplate but with the three-line "J. J. DUCK/ ANYTHING WIRELESS / TOLEDO, OHIO" replacing the two-line "CLAPP-EASTHAM CO. / CAMBRIDGE, MASS." In 1914, the Wm. B. Duck Co., successor to the J. J. Duck Co., offered the horizontal Clapp-Eastham detector as the "Type E Universal Ferron." Nichols Electric Co. (1915), J. H. Bunnell & Co. (1916), and, undoubtedly, other major companies also sold the Clapp-Eastham Universal Ferron detector. Existing examples of this detector vary considerably in the hue and veining of their "Holland Blue" marble bases.

DB, Westinghouse

The Westinghouse Type DB detector (Figs. 321A and B) was the best-selling dual detector stand marketed. Both of its detectors (catwhisker contact on one; Perikon on other) are *horizontal* in the version usually seen. A rare version of the Westinghouse Type DB has two *vertical* detectors (both catwhisker contacts; see illustration in ad). The unsophisticated vertical version, which antedated the horizontal DB detector, apparently was marketed only briefly. Among the several existing *horizontal* DB detectors, collectors find some variations in catwhisker types, terminal labels, and word placement on the metal tags. (See "RCA," pp. 20–23.)

NEW UNIVERSAL FERRON DETECTOR.

The Ferron Detector is so well known among experimenters that extended comment on its merits is unnecessary. It has been listed in our catalog for five years. This crystal is one of the most sensitive yet discovered for the reception of wireless signals. The instrument is sold **with** and **without** the Ferron crystal. The crystals are carefully tested for sensitiveness. Any crystal may, with fusible alloy, listed elsewhere herein, be mounted in the cup. All parts of the metal in the cup may be reached by the contact point which is held by a ball and socket joint, providing a neat and convenient means of adjustment. Suitable pressure is given by a fine coiled spring, the tension of which may be varied by sliding the sleeve in and out of the ball. The base is of beautiful Holland Blue marble 3 x 5½ in., giving great stability. All metal parts are brass, nickel plated.

No. A150 Universal Detector Stand, complete with all parts shown in the illustration $3.25

Additional stationary cups to fit holder, each .... .15
Fusible metal, per ounce .... .30
Shipping weight, 3 lbs.

No. A6054 "Universal" Type Ferron Detector, complete with Ferron Crystal $5.00
Shipping weight, 3 lbs.

Wm. B. Duck Co., *Catalog No. 11* (1918), p. 39

Type DB Crystal Detector

The type DB Crystal Detector is provided with two crystals, either of which may be used by merely turning the two point dial switch to the desired position. The movable contact point is so secured that any part on the crystal can be reached. The metal parts of the Detector are mounted on a moulded black composition base in which is placed a mica stopping condenser. Three terminals are brought out for connections.

FIG. 6 – TYPE DB CRYSTAL DETECTOR

Crystal Detector

Style No.	Description	Dimensions Height	Width	Length
307216	Crystal Detector complete with two crystals	4¼″	5″	5″

(Order by Style Number)

Westinghouse Elec. & Mfg. Co., *Radio Circular 1641* (circa 1921), p. 4

This type "DB" detector differs vastly from the more commom DB. This version is both rare and early.

DeForest D-101, D-101-b, UD-100

The DeForest D-101, a well-known enclosed detector stand, appeared in two versions: an *early* style with a thumb-screw-adjustable clamp-holder for the crystal and metal binding posts of the retention type; and a *later* style with a winged, spring-metal crystal cup and with binding posts of the black molded type. The lower part of the bottom metal holder for the glass envelope is recessed in the later style but remains flush (not contoured) in the early style detector. A less common variation, Type D-101-b, differs from the *later* style, vertical D-101 only in the *horizontal* orientation of its enclosed detector—the type used on the DeForest Everyman crystal set (Fig. 42). In Type UD-100, the *early* style D-101 detector is mounted on a unit panel for use in interpanel sets. The BC-14A crystal receiver (Fig. 19), which was used by the Signal Corps of the U. S. Army during World War I, was equipped with the early-style D-101 detector (Fig. 224A, right). Because the name does not appear on the DeForest D-101, this detector is identified only by the characteristic appearance of its two versions (Fig. 224A).

TYPE D-101 WEATHER AND DUST PROOF

GALENA CRYSTAL DETECTOR

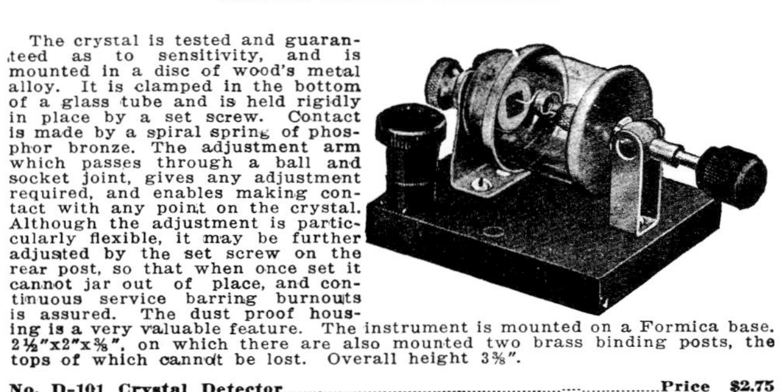

No. D-101 Crystal Detector.......... **Price $2.75**

Shipping Weight 1 lb.

Wm. B. Duck Co., (*Catalog No. 16* (1921), p. 150

Note that the ad description of retention-type binding posts is *incorrect* for the horizontal detector (applies only to the *early* vertical D-101 detector).

Duck, J. J. Ferron; Duck, Wm. B. Type E Universal Ferron (see *Clapp-Eastham Universal Ferron*)

E. I. Baby; RASCO Baby

Although very small crystal detectors were unusual before the 1920's, the E. I. Baby was marketed in 1915 and the RASCO Baby in 1921. These two miniature detectors (Figs. 230, 277, 331, 332) have the same type and sized base, and identical detector-arm supports. The right-angle

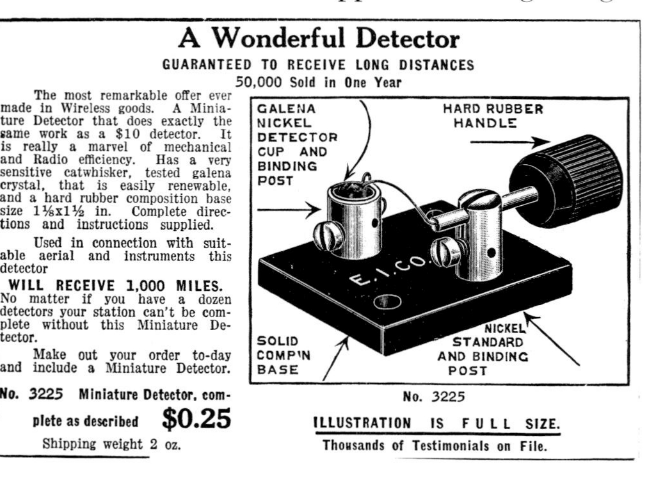

Electro Importing Co., *Catalog No. 16* (1916), p. 19

THE RASCO "BABY"

The Rasco "Baby"

Here it is boys! The smallest and most efficient detector in the world—as well as the cheapest. Our illustration is full size, and while the various details can be seen at a glance, we feel so enthusiastic about it that we must tell you all of its good points. First, there is a solid hard rubber composition base, size 1½" x 1⅛". We have not forgotten two holes to screw down the detector.

Then we have the nickle holder and binding post combined which holds the sliding, knurled, hard rubber composition knob. As you see, this knob not only revolves in its holder, but can also be moved back and forward in order to explore each point of the detector crystal.

Next we see the patent nickle detector cup and binding post combined. This is a little marvel all by itself and will not fail to avoke your admiration. No clamps, no soft metal to fuss with. You simply unscrew the knurled cap and insert your crystal into the stand, screw home the cap which leaves a goodly portion of the galena exposed. The contact is perfect, while the crystal can be exchanged quickly in less than three seconds. By slightly unscrewing the cap, the crystal can be changed in position, in order to explore other sensitive spots. The catwhisker is of phosphor bronze and is attached to the horizontal bar by means of a filister head screw. Can be readily exchanged in less than two seconds. Wires can be connected to the binding post in a jiffy. All metal parts are nickle plated, and you will be proud of this little masterpiece.

GALENA PATENT NICKEL DETECTOR CUP AND BINDING POST
SLIDING HARD RUBBER KNOB
NICKEL HOLDER AND BINDING POST
SOLID COMPOSITION BASE
RASCO

Illustration Full Size

No. 1898 Rasco Baby Detector complete with galena crystal, prepaid.............. **50c**

No. 1899, The same but furnished with an additional piece of **tested radiocite crystal, prepaid** **75c**

Ready for distribution June 30th.

Radio News (June 1921), p. 852

bend in the detector arm of one version of the E. I. Baby is rare among American detectors, as is the *laterally directed*, arched catwhisker of the RASCO Baby. These two detectors have crystal cups that differ in appearance, but both have setscrews to hold the mineral.

ERLA Fixed; ERLA Semi-Fixed; Scientific Fixed; Star Fixed

In its first version in 1923 or earlier, the ERLA fixed detector has no label or lid logo. This version appears to be identical to the unlabelled fixed detector marketed by Scientific Research Labs. in 1924 and 1925 (Fig. 282A). The list price of the Scientific ($1.50) was higher, however, than that of the ERLA ($1.00). By 1924, the fixed detector of the Electrical Research Labs. had acquired a distinctive ERLA logo on the top of its removable round metal cap, but the fiber base had no markings. Subsequently, the base was changed from fiber to Bakelite, and an ERLA logo in raised letters appeared on its underside. In addition, by then—and possibly earlier—this detector was sold in a prominently labelled, small matchbox-type carton (Fig. 236). The Star Fixed (Fig. 282B) is very similar to the ERLA Fixed detector. For a short time in 1924, the Electrical Research Labs. marketed the "ERLA Semi-fixed Crystal Rectifier" for panel mounting. It is tubular in form and has a crystal-to-crystal contact.

Electrical Research Laboratories
Dept. C **2515 Michigan Avenue, Chicago**

Present types of detectors are antiquated by the new Erla crystal rectifier. No adjustmentrequired. Proof against jolt or jar. List, $1

Radio News (Sep. 1923), p. 233

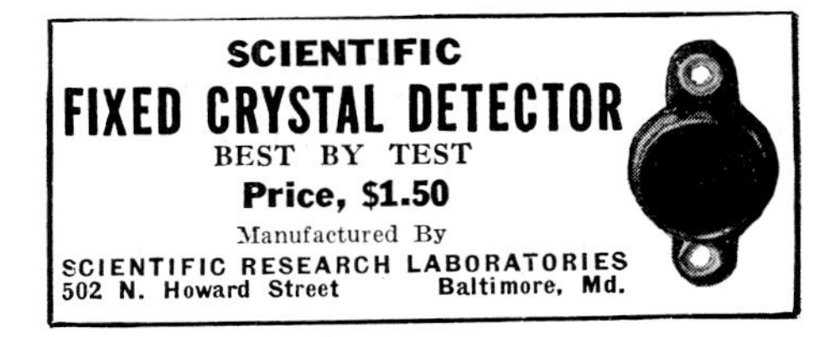

Radio News (Oct. 1924), p. 621

The Scientific Fixed appears identical to the ERLA fixed detector.

Electrical Research Laboratories
Dept. B **2515 Michigan Ave., Chicago**

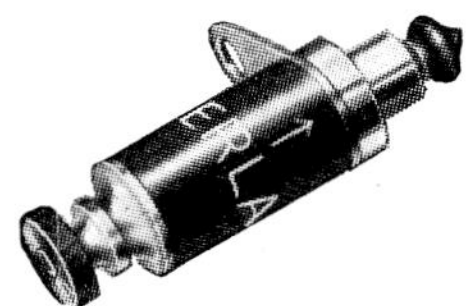

Erla semi-fixed crystal rectifier enables accurate initial adjustment to individual circuit characteristics. No attention required. List $1

Radio Broadcast (Mar. 1924), advertising section

--your confidence

This is my first ad in QST. I want you to buy your radio supplies from me. I want to be known from one end of this country to the other as the man who makes the highest quality radio instruments. I extend prompt and courteous service to each and every customer. Your confidence is the most important object of the policy of the FADA shop.

My Policy

To earn the confidence of my customers by earnestly striving to deserve it; to better my product; to improve my service; and to develop so far as I may, a spirit of friendly and helpful co-operation between the entire FADA organization and the people I serve.

My Equipment

A modern, clean shop in which is installed lathes, drill presses, milling machines, punch presses, saw table and engraving machine. An experimental laboratory and testing station. A high-speed shipping room.

FADA Crystal Detector as illustrated	$3.00
Radiotron UV-200, $5.00; UV-201	6.50
FADA Super-sensitive Galena Crystal	.50
No. 1 Chelsea condenser .001 mfd. mtd.	5.00
No. 2 Chelsea condenser .006 mfd. mtd.	4.50
No. 3 Chelsea condenser .001 mfd. un-mtd.	4.75
No. 4 Chelsea condenser .0006 mfd. un-mtd.	4.25
FADA Control Panel with Auto-jacks	17.50
FADA Two-Stage Amplifier	50.00
FADA Detector and Two-Stage Amplifier	65.00

Watch for announcement of details of FADA instruments. Send me your order for supplies today.

FRANK A. D. ANDREA

1882-B Jerome Avenue, New York City

QST (May 1921), p. 110

FADA 101-A; FADA X101

One of the best-liked early detector stands was the vertical, enclosed FADA Type 101-A (Fig. 237A). A less common variation, sometimes called the X101, differs by having a horizontal orientation (Fig. 238). At first (in 1921), the vertical detectors had only "FADA" engraved on their bases; but by the date of marketing the horizontal X101 in 1922, the labelling had changed to an elliptical, stylized FADA logo positioned between the words "Trade Mark." Frank A. D. Andrea Co. not only sold the Type 101-A directly to customers, but also marketed it through numerous retail dealers. At least one mail-order retailer, the Barawik Co., sold this vertical detector stand without the distinctive engraved FADA logo on its base; identification of such detectors is based solely on appearance (Fig. 237A, right).

Frank A. D. Andrea
1581-E Jerome Ave. New York City

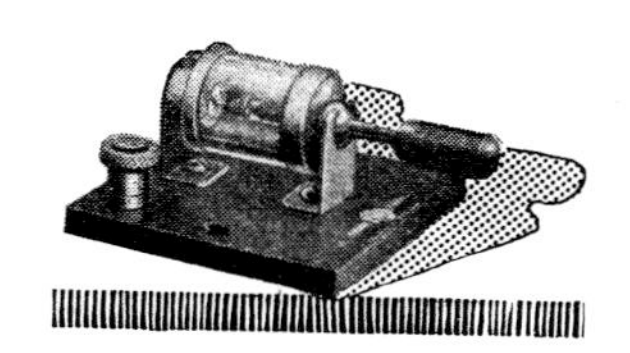

QST (Oct. 1922), p. 3

Grebe RPDA; Grebe RPDB

In 1919, A. H. Grebe & Co. marketed its RPDA "duplex" and RPDB "single" (Fig. 245) crystal detectors. Except for having either one (RPDB) or two (RPDA) detectors, they are basically the same. They differ from other detector stands in that the enclosed detector of both is mounted on a vertical panel instead of being either attached directly to the horizontal base or held by vertical supports.

No. RPDB. Grebe modern detector stand (single) 2.75
Shipping weight 1 pound.

No. RPDA. Grebe modern detector stand (duplex) $9.00
Shipping weight 2 pounds.

F. D. Pitts Co., *Radio Apparatus* Catalog No. 21 (1921), p. 7

Jove; Jumbo Jove

The Jove detector (Fig. 249) has a molded base, while the Jumbo Jove has a marble base. Two styles of catwhiskers (or "contact springs," as they were called in ads) were provided with most Jove detectors. One is a straight spring wire extending horizontally to a position above the crystal holder where it bends downward at a right angle; the other varies only in that it has a spiral coil about midway in the horizontal portion. A highly maneuverable (Ghegan patent) mechanism for the detector arm was used not only on the Jove and the Jumbo Jove, but also on other detectors marketed by J. H. Bunnell & Co. (Figs. 171B, 247).

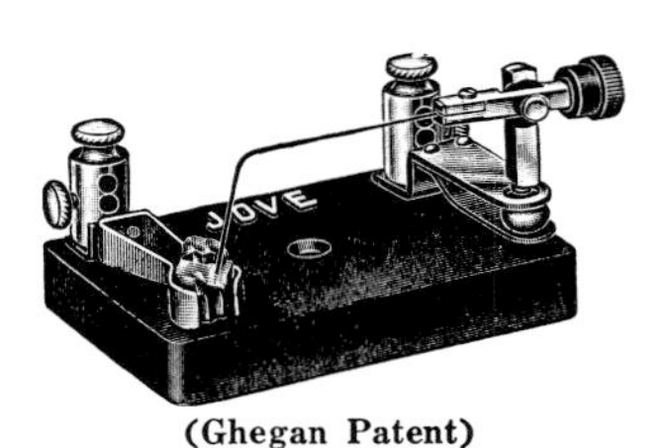

(Ghegan Patent)

BUNNELL

INSTRUMENTS
Always Reliable

Jove Detector

Handiest, Handsomest, Best
Sample by Mail, $1.80
Tested Galena Crystal, 25c

JUMBO-JOVE CRYSTAL DETECTOR HOLDER

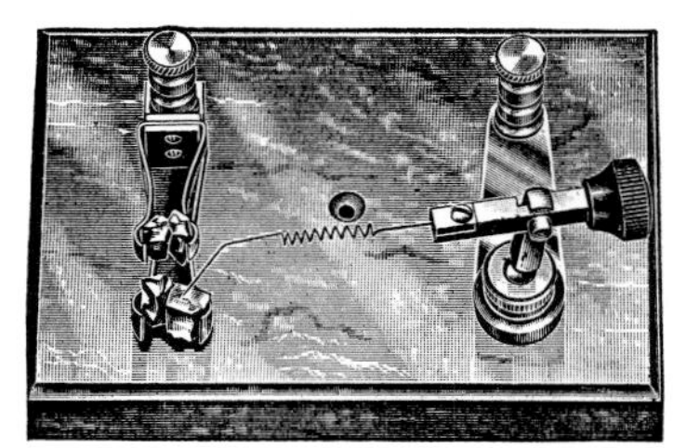

This Holder is mounted on a polished Marble Base, measuring 6 x 4 x ¾ inches. It has Double Binding Posts and a Double Clip for holding crystals, whereby several minerals may be held at one time and comparative tests of their efficiency quickly made. Can be furnished on either Electric Blue or Pure White Marble Base.

It is very attractively designed and is by far not only the best but the handsomest Crystal Holder on the market.

Schedule BA

List No.	Price each
8850	$8.50

Postage weight, 3 pounds

No. 8850

National Radio Supply Co., *Catalogue No. 3* (1920), p. 48

Merit (see *Pacent 31*)

Murdock No. 324; Murdock Silicon

The small Murdock No. 324 detector appeared in two versions. In its early (1916) style, a tapered, spring-coil catwhisker extends horizontally from a vertical adjustable support to a point above the crystal holder; from there, a straight portion of the catwhisker extends downward to establish contact with the crystal (similar to the Philmore catwhisker). This mechanism was replaced in 1921 by a horizontal arm for the smaller catwhisker. The improved maneuverability of the later version (Fig. 261) permits "universal adjustment."

Several early radio supply catalogs, such as those of Julius Andrae & Sons; J. H. Bunnell & Co.; Montgomery Ward & Co.; National Radio Supply; F. D. Pitts; Sears, Roebuck & Co.; and W. M. Welch Scientific Co., listed the Murdock No. 324. In some catalogs and early ads, this detector was called "Chelsea" (Wm. J. Murdock Co. was located in *Chelsea* MA), "Junior," or "Midget" detector stand.

In 1913, prior to the appearance of the No. 324 detector, Murdock marketed its larger, more sophisticated Silicon detector (Fig. 262), available with or without a condenser in its base.

DETECTOR STAND No. 324
324 Detector stand only......$ .75
Catalog No. Price each

National Radio Supply Co., *Catalogue No. 3* (1920), p. 50

Early type of Murdock No. 324 detector.

No. 324 Murdock Crystal Detector Stand only **$1.00**

Julius Andrae & Sons Co., *Radio Supplies* Catalogue R-6 (1923), p. 57

Later type of Murdock No. 324 detector.

Pacent 30; Pacent 31; Merit

The petite Pacent No. 30 detector (Fig. 265), which appeared in 1922, differs in only a few regards from the later-marketed Pacent No. 31 (Fig. 266). They are the same size, and both have round Bakelite bases upon which very similar, raised-design Pacent logos appear directly across from the terminals. On the base of the earlier-marketed detector in raised letters, "NUMBER 30" is displayed to the left of the logo, and "DETECTOR STAND" is at the right. The top holder of the glass envelope of the No. 30 detector is made of Bakelite; a raised-design Pacent logo is displayed directly above the logo on the base. The top of

More New Pacent Radio Essentials

Pacent Crystal Detector

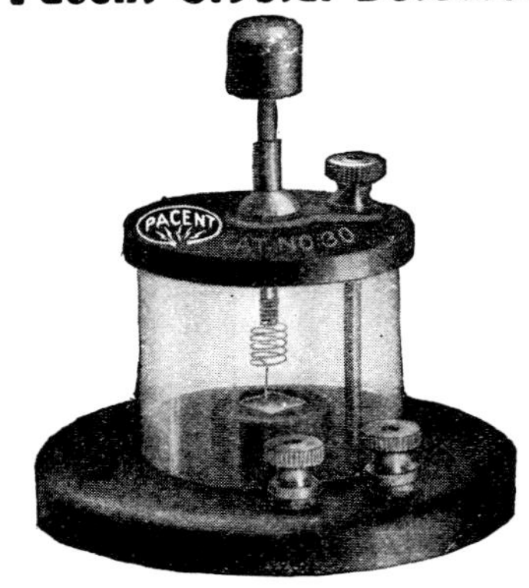

The Pacent Crystal Detector is perfect in construction and reasonable in price. It is dust-proof, rust-proof and fool-proof. Has moulded top and base with glass cover. Crystal instantly accessible. All parts heavily nickeled and highly polished. A truly beautiful instrument incorporating every feature a good detector should have. Pacent-made and Pacent-stamped.

Catalog No. 30A Crystal Detector. Price without crystal $1.50

Catalog No. 30 Crystal Detector. Price with crystal $1.75

Pacent Electric Company, Incorporated
LOUIS GERARD PACENT, PRESIDENT
150 Nassau Street New York City
MEMBER OF RADIO SECTION, ASSOCIATED MANUFACTURERS OF ELECTRICAL SUPPLIES

Radio News (June 1922), p. 1171

the No. 31 detector is nickel-plated metal; an engraved Pacent logo is located straight above the logo on the base. Inscribed on the top are: "No." at the right of the logo, "31" at its left, and "DETECTOR STAND" at the opposite side (directly above the terminals).

It is hard to convince some people that xtal detectors actually use cat whiskers so it is necessary to produce the proof that one manufacturer even goes so far as to number the cats. The illustration shows that this particular cat whisker came from "Cat No. 30."

QST (Feb. 1923), p. 68

The Merit crystal detector stand (Fig. 260) is identical to the Pacent No. 31 except that the Pacent logo is absent from its base (a faintly discernible circular imprint remains), and the word "MERIT" is inscribed on the top in place of the Pacent logo and "No. 31." The words "DETECTOR STAND" remain in the same location.

Why Pacent Electric Co. changed the name of the "No. 31" detector to "Merit" is unknown. Possibly it was produced for another firm. The Merit detector has been seen on at least one unidentified, presumably factory-made crystal receiver (Fig. 173) and on a few, obviously home-brew crystal sets.

Philmore: Open Type; Unmounted; Glass Enclosed Fixed; Fixed

The early Philmore detectors were marketed in colorful cardboard boxes. Collectors frequently find these detectors in the original boxes, which provide important information. The catalog number (Cat. No.) and company address on these cartons indicate the approximate vintage (see "Philmore Manufacturing Company"). A three-digit Cat. No. was used in early production, and a four-digit number was used later.

Until at least 1926, the early open detectors sold by Philmore had the trade name "Ajax." For the *unmounted* open detector, the Ajax box, either yellow (probably earlier) or blue, carries Cat. No. 101 as does the red-and-black box of the subsequent unmounted Philmore detector. Some of the earliest unmounted detectors are identified by Cat. No. 308 instead of 101. On the box for later unmounted Philmore detectors, Cat. No. 7010 is used. The *mounted* open Philmore/Ajax detectors first appeared with Cat. No. 310 on their maroon-and-green box (Fig. 268), and later had Cat. No. 7003 on the red-and-black box (Fig. 269).

The early "Glass Enclosed Fixed Detector," which is actually only *semi-fixed* (the type used on the Supertone crystal set), was assigned Cat. No. 309; the later version with a plastic enclosure has Cat. No. 7008 (Fig. 270). For its sealed "Fixed Detector," (Fig. 271) Philmore first used Cat. No. 100 on the maroon-and-green (then, red-and-black) box, and later, No. 7002 (red-and-black box).

RADIUM JEWELL DETECTOR

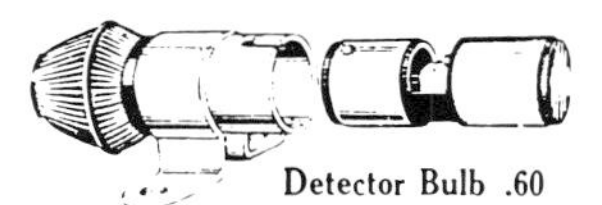

Detector Bulb .60
Bulb Holder .40

Requires no adjusting. Only occasional tuning, which can be done in one second.

No battery used. Can be applied to any receiving set. Takes the place of crystal detectors. No more fiddling around for sensitive spots; just pick up the phones and listen. Worth several times the price asked for this new invention. If your dealer can't supply, we will mail direct.

Dealers write for proposition.

Newark Electrical Supply Distributor for New Jersey

STAR MFG. CO., Newark, N. J.

Radio News (Mar. 1922), p. 882

STAR CRYSTAL DETECTOR

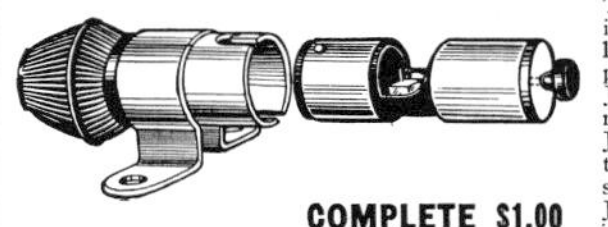

COMPLETE $1.00

This is a small high grade Detector. Adjustment is easily and quickly made. Movable parts are light and perfectly balanced. Does not jar from position. Fits the same socket as the Radium Jewell STAR. Comes complete with crystal mounted. You can change crystal. Star and Jewell make a wonderful combination. Use Star to locate party then slip in Jewell and you are set for the evening. Both same price. Star 60c, Jewel 60c, Holder 40c. Jewell was advertised in March issue.

If your dealer can't supply we will mail direct.

Newark Electrical Supply Co., Distributors for New Jersey

STAR MFG. CO., Newark, N. J.

Radio News (Apr.-May 1922), p. 1039

Notice the slight difference in these two detectors: the "Star" has a small adjustment knob on its left that the "Radium Jewell" does not have.

Radium Jewell; Star Semi-fixed

Throughout most of 1922, the name of the semi-fixed detector in all Star Manufacturing Company ads was "RADIUM JEWE*LL*" (double L). Ads appearing in December 1922 and during 1923, however, listed "RADIUM JEWE*L*" (single L). Although these later ads probably indicate a correction of spelling, the earlier "JEWELL" is used in the table of this book because of its repeated appearance in earliest ads. The manufacturer also marketed the STAR with a combination crystal, "set by turn of tube." This glass-tubular detector "fits the same socket as the Radium Jewel."

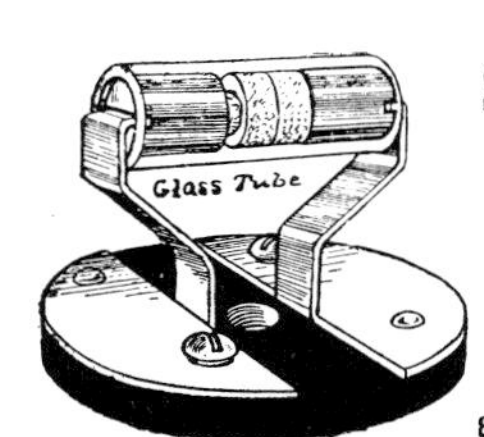

Star Detector

Combination Crystal
Set by turn of tube.
Volume and Distance.

List Price, 50c.

STAR MFG. CO.,
868 Bergen St., Newark, N. J.

Radio News (Dec. 1923), p. 800

This "Star" detector appears very different from the one with the same name shown in the *Radio News* ad from the April-May 1922 issue.

Big Improvements for Your Set At Small Cost

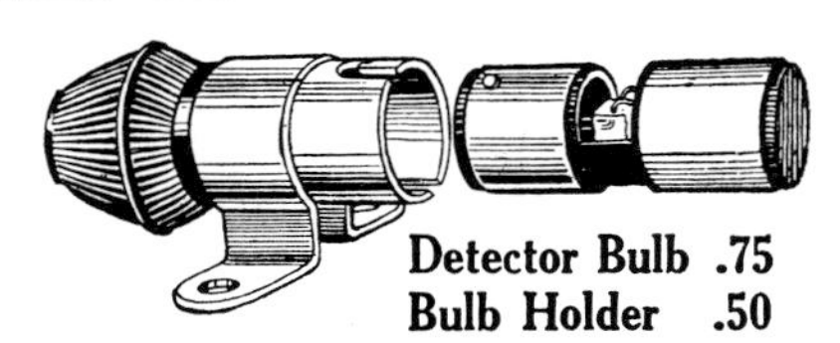

Detector Bulb .75
Bulb Holder .50

A New Kind of Detector

RADIUM JEWEL DETECTOR

IMPROVED crystal detector, yet it requires no battery and no very little adjusting. Extremely sensitive. Occasional tuning is the only care required.

Steps up the efficiency of your detector set several times. Saves fuss and inconvenience. Worth a great deal more to you but you can buy it from most any dealer complete with holder for $1.25.

AEROPLUG—The Socket Aerial Eliminates Lightning and Static

MAKES a powerful aerial of your electric lighting circuit, simply by screwing it into the ordinary lamp socket. Uses no current. NO danger. Can't get out of order. A good aerial in any kind of weather because lightning and static are absolutely banished. Made extraordinarily well.

Used for Crystal Sets Only When Close to Stations $1.50

If your dealer can't supply, we will mail direct

STAR MANUFACTURING CO. 868 BERGEN STREET NEWARK, N. J.

Radio News (Dec. 1922), p. 1196

Note change in spelling: "Jewell" (March, April-May 1922 ads) to "Jewel" (Dec. 1922 ad).

RASCO Baby (see *E. I. Baby*)

Scientific Fixed; Star Fixed (see *ERLA Fixed*)

Star (see *Radium Jewell*)

Unknown Manufacturers

A sizable number of commercial crystal detectors did not have labels. Some of these, such as the DeForest D-101, have distinct identifying characteristics. The identity of several others can be discerned from illustrations and descriptions in early ads or from the labelled original boxes in which some have been kept. Nonetheless, many factory-made detector stands remain unidentified (Figs. 287–318), even though, in some instances, several examples of identical or very similar detectors have been found. Such a detector is categorized as "unknown manufacturer" (and not included in the table of this book unless the trade name is known). Several detectors in this category are miniature or small inexpensive detectors; but they vary widely in size, type, quality, and apparent dates of manufacture. Many unidentified detectors have some fascinating feature such as a swirl-type, or clock-spring, catwhisker (Figs. 297–299).

Crystal Detectors and Crystals

Cat. No.	DESCRIPTION	Price
195	Crystal Detector for use with either mounted or unmounted Crystals	25c ea.
196	Telurium and Zincite Crystal Detector	25c each
197	Mounted Crystals "Galena", "Pyrite", "Silicon"	10c each
198	Genuine "B" Metal Crystals	25c each
199	Fixed Crystal Detectors	25c each
200	Mountings for Fixed Crystal Detectors	10c each

Cat. No. 195

Cat. No. 197

Cat. No. 198

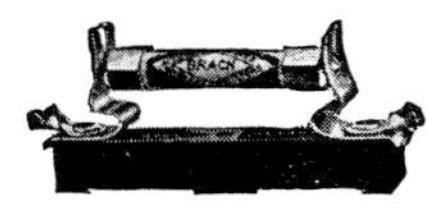

Cat. No. 199 & 200

Page 17

Kresge's *Radio Buyer's Catalog* (1925-1926), p. 17

Crystal Detectors

Dimensions for base or case = length x width x height (length is the larger and width, the smaller of the two horizontal measurements). Consult crystal detector table for additional details including date of manufacture. See text for discussion of some special features.

Fig. 210. Aerophone (Essex Specialty Co.); 3 1/4″ x 1 3/4″ x 1/2″. Al Reymann collection and photo.

Fig. 211. AMCO Paragon No. 45 (Adams-Morgan Co.); 2 1/2″ x 1 1/2″ x 1/2″.

Fig. 212. Bestone (Hyman & Co., Henry); 2 1/2″ x 2″ x 3/8″.

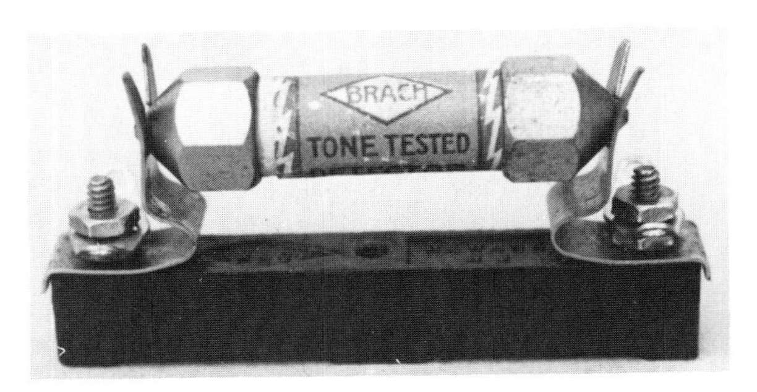

Fig. 213. Brach Fixed (Brach Mfg. Co., L. S.); base = 2 1/2″ x 5/16″ x 3/8″; detector = 1/2″ diam., 1 7/8″ length.

Fig. 214. Brownlie Vernier (Brownlie & Co., Roland); detector = 1/2″ diam., 1 3/8″ length (tubular portion).

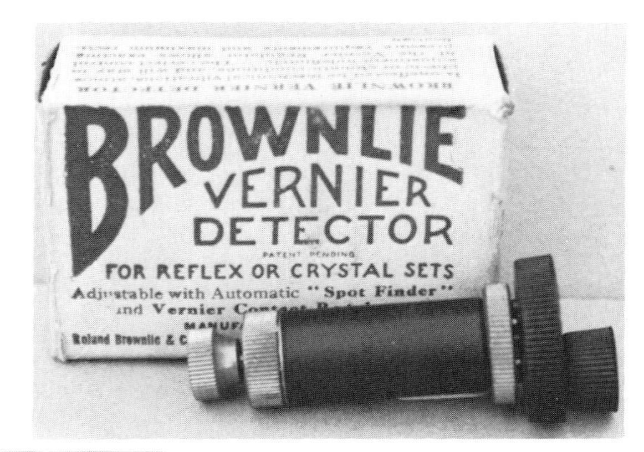

Fig. 215. Carborundum Detector No. 30 (Carborundum Co.); 9/16″ diam., 2″ length (tubular portion). See Figs. 216A and B.

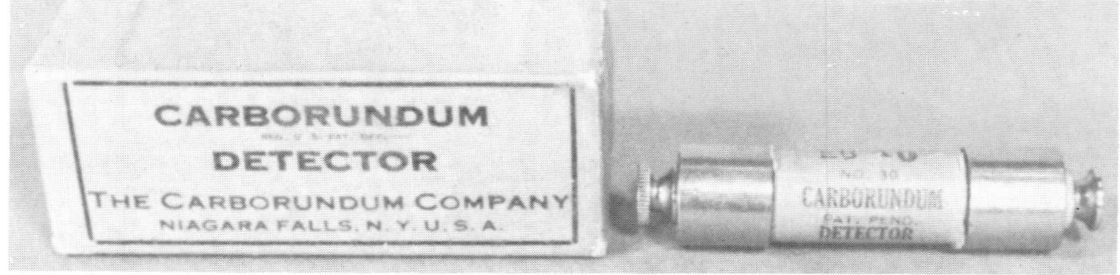

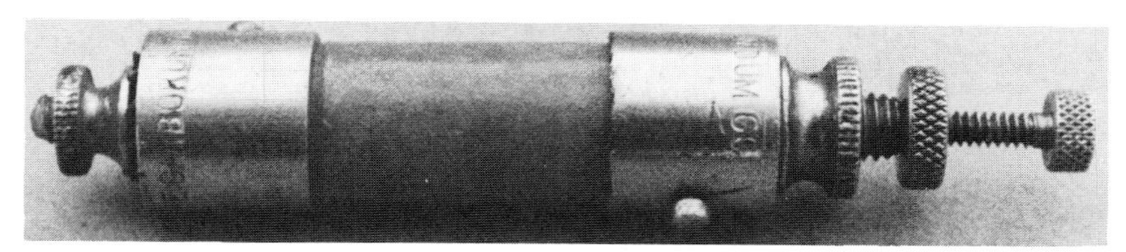

Figs. 216A and B. Carborundum Detector No. 30 Improved, Adjustable (Carborundum Co.); 9/16″ diam., 1 7/8″ length (tubular portion). Fig. 216B shows both the fixed (lower) and the adjustable (upper) Carborundum Detector No. 30 with original boxes and the instruction sheet for "the improved (adjustable) Carborundum Detector." (Fig. 216A, previously Frank Falkner collection.)

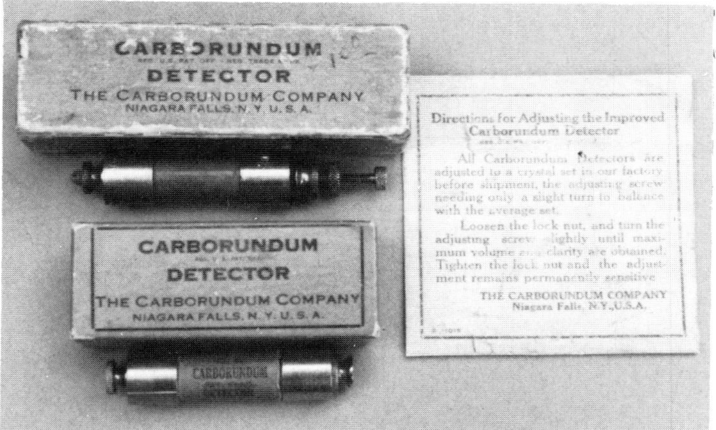

Fig. 217. Champion (Champion Radio Products Co.); 3 3/8″ x 2″ x 3/8″.

Figs. 218A and B. Carborundum Stabilizing Detector Unit (Carborundum Co.); 3 3/8″ x 2 7/16″ x 3/8″. Front and rear views of detector unit shown in Fig. 218B; battery ("C" cell) is absent from the holder (extending from sides of lower part of unit on left).

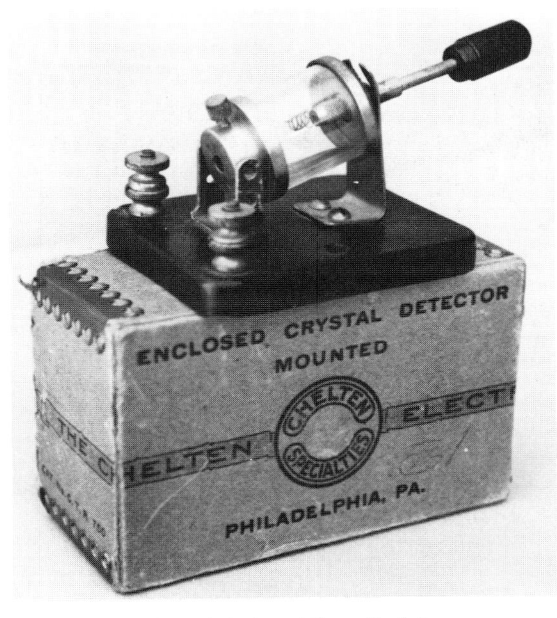

Fig. 219. Chelten (Chelten Elec. Co.); base = 2 5/8″ x 2 1/8″ x 5/16″, slant detector ht. = 1 1/4″ to 1 5/8″.

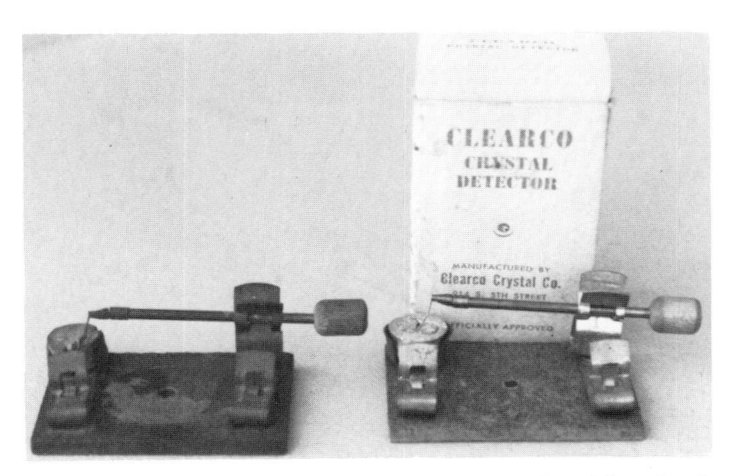

Fig. 220. Clearco (Clearco Crystal Co.); 2″ x 1 5/8″ x 1/8″. A decal label appears on the mid-portion of the base in the earlier version (left) but is not on the later detector, which was sold in a labelled box (right).

Fig. 221. Cleartone Radio Concert Model UD-1432 (RCA; Westinghouse Elec. & Mfg. Co.); 3 1/4″ x 2″ x 1/2″. See Figs. 222, 323-325.

Fig. 222. Crosley (Crosley Mfg. Co.; Wireless Specialty Apparatus Co.) 3 1/2″ x 2″ x 3/8″. See Figs. 323-325.

Fig. 223. Crystaloi Type O (Turney Co., Eugene T.); base = 2 1/4″ diam., 1/2″ thick (legs, 1/4″ ht.), detector = 1 1/4″ diam. Type O and A detectors probably are identical.

Figs. 224A and B. In A, DeForest D-101 (DeForest Radio Tel. & Tel. Co.); 2 1/2″ x 2″ x 3/8″. Earlier version on right. In B, DeForest D-101-b, horizontal detector; 2 1/2″ x 2″ x 3/8″. Unlike the vertical D-101 detector, the D-101-b has "DeForest" on its base in raised letters.

Figs. 225A and B. DeForest, Cam Adjustable (DeForest Radio Tel. & Tel. Co.); 2 1/4″ x 1 1/8″ x 1/4″. Cam shown in Fig. 225B. This detector was used in reflex radio circuits of DeForest D-7, D-7A, D-10, and D-12.

Figs. 226A and B. Doron Universal (Doron Bros. Electrical Co.); 3 1/2″ x 2″ x 5/8″. Views from both sides. May be used as either a catwhisker-type or Perikon detector. Don Patterson photo.

Fig. 227. Dubilier (Dubilier Condenser & Radio Corp.); 3 3/4″ x 1 1/8″ x 3/16″.

Fig. 228. Duck Type G (Duck Co., Wm. B.); 3 3/8″ x 2 1/8″ x 5/8″. Base is a replacement.

Fig. 229. Edgcomb-Pyle (Edgcomb-Pyle Wireless Mfg. Co.); 5″ x 3 3/4″ x 3/4″.

Fig. 230. E. I. Baby (Electro Importing Co.); 1 1/2″ x 1 1/8″ x 3/16″. The *usual version* of the E. I. Baby has a straight detector arm with laterally directed catwhisker, appearing identical to the RASCO Baby (Fig. 277)

Fig. 231. E. I. Bare Point Electrolytic (Electro Importing Co.); 3″ x 3″ x 1/4″.

Fig. 232. Electro Commercial (Electro Importing Co.); 4 7/16″ x 3″ x 1/2″.

Fig. 233. Electro Radiocite (Electro Importing Co.); 4 1/4″ x 2 1/2″ x 7/16″. Sold with a 3/8″ felt sub-base as shown.

Fig. 234. Electro Universal (Electro Importing Co.); 2 3/8″ x 2″ x 3/8″. Base is a replacement.

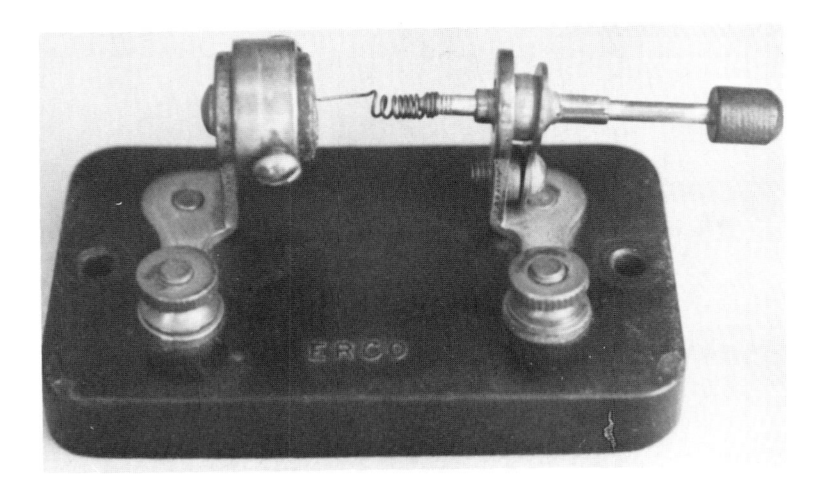

Fig. 235. ERCO (unknown mfr.); 2 3/4″ x 1 5/8″ x 3/8″. The handle knob is larger on some ERCO detectors.

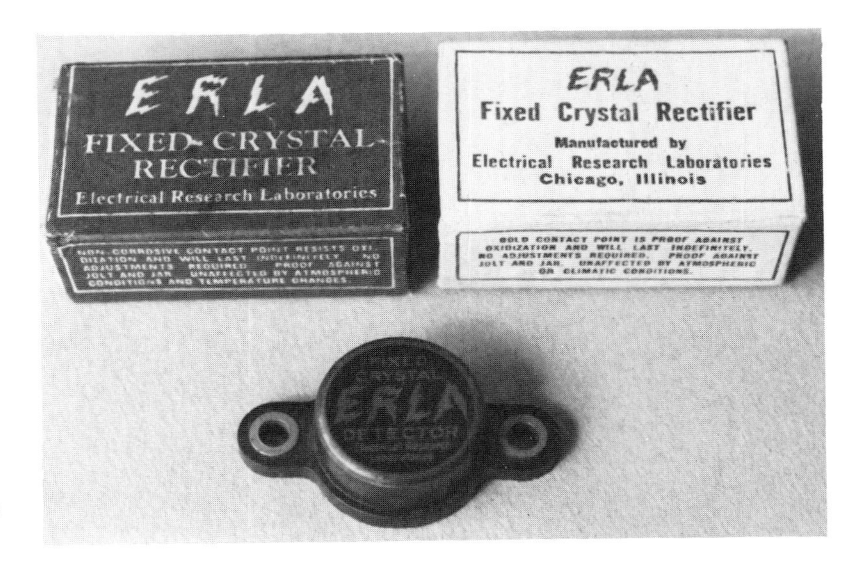

Fig. 236. ERLA Fixed, with two types of boxes (Electrical Research Labs.); base = 7/8″ diam. (1 5/8″ broad part), 1/8″ ht. See Figs. 282A and B for similar detectors.

Figs. 237A and B. FADA Type 101-A, Vertical (Andrea, Frank A. D.); 2 1/2″ x 2 1/2″ x 1/4″. Detector on right without "FADA" logo probably was made for Barawik Co. Date (1920) shown on sticker on base of detector on left is incorrect as this logo type was introduced in 1922 (see text). Fig. 237B is original box with FADA 101-A inside. (Fig. 237B, Al Reymann collection and photo.)

Fig. 238. FADA Type X101, Horizontal (Andrea, Frank A. D.); 2 1/2″ x 2 1/2″ x 1/4″. Al Reymann collection and photo.

Fig. 239. Federal No. 17 (Federal Tel. & Tel. Co.); 3 5/8″ x 3 3/8″ x 5/8″. Woody Wilson collection (previously) and photo.

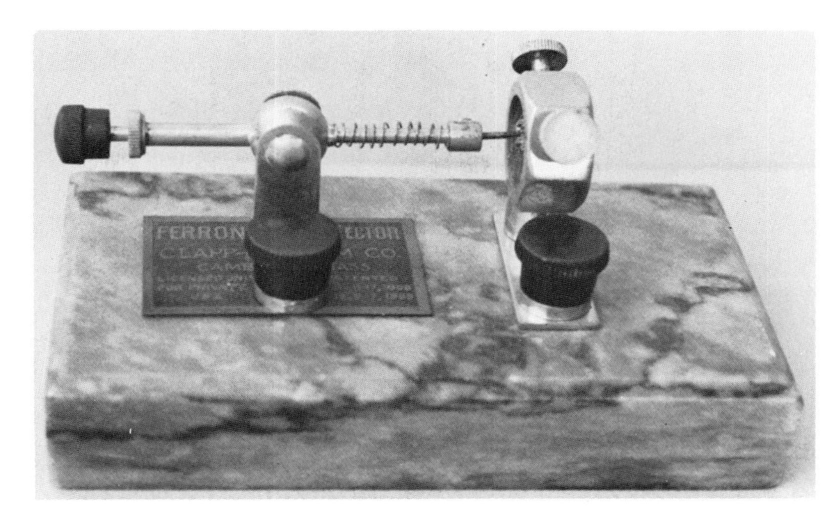

Fig. 240. (Universal) Ferron (Clapp-Eastham Co.); 5 1/2″ x 3″ x 1″.

Fig. 241. (Universal) Ferron, unlabelled (Clapp-Eastham Co.); 5 1/2″ x 3″ x 1″. This detector, without the metal-tag label of the manufacturer, was sold by some retailers.

Fig. 242. Foote Fixedetector (Foote Mineral Co.); 7/16″ diam., 1 1/8″ ht. to top of terminal. "Fixedetector" fits conventional crystal cup as illustrated on box label. See also Fig. 319.

Fig. 243. Firth Perikon (Firth & Co., John A.); 4″ x 2 1/4″ x 3/8″.

Figs. 244A, B, and C. In A, Ferron (Clapp-Eastham Co.); in B and C, Ferron (Duck Co., J. J.; Clapp-Eastham Co.); 5 1/2" x 3" x 7/8". (Figs. 244 B and C, Woody Wilson collection and photos.)

Fig. 245. Grebe RPDB (Grebe & Co., A. H.); base = 2 1/2" x 2 1/2" x 1/4", vertical panel = 2" (ht.) x 1 3/4" x 3/8".

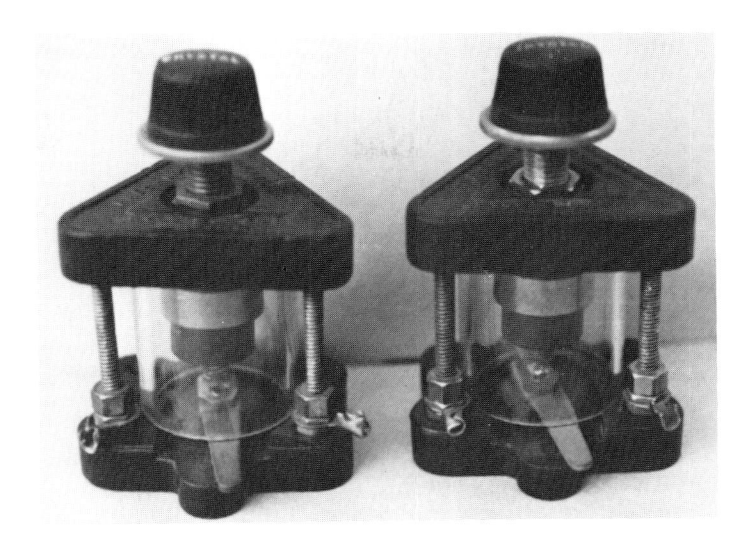

Fig. 246. Freshman Double Adjustable (Freshman Co., Chas.); 2″ x 1 3/4″ x 3/8″.

Fig. 247. Ghegan-patent detector (Bunnell & Co., J. H.?); 3 7/8″ x 2″ x 3/8″. Probably modified from Jove detector (Fig. 249).

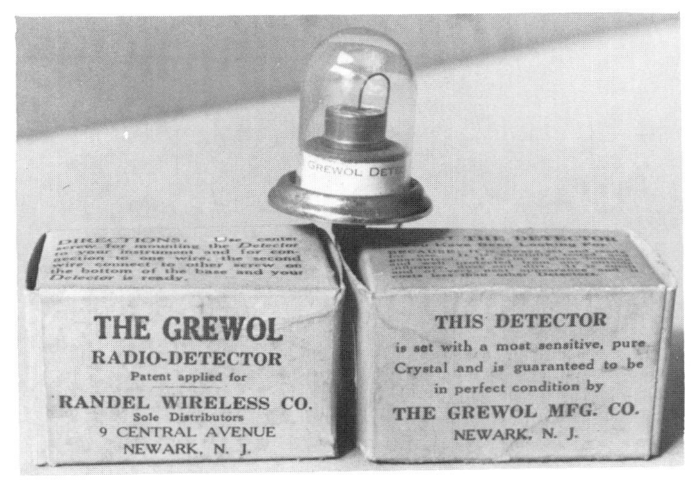

Fig. 248A and B. Grewol (Grewol Mfg. Co.); base = 1 1/4″ diam.; ht. to top of dome = 1 1/2″. In A, early detector and box (2 views) while patent was pending. In B, detector at left ("Pat. May 15, 1923") and its box after patent issued. "Tested" date is stamped on the end of each box.

Fig. 249. Jove (Bunnell & Co., J. H.); 3 1/4″ x 2 5/16″ x 1/2″. Detector was sold with both catwhisker types shown.

Fig. 250. K-3 (unknown mfr.); 2 1/4″ x 1″ x 1/8″.

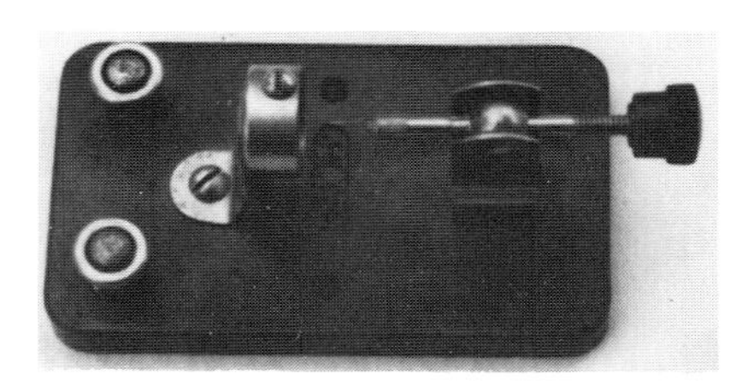

Fig. 251. J. K. Corp. (J. K. Corp.); 3 1/4″ x 2″ x 1/2″. Al Reymann collection and photo.

Fig. 252. Kennedy (Kennedy Co., Colin B.?); 3 3/8″ x 2″ x 1/2″.

Fig. 253. Kilbourne & Clark /K & C (Kilbourne & Clark Mfg. Co.); 2 3/4″ x 1 3/4″ x 3/16″. Detector as originally sold was unmounted, but the mounted version also was marketed.

Fig. 254. McCarty Micrometer (McCarty Wireless Telephone Co.); 6″ x 3 1/8″ x 3/4″. Red hard-rubber base. Ed Sharpe collection.

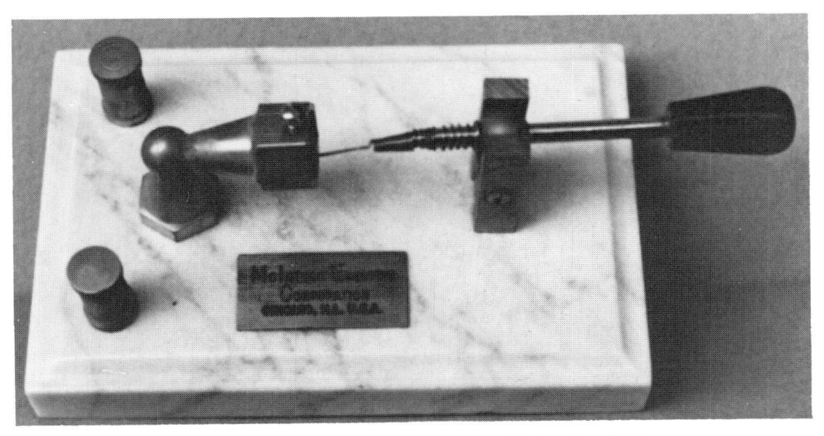

Fig. 255. McIntosh (McIntosh Elec. Corp.); 4 1/2″ x 3″ x 3/4″.

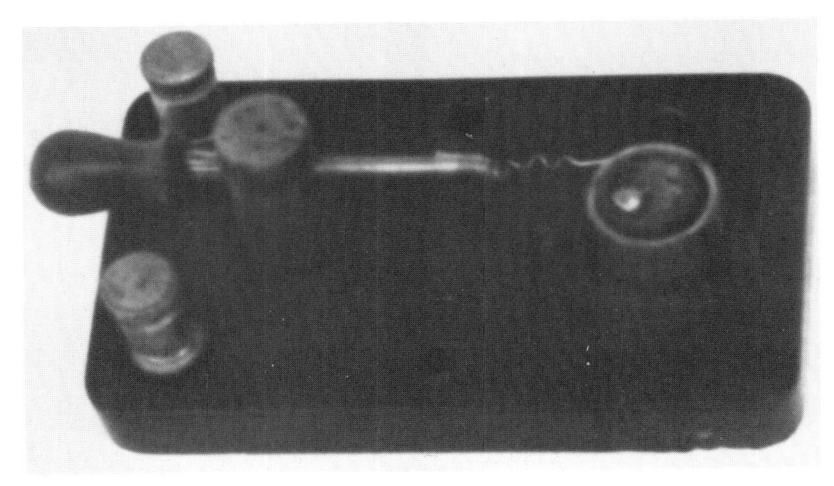

Fig. 256. Magnus (Magnus Elec. Co.); 3 3/8″ x 2″ x 3/8″. Same type detector was used for the Duncan Simple (Fig. 44) and Victor (Fig. 179) crystal sets. Don Patterson photo.

Fig. 257. MAR-CO (Martin Copeland Co.); 2″ x 1″ x 3/16″.

Fig. 258. MESCO Perikon (Manhattan Electrical Supply Co.); 5″ x 3″ x 3/8″. Wood base.

Fig. 259. Massie-type carbon block (not commercial) 2 1/2″ x 2 1/2″ x 3/8″. Needle across carbon blocks.

Fig. 260. Merit, left; Pacent No 31, right (Pacent Elec. Co.); both = 2″ diam., 1/4″ ht. See Fig. 266.

Fig. 261. Murdock No. 324, later version (Murdock Co., Wm. J.); 2 1/2″ x 1 1/2″ x 1/2″.

Fig. 262. Murdock Silicon with Condenser (Murdock Co., Wm. J.); 4 1/2″ x 2″ x 1/2″.

Figs. 263A and B. NAECO (Radio Surplus Corp.); 2 1/4″ x 1 3/8″ x 1/4″. The joint for the detector arm is unusual—a rotating rod (instead of a ball) in the socket. The catwhisker is mounted on the arm as an "eccentric." (Fig. 263A, Al Reymann collection and photo; Fig. 263B, Ed Sharpe collection.)

Fig. 264. Parkin (Parkin Mfg. Co.); 2 1/2″ x 1 1/2″ x 1/2″.

Fig. 265. Pacent No. 30 (Pacent Elec. Co.); 2″ diam., 1/4″ ht. Viewed in three positions.

Fig. 266. Pacent No. 31 (Pacent Elec. Co.); 2″ diam., 1/4″ ht. Viewed in three positions. See Fig. 260.

Fig. 267. Peppy Pal (Radio-Ore Labs.); 3 1/2″ x 1 3/4″ x 3/8″ wood base. Although the label of the box as shown has "PEPPY PAL / CRYSTAL ***SET***," the wrapped and sealed item inside was this detector. The small size of the box (4″ x 2 1/2″ x 1 1/2″), which is just right to contain this detector, suggests that the word "set" instead of "detector" on the label is an error.

Fig. 268. Philmore Open Type, Cat. No. 310 (Philmore Mfg. Co.); 1 1/2″ x 1 1/4″ x 1/8″ (fiber base). An earlier version of the No. 310 had a larger *molded* base as shown in detector in illustration on side of box in Figs. 268, 269; a later version of No. 310 is same as No. 7003 (Fig. 269).

Fig. 269. Philmore Open Type, Cat. No. 7003 (Philmore Mfg. Co.); 1 1/2″ x 1 1/4″ x 1/8″. Later than No. 310. (The boxed, *unmounted*, open-type Philmore detector had Cat. No. 101 or No. 308, early, but No. 7010, later.)

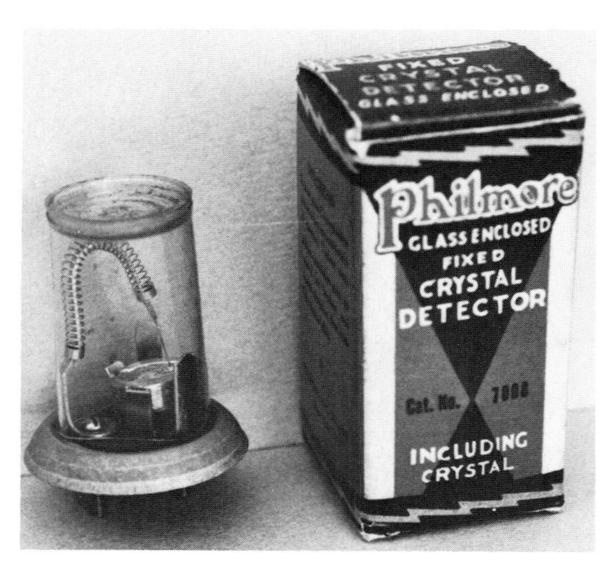

Fig. 270. Philmore "Glass Enclosed Fixed," Cat. No. 7008 (Philmore Mfg. Co.); base = 1 3/8″ diam., 1/4″ ht., enclosure = 7/8″ diam., 1 1/2″ ht. Earlier version, *Cat. No. 309*, is glass-enclosed; the carton for this later item, Cat. No. 7008, continues the designation of "glass enclosed," even though the enclosure is clear plastic. These detectors are called "fixed" but are actually semi-fixed.

Fig. 271. Philmore Fixed, Cat. No. 100 (Philmore Mfg. Co.); 1″ diam., 1/2″ ht. Boxes viewed in three positions. Unlike the "Glass Enclosed Fixed" (Fig. 270), the sealed detector shown here is, indeed, fixed.

Fig. 272. Pur-a-tone (Brach Mfg. Co., L. S.); 4 1/2″ x 2″ x 1/2″. Al Reymann collection and photo.

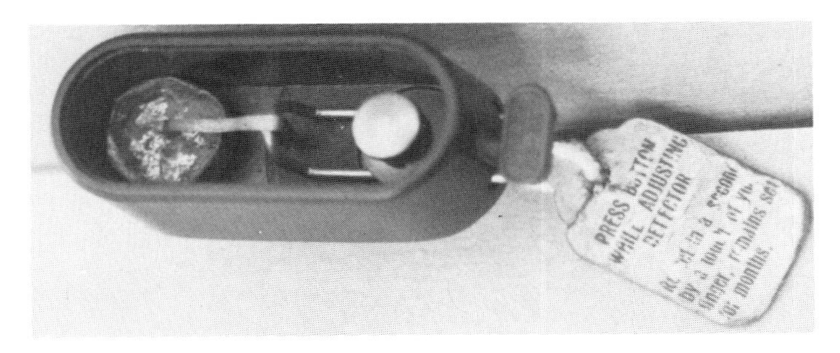

Fig. 273. Puretone Adjustable (Nilsson, O. F.); 1 1/2″ x 5/8″ x 5/8″. At the top of detector is a protective glass window-cover. The MacKay Marine crystal receiver (Figs 82A and B) has a very similar detector.

Fig. 274. Pyratek Fixed (Erisman Labs.); base = 3 1/16″ x 5/16″ x 3/16″; detector = 1/2″ diam., 1 5/8″ length.

Fig. 275. Radio Service & Mfg. Co. Type S-20 (Radio Service & Mfg. Co.); base = 2″ x 2″ x 3/8″.

Fig. 276. Radiotector (Gilbert Co., A. C.); 5″ x 4″ x 3″. Les Rayner collection.

Fig. 277. RASCO Baby (Radio Specialty Co.); 1 1/2" x 1 1/8" x 3/16". SeeE. I. Baby (Fig. 230).

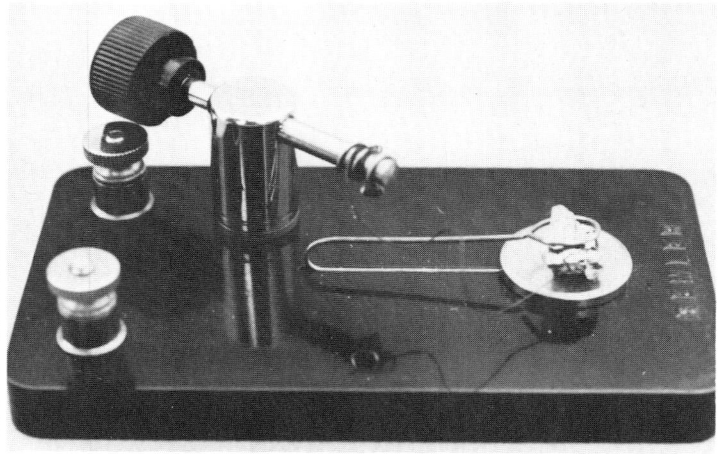

Figs. 278A and B. Remler No. 50. (Remler Radio Mfg. Co.); 3 7/8″ x 2 1/8″ x 3/8″. Unusual spring-wire holder for crystal.

Fig. 279. Redden's (Redden, A. H.); 3 1/2″ x 2″ x 1/2″. Al Reymann collection and photo.

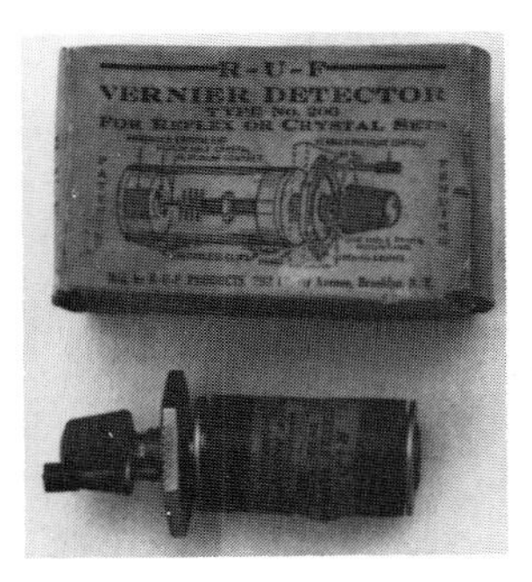

Fig. 280. R-U-F Vernier Type 200 (R-U-F Products); detector = 3/4″ diam., 1 1/2″ length (tubular portion). Al Reymann collection and photo.

Fig. 281. R-W self-adjusting (R-W Mfg. Co.); 3 1/16″ x 1 1/8″ x 3/16″.

Figs. 282A and B. In A, Scientific Fixed (Scientific Research Labs.); in B, Star Fixed (Star Crystal Co.); both = 7/8″ diam. (1 5/8″ broad part), 1/8″ ht. These detectors are similar to the ERLA Fixed (Fig. 236).

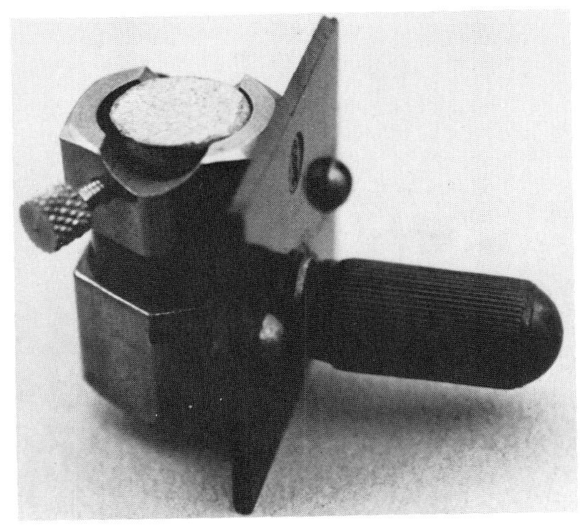

Fig. 283. Shamrock-Harkness, Panel Mount (Shamrock Mfg. Co.); Bakelite for panel mount = 1 3/8″ x 1 1/4″ x 1/16″.

Fig. 284. Signal Standard Galena No. 41 (Signal Elec. Mfg. Co.); 3 7/16″ x 1 15/16″ x 1/2″. Two variations of terminals and crystal cups are shown.

Fig. 285. Stellar Fixed (Stellar Mfg. Co.); 1" diam., 5/8" ht.

Fig. 286. Tillman (Tillman Products Co.); 3 1/2" x 2" x 1/2".

Fig. 287. Unknown mfr.; 3" x 3" x 1/4".

Fig. 288. Unknown mfr. (Perikon detector); 4 3/4" x 2 1/4" x 1/2".

Fig. 289. Unknown mfr. (Perikon detector); 2 1/2" x 1" x 3/16". Probably Radio/Radjo Permanent (Harris Lab.), which appears the same as the Harco detector.

Fig. 290. Unknown mfr. (homebrew?); 2 5/8" x 1" x 1/16". Replacement arm and handle knob.

Fig. 291. Unknown mfr. (Duck Co., J. J.?); 5″ x 3″ x 3/4″. Replacement base.

Fig. 292. Unknown mfr.; 3 1/4″ x 2″ x 3/8″. Replacement arm and handle knob.

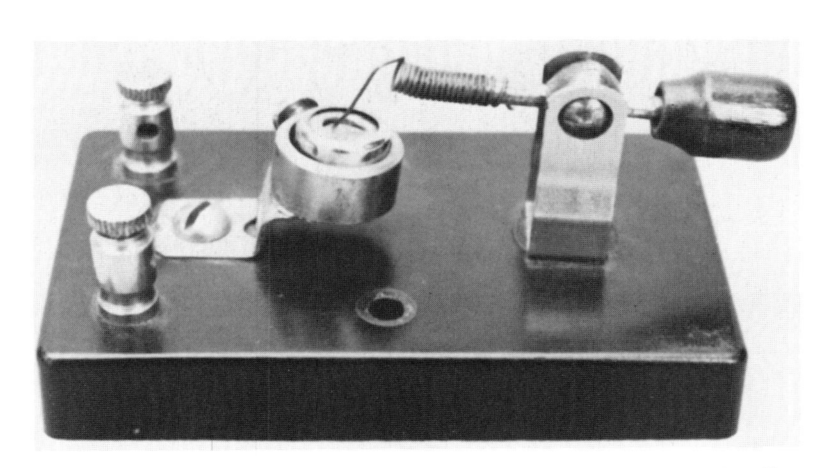

Fig. 293. Unknown mfr.; 3 1/4″ x 2″ x 1/2″. Replacement arm and handle knob.

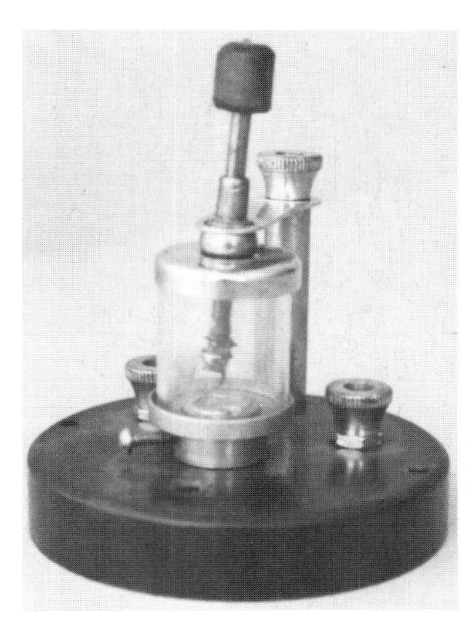

Fig. 294. Unknown mfr.; 2 3/4″ diam., 1/2″ ht.

Fig. 295. Unknown mfr.; 2″ x 1 1/8″ x 1/4″.

Fig. 296. Unknown mfr.; left = 2″ x 1 1/8″ x 1/4″, right = 1 1/2″ x 1 1/4″ x 1/8″.

Fig. 297. Unknown mfr. (swirl-catwhisker type); both = 1 1/2″ x 1 1/4″ x 1/8″.

Fig. 298. Unknown mfr.; all = 1 1/2″ x 1 1/4″ x 1/8″. Detector style differs from Fig. 297 but same swirl-type catwhisker.

Fig. 299. Seven miniature detectors (unknown mfrs.); dimensions as listed for Figs. 296-298.

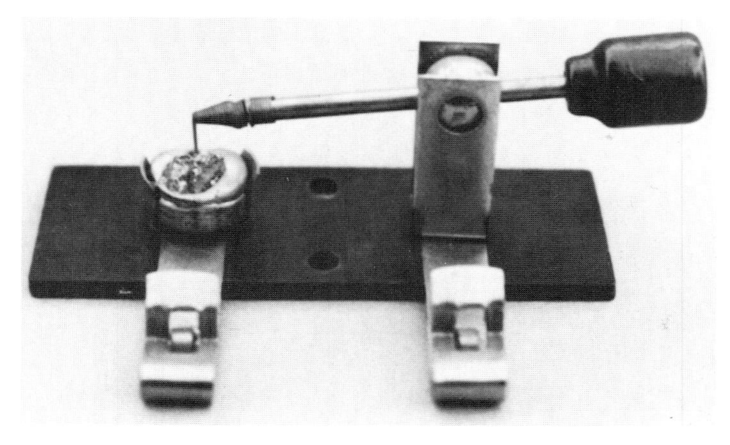

Fig. 300. Unknown mfr.; 2 3/4″ x 1″ x 1/8″. Philmore-type components but arm length (2 1/8″) is greater than usual (1 1/2″) for Philmore detectors (Fig. 269).

Fig. 301. Unknown mfr.; 1/2″ diam., 3/4″ length. Small capsule with its fixed detector fits special holder.

Fig. 302. Unknown mfr.; 3 7/8″ x 1 7/8″ x 3/8″.

Fig. 303. Unknown mfr.; 2 7/8″ x 2″ x 1/2″.

Fig. 304. Unknown mfr.; 2 3/4″ diam., 1/2″ ht.

Fig. 305. Unknown mfr.; 2 7/8″ x 2 1/2″ x 3/8″.

Fig. 306. Unknown mfr. (probably sold by Sears, Roebuck, & Co.); wood base = 3″ x 1 5/8″ x 3/8″.

Fig. 307. Unknown mfr. (Perikon detector); 3 1/2″ x 1 5/16″ x 3/8″. See Figs. 289 and 309.

Fig. 308. Unknown mfr. (Electrolytic detector); 3 1/2″ x 2 1/2″ x 3/16″.

Fig. 309. Unknown mfr. (Perikon detector); 2″ x 1″ x 3/16″. Probably Whiz Bang! (Federal Radio Products Co.). See Figs. 289 and 307.

Fig. 310. Unknown mfr.; 4″ x 2 3/8″ x 3/8″. Arched yoke-arm support with pivot pins.

Fig. 311. Unknown mfr.; marble base = 3 3/4″ x 3″ x 5/8″. Arched yoke-arm support with pivot pins.

Fig. 312. Unknown mfr.; marble base = 5 5/16″ x 4″ x 1 1/8″, detector = 3 1/8″ ht. above base. See Fig. 332.

Fig. 313. Unknown mfr.; 3″ x 2″ x 11/16″.

Fig. 314. Unknown mfr.; 2 7/8″ x 2″ x 3/8″.

Fig. 315. Unknown mfr.; 2 3/4″ diam., 3/8″ ht.

Fig. 316. Unknown mfr.; 3 3/8″ x 2 1/16″ x 3/8″. Replacement base.

Fig. 317. Unknown mfr.; 4″ x 2 5/16″ x 5/16″; multi-mineral cup.

Fig. 318. Unknown mfr.; marble base = 5″ x 3″ x 3/4″, detector = 2″ ht. above base.

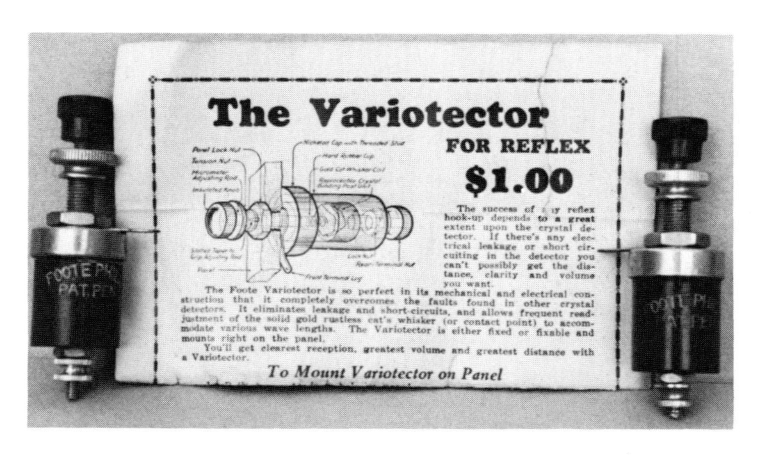

Fig. 319. Variotector (Foote Radio Corp.); 3/4″ diam., 1 1/16″ length (tubular portion). See Fig. 242.

Fig. 320. Western Wireless (Western Wireless); 4″ x 2″ x 1/2″. Replacement knob.

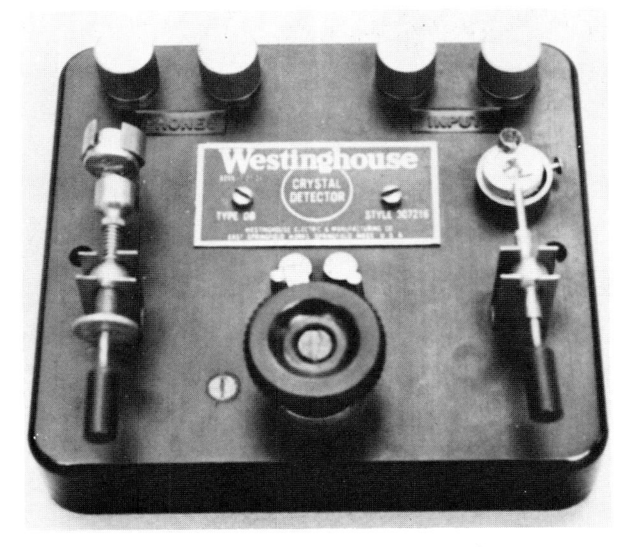

Figs. 321A and B. Westinghouse Type DB (Westinghouse Elec. & Mfg. Co.; RCA); 4 1/2″ x 4 1/4″ x 7/8″. The detector in Fig. 321A has minor variations (in catwhisker, type of terminal labels, word placement on metal tag) from the one in Fig. 321B. Also see ad on p. 112 for an illustration of an earlier, quite different Type DB detector.

Fig. 322. WMC (Westfield Machine Co.); 2 1/2″ x 1 1/2″ x 3/16″. Sold as "Wizard" by Montgomery Ward & Co.

Fig. 323. WSA / SORSINC (Wireless Specialty Apparatus Co.; Ship Owners Radio Service, Inc.); 3 1/2″ x 2″ x 3/8″. See Figs. 222, 324, 325.

Fig. 324. WSA / Westmore-Savage Co. (Wireless Specialty Apparatus Co.; Westmore-Savage Co.); 3 1/2″ x 2″ x 3/8″. See Figs. 222, 323, 325.

Fig. 325. WSA / SORSINC, left; WSA / Westmore-Savage Co., middle; WSA /Crosley, right. See Figs. 222, 323, 324; refer to discussion in text on detectors of Wireless Specialty Apparatus Co.

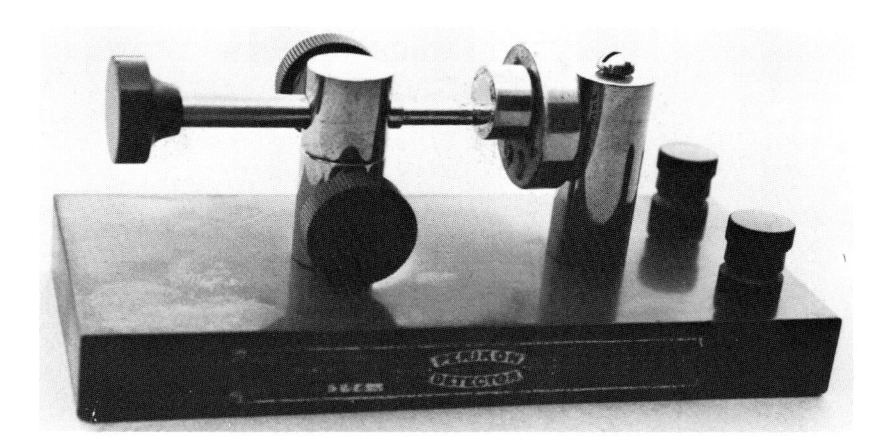

Fig. 326. WSA Perikon Type IP-162A (Wireless Specialty Apparatus Co.); 5 1/2″ x 2 1/2″ x 5/8″. Ed Sharpe collection.

Fig. 327. WSA Perikon Detector P-520 (Wireless Specialty Apparatus Co.); 4 3/8″ x 3 3/16″ x 1/2″.

Fig. 328. WSA Triple Detector / Lowenstein Type SE 183-A (Wireless Specialty Apparatus Co.; Lowenstein Radio Co.); 5 5/8″ x 5 5/8″ x 5/16″. Ed Sharpe collection.

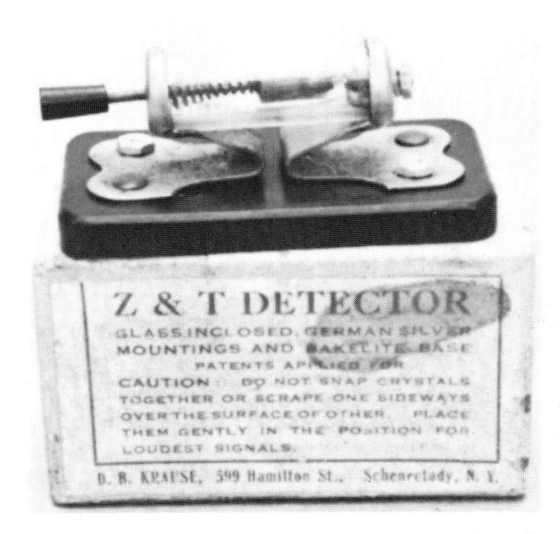

Fig. 329. Z & T (Krause, D. B.); 1 5/8″ x 1″ x 3/16″.

Fig. 330. Zinc & Tellurium (Tool & Device Corp.); 2″ x 1″ x 1/4″.

Fig. 331. *SMALL WONDERS*: left = E. I. Baby (Electro Importing Co.), right = RASCO Baby (Radio Specialty Co.); both = 1 1/2″ x 1 1/8″ x 3/16″.

Fig. 332. *THE LARGE AND THE SMALL OF IT*: back = unknown mfr., marble base, 5 15/16″ x 4″ x 1 1/8″; front left = E. I. Baby; front right = RASCO Baby; both 1 1/2″ x 1 1/8″ x 3/16″. See Figs. 230, 277, 312, 331.

Crystal Packaging Differences and Details

From 1910 to 1919, crystals (galena, silicon, carborundum, iron pyrites, bornite, molybdenum, zincite, and others) were generally sold unmounted, usually in bulk quantities, and by the ounce or by the piece. A purchaser acquired enough mineral to produce several smaller pieces for use in a detector that, at that time, often had a holder designed for an unmounted crystal. Distributors gave little attention to packaging these inexpensive minerals, although by 1918 the Electro Importing Company was using labelled cartons for its unmounted galena (rectangular box) and unmounted Radiocite (round box).

By 1920, minerals for use in detectors were often mounted in Wood's metal (an alloy melting at a low temperature) to fit conveniently in the setscrew-type crystal cups. These mounted crystals were usually sold individually packaged. For sales allure, the nature of the packaging was probably almost as important as the quality of the item within. Those colorful labelled containers for crystals continue to entice present-day collectors.

Types of Packaging

Tins

Newman-Stern Co. was one of the first companies to use an attractive substantial container to market its minerals: Arlington-Tested NAA Galena or Silicon (Fig. 343). In early 1920, its illustrated ads showed a labelled, cube-shaped, lidded tin container. By 1923, the shape of the lidded tin box had changed to cylindrical ("round"). This *high-profile*, round, tin container (and its contents, of course) continued to be marketed un-

The Famous
ARLINGTON
TESTED CRYSTALS
The Acknowledged Standard in Radio Detector Minerals.

It has been said by advanced radio physicists that vacuum tubes are no more sensitive than "very good" spots on galena. Hundreds of testimonials now in our files testify to the probability of such a statement. These letters come from users of Arlington Tested Crystals.
Arlington Tested Minerals have been on the market seven years and are sold by reputable dealers everywhere. Each crystal is individually tested and boxed in a lithographed metal container. Imitations are available but you can be assured of the original genuine crystals by looking for the signature "J.S. Newman" on each box.
NAA Tested Galena 35c
NAA Tested Silicon 35c

THE NEWMAN-STERN CO.
1874-76 EAST SIXTH STREET. CLEVELAND, OHIO

QST (Oct. 1920), p. 65

til at least 1934—as a mail-order catalog item, by then.

Many crystal manufacturers used *low-profile*, round, "pill-box-type," lidded, tin containers (Figs. 333A, 334, 341, 345, 348). Some trade names of crystals packaged in this style are: B-Metal Loud Talking, Foote's Radio Triplets, HARCO Culina, HARCO Silvertone, E-Z, Philmore Hi-Volt, Philmore "X"TRA Loud, and Star.

Radio News (Aug. 1923), p. 216

RADIOCITE

TESTED FOR SENSITIVITY

The most wonderful of all Wireless Crystals

Use Radiocite in your detector and then forget it

The Wireless Crystal that DON'T Jar Out

RADIOCITE is the most wonderful of all radio crystals. It is more sensitive than Galena and far more sensitive than ANY other crystal or mineral. RADIOCITE is a specially selected grade of a rare crystal chemically treated by our own secret process.

The mineral that looks like liquid gold. It has a highly, wonderfully polished surface giving it a perfectly burnished appearance. This crystal is now in use by several governments, and is conceded to be the most satisfactory of all. It is used with a medium stiff phosphor bronze spring, or with a stiff silver wire, about No. 30 B. & S. Gauge. One of the important features of RADIOCITE is that it does not jar out easily. Each crystal is **tested out individually for sensitivity** and guaranteed. RADIOCITE comes packed separately in a box, wrapped in tin-foil. Full directions for use accompany it. RADIOCITE can be mounted like any other crystal; it may be clamped between springs, but it is best to set it in **Hugonium** soft metal. Money refunded if our claims are not substantiated.

Shipping weight 2 oz.

No. EK3939 Generous piece of tested RADIOCITE. Prepaid $0.50

The one up-to-date mineral which every amateur must have.

Cleveland, Ohio,

Electro Importing Co.
233 Fulton St.,
New York.

Gentlemen:—

Your piece of radiocite received in excellent condition and am glad to inform you that it is without doubt the best mineral ever put on the market. It has any silicon or galena beat forty different ways and back again. I have tried it out on an indoor set consisting of a piece of bare copper wire 20 feet long, a gas pipe ground, a forty cent detector and a pair of 2000 Ohm phones. This set was used merely for the purpose of testing Radiocite and the results obtained "knocked me off my feet." I have not yet tried it on my big set but if it works as good as it did on the small set—why, I'll have **"some"** set.

Yours truly,
L. PLACEK,
316 W. 84th St.,
Cleveland, O.

Electro Importing Co., *Catalog No. 20* (1918), p. 38

Super-Sensitiveness!

The crystal is the "bull's-eye" of your crystal receiving set. Unless it is supersensitive you are wasting time and entertainment and cannot "hit" the combination for best results. Insist upon the genuine original Arlington Tested "NAA" Detector minerals They are carefully selected from bulk stock, individually tested and guaranteed super-sensitive. Galena, Goldite or Silicon, price for crystal, 25¢. Same mounted in brass cup, 40¢. Obtainable at your dealers or sent direct (post-paid) on receipt of price.

The Newman-Stern Co

Newman-Stern Building Cleveland

QST (Jan. 1923), p. 134

Radio News (Nov. 1923), p. 643

Boxes: Wood, Cardboard

Round, low-profile wood boxes were used for several crystals (Figs. 340, 347A and B). Some trade names of minerals packaged in wood boxes are: Gotham Galena, Harris Galena, Rosite, and San. The Westinghouse Spare Crystal, for detectors of the Aeriola Jr. crystal set or DB detector stand, was sold in a smaller, high-profile wood box (Fig. 349).

Small cardboard boxes were a common package form for crystals (Figs. 333A and B, 335, 342A and B, 344, 346, 350A and B). A few trade names of crystals sold in this type of container are: A-1, American, Brownlie Quick Contact, Million Point Mineral (M. P. M.), Parkin, Philmore (see comments in next section), Powertone, T. N. T. Low Loss Galena, T. N. T. High Power, and T. N. T. Steel Galena.

Card-Mounted Envelopes; Envelopes

Some crystals were packaged in small envelopes attached to larger cards bearing the trade name and other information. A prominent user of this type of packaging was the Philmore Manufacturing Company (Fig. 346). Previously, Philmore had marketed its crystals in boxes (see above) of two types: a yellow box (believed to have been used before the card-mounted style) and a red-and-black box (definitely used at a later date) labelled Cat. No. 7004. A few manufacturers sold their mounted crystals (e.g., Clearco; see Fig. 338) in small labelled envelopes unattached to cards.

Special Case

One of the more elaborate styles of crystal packaging was used by the Keystone Products Company for its "Atomite." An ad in *Radio News* for December, 1924, entitled "An Ideal Gift" shows a mounted crystal glowingly reposing in a hinged, lidded, "plush lined, leatherette novelty (jeweller's type) case." This ad states that the crystal "sparkles in near platinum mount." The packaging and claims were undoubtedly intended to justify the list price of $1.00, gold catwhisker included. Without the special case (or gold catwhisker), the Atomite sold for $0.50.

Radio News (Dec. 1924), p. 1044

Pacific & Atlantic

To while away her spare moments while waiting to keep appointments, Miss Addie Rolf of New York need merely to open her leather card case and listen in on what the wild ether waves are saying. But who would keep Miss Rolf waiting?

Popular Radio (Feb. 1926), p. 125

Crystals and their Containers

Consult tables for further details, including date of manufacture. See text for discussions of some special features.

Figs. 333A and B. In A, A-1 (California Radio Minerals); E-Z (E-Z Crystal Co.); Powertone (Powertone Radio Products Co.); Shur-Hot (Non-Skid Crystal Mfg. Co.). In B, American (American Radio Mfg. Co.)

Fig. 334. B-Metal Loud Talking (B-Metal Refining Co.).

Fig. 335. Right = Brownlie Quick Contact (Brownlie, R.); left = Mounted Galena (unknown mfr.).

Fig. 336. Dr. Cecils (unknown mfr.; British origin).

Fig. 337. Carrs Banketite (Carrs; British origin).

Fig. 338. Clearco (Clearco Crystal Co.). The Clearco mounted crystals were available with either a straight-type or flange-type rim.

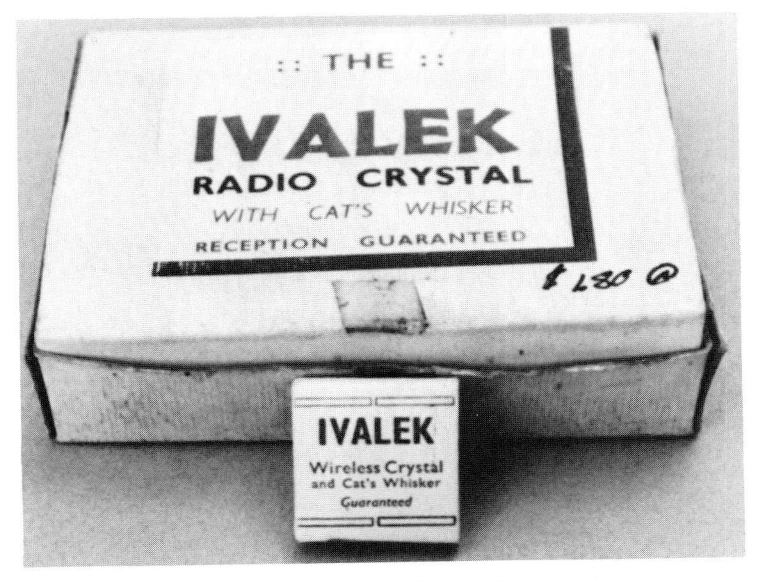

Fig. 339. Ivalek (unknown mfr.; British origin).

Fig. 340. Left = Tested Galena (Harris Lab.); middle = unknown mfr.; right = "W" (unknown mfr.).

Fig. 341. HARCO Silvertone (Harcourt Radio Co.?).

Figs. 342A and B. M. P. M. / Million Point Mineral (M. P. M. Sales Co.).

Fig. 343. NAA Galena (Newman-Stern Co.).

Fig. 344. Parkin (Parkin Mfg. Co.).

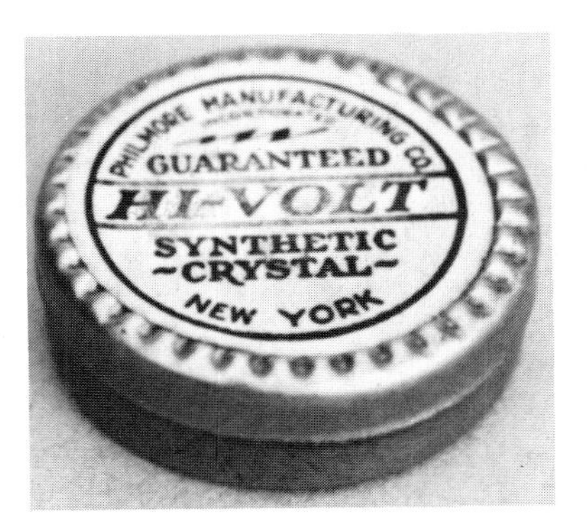

Fig. 345. Philmore Hi-Volt Crystal (Philmore Mfg. Co.).

Fig. 346. Philmore Meter Tested; Hi-Volt; Philmore Crystal; Philmore Cat. No. 7004 (Philmore Mfg. Co.).

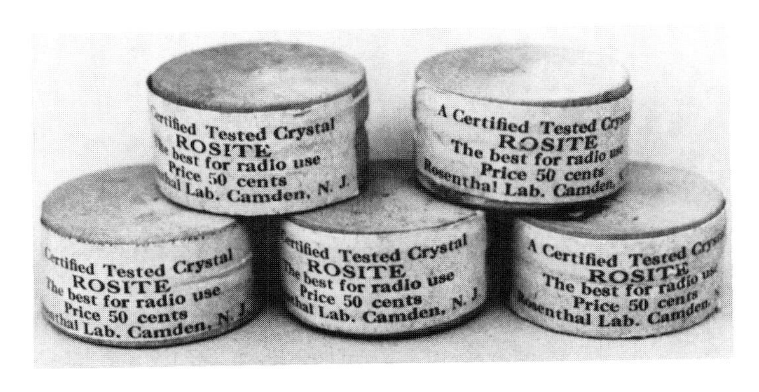

Figs. 347A and B. In A, Rosite (Rosenthal Lab.). In B, San (San Co.).

Fig. 348. Star Galena (Star Crystal Co.).

Fig. 349. Westinghouse Spare Crystals (Westinghouse Elec. & Mfg. Co.; RCA); for use on detector of Aeriola Jr. crystal set and the DB detector.

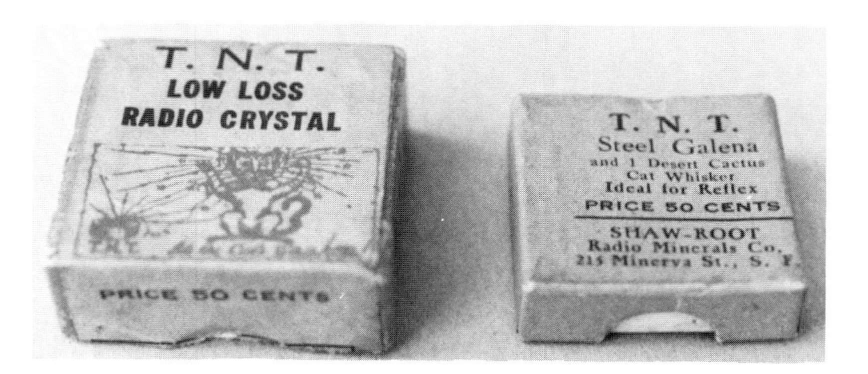

Figs. 350A and B. T. N. T. Low Loss (T. N. T. Products Co.); T. N. T. Steel Galena (Shaw-Root Radio Mineral Co.). A "Desert Cactus Cat Whisker" was supplied with the T. N. T. Steel Galena mounted crystal (box on right).

Fig. 351. Some of the commercially available early crystals.

THIS WONDERFUL DETECTOR

Are you interested enough in wireles to spend this small sum for something you cannot do without? Needs no adjusting--just snap it in your hook-up and receive--very convenient. Comes in four styles — GALENA — SILICON — IRON PYRITES and CARBORUNDUM. You

BOYS!!!

FOR 25c ONLY

Including Postage

can have five of them for half the cost of any one other detector, and you have the advantage of experimenting with four crystals and using the mineral most suited to your apparatus. FIVE WRIGHT DETECTORS—any selection $1.00 or 25c each. Act Quickly—write at once before you forget.

THE WRIGHT RADIO CO. (No Stamps Accepted)
P.O. BOX 534, CITY HALL STATION, NEW YORK, N. Y.

Enclosed is....................for which please send me:

........Galena Detector ....Silicon Detector

........Iron Pyrites Detector ....Carborundum Detector

Name .... Address ....

QST (July 1921), p. 100

Tell them that you saw it in RADIO

Radio (Nov. 1924), p. 88

Table Talk

Entries in the three tables are arranged alphabetically, primarily by the names of manufacturers as shown in the first column. Secondly, items in the next column (trade names of crystal sets, detectors, or crystals) are also listed in alphabetic order relative to the first column; each such item is cross-referenced to manufacturer. This form of duplicate listing permits readers to locate an entry if either the manufacturer or item trade name is known. For an item with undetermined manufacturer, only a single entry exists, alphabetized by the item trade name; in the first column, "unknown mfr." appears in parentheses.

Items listed in the main section of each table were marketed in the arbitrarily defined early (before or by 1929) or later (1930–1939) periods. An addendum, a continuation of each main table, lists some more recent (1940–1955) items.

In these tables, the designation "manufacturer" has a meaning that extends beyond the strict interpretation of the term. Included in this designation, as a small fraction of the total, are companies that used their own labels and trade names for products supplied to them by the producers. For some such products, this extramural origin is quite apparent and, therefore, specified in the tables (e.g., "Liberty" trade name for the crystal set made by Gundlach-Manhattan Optical Co.). In several cases, however, these occurrences are less obvious, probably because the crystal sets, detectors, or crystals were produced primarily or exclusively for the contracting concern with little or no discernible evidence of source.

Many companies using their own labels for crystal set items not made within their organization were mail-order houses; most also had direct-sales outlets. They usually advertised widely and distributed catalogs or brochures with detailed descriptions and good illustrations of their products. Some of these companies had enviable reputations for quality; their prestigious labels identify crystal set items that are still coveted by collectors. Among these organizations are: Manhattan Electrical Supply Co. (MESCO); J. J. Duck Co. and its successor, Wm. B. Duck Co.; Sears, Roebuck & Co.; and Montgomery Ward & Co.

The tables on crystal sets and detectors include a small number of items that have *electrolytic* detectors. Although the electrolytic detector is beyond the scope of these tables, the few entries with this item are representative of devices with functions and appearances approximating those of *crystal* sets and detectors.

In the detector table, all pre-1922 items listed are *adjustable* crystal detectors. Listings for later dates include several fixed and semi-fixed detectors. Although these last two types of detectors were used in some crystal sets, the fixed detectors were used especially in reflex circuits of vacuum-tube radios.

The dates and sales prices given in the tables are derived from the earliest references found. Discovery of other information could indicate that some dates of origin extend further back.

For interpretation of several types of tabular entries, readers should become familiar with the "Guide to Abbreviations Used In Tables," which precedes each of the three tables. That section also gives details of such factors as standards used for size, orientation of detector, and year ranges of the items.

Table Highlights

As might be expected, the majority of the listed manufacturers were concentrated in the most populous states. Seven states (New York, Illinois, Ohio, Pennsylvania, New Jersey, California, and Michigan) produced 58% of the 573 trade-name crystal sets; ten cities in these states (New York City, Chicago, Pittsburgh, Newark, Toledo, Detroit, Cincinnati, Los Angeles, Philadelphia, and San Francisco) made 53% of them. Six states (New York, Illinois, Pennsylvania, New Jersey, California, and Massachusetts) were the source of 72% of the 341 trade-name crystal detectors; seven cities in these states (New York City, Chicago, Pittsburgh, Newark, Boston, San Francisco, and Los Angeles) were the place of origin for 58% of them. Six states (New York, California, Illinois, Missouri, New Jersey, and Pennsylvania) produced 68% of the 207 trade-name crystals; eight cities in these states (New York City, Chicago, Los Angeles, Kansas City, Philadelphia, San Francisco, Newark, and Pittsburgh) were the source of 50% of them.

Total numbers of listed trade-name items, companies, and states of origin are given below. The numbers in the column for companies do not include unknown manufacturers (21 for crystal sets, 3 for crystal detectors, and 14 for crystals). Items for which neither the trade name nor the manufacturer is known are omitted from the three tables, but several such articles are shown in the published photographs. Numerous unidentified, commercially produced crystal detectors have been seen—many more than for crystal sets or crystals.

	TRADE-NAME ITEMS (No.)	ORIGINATING COMPANIES (No.)	ORIGINATING STATES (No.)
Crystal sets	573	379	24
Crystal detectors	341	232	20
Crystals	207	164	20

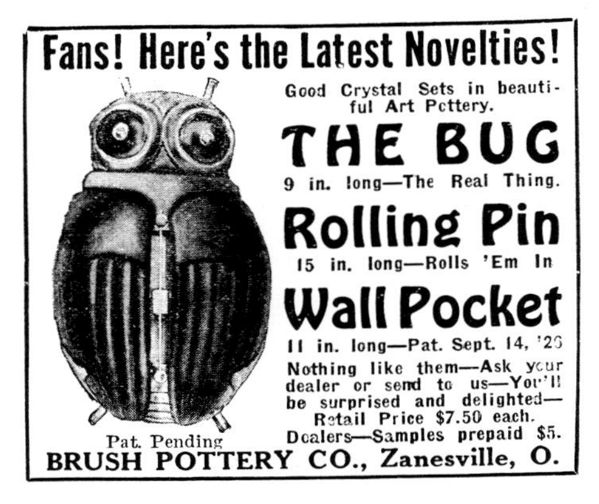

Radio News (Dec. 1926), p. 767

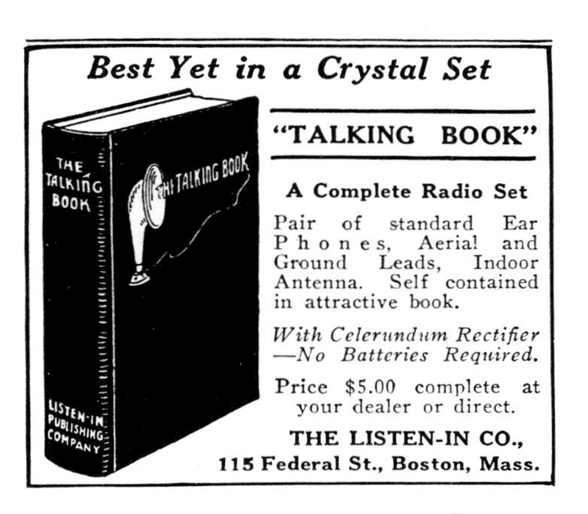

Radio News (Oct. 1926), p. 443

CRYSTAL SETS TABLE

GUIDE TO ABBREVIATIONS USED IN TABLE FOR CRYSTAL SETS:

CASE:

TYPE
- L Lidded Box
- U Unlidded Box
- C Coil, Slide Coil, or Loose Coupler
- B Breadboard or Base & Coil
- T Thin Case
- O "Open"—No Case
- M Miscellaneous Other
- X Unknown

MATERIAL
- W Wood
- M Metal
- P Plastic, Bakelite, Hard Rubber, or Celluloid
- F Fiberboard, Cardboard, or Leatherette
- O Other Miscellaneous
- X Unknown

SIZE*
- M Miniature
- S Small
- I Intermediate
- L Large
- X Unknown

*Size Baselines: (See photo captions of these sets for their dimensions.)

Miniature Philmore Little Wonder or smaller
Small LEMCO No. 340
Intermediate Aeriola Jr.
Large DeForest Everyman DT-600 or larger

DETECTOR:

STATE OF ENCLOSURE
- O Open
- E Enclosed
- X Unknown

ORIENTATION*
- H Horizontal
- V Vertical
- X Unknown

CONTACT TYPE
- C Catwhisker or Other Metal Point
- P Perikon (Mineral-to-mineral)
- E Electrolytic Type
- T Thumbwheel Type
- M Miscellaneous (Fixed, Cylinder or Cartridge Type; Diode)
- X Unknown

*Orientation of detector refers to whether the rod/arm is parallel ("horizontal") or at right angle ("vertical") to the underlying detector base.

PRICE: Earliest retail price found; X unknown.

YEAR:*

MAIN TABLE
- E Early (Before or by 1929)
- L Later (1930–1939)

ADDENDUM TABLE
- R Recent (1940–1955)

*If year is unknown or uncertain, one of the above letters is used for an approximation, based on available evidence; when year is listed, it is for the earliest reference found.

REFERENCES:

- R *Radio News*
- Q *QST*
- M Other Magazine(s)
- C Catalogs
- V Viewed by Author
- O Other Source(s)

MANUFACTURER	SET NAME/CASE; DETECTOR	PRICE	YEAR	REF.
(See Edgcomb-Pyle Wireless Mfg. Co.)	A-1 /BWI;OHC	$5.00	1914	C
	A-2 /BWI;OHC	$11.50	1914	C
	A-3 /BWI;OHC	$14.50	1914	C
	A-4 /BWI;OHC	$13.50	1914	C
	A-5 /BWI;OHC	$17.50	1914	C
	A-6 /BWI;OHC	$21.00	1914	C
	A-7 /UWI;OHC	$25.00	1914	C
(See Duck Co., Wm. B.)	A505 /BWI;OVC	$47.50	1914	C
	A1002 /BWI;OVC	$8.80	1914	C
(See Wireless Equipment Co.)	ABC /UWI;OHC	$24.50	1920	RQ
(See Jewett Mfg. Co.)	ABC /UWI;OHC	$40.00	1922	R
A-C Electrical Mfg. Co., Dayton OH (Called A-C Dayton Co. in 1928)	A-C Dayton /UWI;OHC	$5.00	1923	VO
(See Everett Radio Co.; Pacific Screw Co.)	Acme /XXX;XXX	$3.50	1925	O
(See Sears, Roebuck & Co.)	Acme Mounted:			
	Single-slide Tuner /BWI;OHC & OHP	$5.50	1917	C
	Double-slide Tuner /BWI;OHC & OHP	$7.00	1917	C
(See Ajax Elec. Specialty Co.)	ACS /UWI;OHC	$5.00	1925	RO
Adams-Morgan Co., Upper Montclair NJ (AMCO)	Type RA /BWS;OVC	$4.45	1913	C
	Type RB /BWI;OVC	$6.60	1913	C
	Type RC /BWI;OVC	$8.80	1913	CVO
	Type RD /UWL;OVC	$20.00	1913	C
	Type RE /UWL;OVC	$45.00	1913	C
	Double Crystal, The Arlington /UWI;OHC & OHC	X	E	O
Aeolus Corp., NY NY (see also Benson Engineering Co., mfr. in 1924)	Radiolette /MMS;EHM	$4.00* (*price in 1924)	1925	RO
Aerex Radiophone Corp., NY NY	King of the Air /LWI;OHC	X	1922	O
Aerial Elec. Co., Chicago IL	Crystal Mystery /LWS;OHC	$10.00	1925	RO
	Crystal King /XXX;XXX	$8.00	1926	RO
	Combination King-1 /TPI;OVC	$10.95	1926	RO
	Combination King-3 /XXX;XXX	$19.95	1926	RO
(See RCA; Westinghouse Elec. & Mfg. Co.)	Aeriola Jr. (Model RE) /LWI;OHP	$25.00	1921	RQMCVO

MANUFACTURER	SET NAME/CASE; DETECTOR	PRICE	YEAR	REF.
(See Crain Bros. Radio Shoppe)	Aeriola X /LWS;OHC	$6.00	1925	RVO
(See Essex Specialty Co.)	Aerophone /XXX;OHC	X	E	V
Aerovox Wireless Corp., NY NY	Aerovox /XXX;XXX	$16.00	1923	O
Aetna, Hartford CT	Aetna /LWI;OHC	X	E	O
(See Clapp-Eastham Co.)	AF /UWL;XXX	$550.00	1914	O
(See Custer)	Air Bug /CWM;OHC	X	E	V
(See Montgomery Ward & Co.)	Airline Jr. /UWI;EHC	$7.40	1923	C
	Airline /UWI;EHC or OHC	$3.95	1925	C
(See Detroit Radio Co.)	Airmuse /XXX;XXX	$6.00	1923	O
(See Bowman; Radio Equipment Co.; Telephone Maintenance Co.)	Airophone /UWI;OHC	$20.00	1922	RCO
(See Brownie Mfg. Co.; Rivero & Co.)	(Brownie) Airphone /TMS;OHC	$7.50	1924	RVO
(See National Airphone Co.)	(National) Airphone G /TPS;EVT	$12.50	1922	RMO
(See Radio Exchange & Supply Co.)	Airscout /XXX;XXX	X	1923	O
(unknown mfr.)	Air-Scout Senior/UFS;OHP	X	E	V
(See Philmore Mfg. Co.)	Ajax (early version of Supertone?) /XXX;XXX	X	1925	O
Ajax Elec. Co., Cambridge MA	Ajax /XXX;XXX	X	1925	O
Ajax Elec. Specialty Co., St. Louis MO	Crystal Set /XXX;XXX	X	1924	O
	ACS /UWI;OHC	$5.00	1925	RO
	CST Jr. /CWI;OHC	$3.00	1925	RO
	VS&L /UWS;OHC	$2.50	1926	RMVO
(See Edelman, P. E.)	Aladdin Multitone:			
	Large Set /UWI;OHC	$22.00	1922	R
	Smaller Set /UWS;OHC	$15.00	1922	R
Alden Wireless Co., Brockton MA	Alden /BWI;OHP & OHP	X	1910	M
Aldine Radio & Mfg. Co., Chicago IL	Micr-o-phone /CWI;OHC	$10.00	1922	R
American Can Co., Chicago IL	Skeezix Radio Toy /MMS;OHC	X	E	O

MANUFACTURER	SET NAME/CASE; DETECTOR	PRICE	YEAR	REF.
American Electro Technical Appliance Co., NY NY (later called Radio Service & Mfg. Co.)	Little Wonder Portable Type S-8 /LWS;OHC	$6.00	1920	RMO
American Leader Products Co., Chicago IL	American Leader Pocket Radio /UWS;EHM	X	L	VO
American Radio & Research Corp. (AMRAD), Medford Hillside MA & NY NY	AMRAD 2331 Type E /UWS;EVC	$23.50	1920	RQCO
	AMRAD 2575 /UWS;OVC	$21.50	1922	RO
	Crystal /XXX;XXX	$10.00	1922	O
American Radio Corp., Indianapolis IN	American /CWS;OHC	$4.00	E	O
American Specialty Mfg. Co., Bridgeport CT	Regal Jr. /UWI;OHC	$20.00	1922	RO
(See California Radio Co.)	Angeleno /XXX;XXX	X	E	O
(See General Elec. Co.; RCA)	AR 1300 /UMI;OHP	$50.00	1922	RQMCVO
(See RCA; Wireless Specialty Apparatus Co.)	AR 1375, Concert Radiola /LMI;OHC	$40.00	1922	RMCO
	AR 1382, Concert Radiola /UWI;OHC	$70.00	1922	C
(See Curry & Coutillier Labs.)	(Radio) Argentite /XXX;XXX	X	1923	R
ARJO Radio Products Co. (address unknown)	Simple-X /CWI;OVC	$4.75	E	VO
(See Adams Morgan Co.)	(Double Crystal,) (The) Arlington /UWI;OHC & OHC	X	E	O
(See Duck Co., Wm. B.)	Arlington (R 40) /BWI;OHC	$17.50	1916	C
(See Hecco Elec. Co.)	Arlington /BWI;OHC	$7.50	E	O
(See Wilkins Radio Co.)	(Crystal) Artay /XXX;XXX	$5.00	1926	O
(See Clapp-Eastham Co.)	ASA /UWL;XXX	X	1916	O
(See Buscher Co., C. A.)	Atchison /UWI;OHC	$7.50	1923	C
Atlantic Instrument Co., NY NY	Atlantic Jr. /UWI;OHC	$18.00	1922	RQ
Atwater Kent Mfg. Co., Philadelphia PA	Atwater Kent /BWI;OHC	X	1923	VO

MANUFACTURER	SET NAME/CASE; DETECTOR		PRICE	YEAR	REF.
Audiola Radio Co., Chicago IL (see also Splendid Radio Co.)	Audiola Model A, No. 1 /UWS;OHC (same as Splendid)		$10.00	1922	MVO
	Midget /XXX;XXX		X	1925	O
Audio-Tone Radio Co., NY NY	Krystalceptor /XXX;XXX		X	1925	O
(See Lawsam Elec. Co.)	(Lawsam) Baby /UWS;OHC		$17.50	1922	RQV
(See Cannon Co., C. F.)	(Cannon Ball) Baby Grand /BMS;OHC		$1.50* (*$0.88 in 1934)	L	CV
(See Beaver Machine & Tool Co.)	Baby Grand Vest Pocket /TWM or TPM;OHC		$7.50	1923	MCVO
(See Ray-Di-Co Organization)	Baby Wonder /LWI;OHC		$35.40	1922	R
Baldor Elec. Co., St. Louis MO	Ray Tuner /LWS;OHC		X	E	O
B & C Rad. Comm. Co., Boston MA	B & C /UWI;OHC		X	E	VO
Banister & Pollard, Newark NJ	Crystal /BWI;OHC		$7.00	1922	RM
(See Briggs & Stratton Co.)	BASCO 70300 /UWI;EHC		$12.00	1922	RQCO
(See Simplex Electrical Lab.; Simplex Radio Co.)	Basub /UWI;OHC		X	E	VO
Baughman Radio Engineering Co., (address unknown)	King Tut /XXX;XXX		X	1924	O
(See DeForest Radio Tel. & Tel. Co.; General Radio Co.; Liberty Elec. Corp.)	BC-14A (also called "Type SCR-54," "Type SCR 54-A") /LWL;EVC		X* (*$4.95 in 1931 as Army Surplus item)	1918	CVO
Beaver Machine & Tool Co., NY NY, Sales; Forest Hill (Newark) NJ, Mfg./Lab.	Baby Grand Vest Pocket / TWM or TPM;OHC		$7.50	1923	MCVO
(See Sears, Roebuck & Co.; Bowman & Co., A. W.)	Beginners' Sets (6 variations)	From: To:	$8.57 $26.81	1923	C
	Beginner's Unit /O--;OHC (called Bowman Type 1A-120 by mfr.)		$7.95	1921	C
Benson Engineering Co., mfr., Benson Melody Co., distributor, both Chicago IL (see also Aeolus Corp., mfr. in 1925)	Radiolette /MMS;EHM		$4.00	1924	RO
Berman Sporting Goods Co., St. Paul MN	Crystal /XXX;XXX		X	1925	O
Berstan Radio Products Co., Springfield MA	Mysto-C /XXX;XXX		$3.00	1924	O

MANUFACTURER	SET NAME/CASE; DETECTOR	PRICE	YEAR	REF.
(See Hyman & Co., Henry)	Bestone /UWI;EHC (variations with 1 or 2 sets of tap switches)	$22.50	1923	RO
Bethlehem Radio Corp., Bethlehem PA	Crystal-Dyne /MMS;EVC	X	E	VO
	World /TMS;EVC	X	1929	VO
Betta-Tone Radio Co., San Francisco CA	Betta-Tone /LWS;OHC	X	1924	O
(See Martian Mfg. Co.; White Mfg. Co.)	(Martian) Big-4 /MMS;OHC	$7.50	1923	RVO
Bird Elec./Bird Radio Corp., Cincinnati OH	Bird /UWS;EVM	$6.00	1925	RO
(See Philmore Mfg. Co.)	Blackbird /UFI,UMI;EVC	$3.00	1931	CVO
(See National Industries)	Black Diamond /XXX;XXX	$2.50	1923	O
Blair Co., Mitchell, Chicago IL, distributor (own label) (see Martian Mfg. Co. and White Mfg. Co, mfrs.)	Blairco "4" /MMS;OHC (called "Martian Big 4" by mfrs.)	$7.50	1923	M
(See Perry & Sons, Frank B.)	(Radio) Blinker /UMS;EVC	$12.00	1922	RO
(See Clapp-Eastham Co.)	Blitzen /UWI;OHC	$33.00	1913	CO
(See Brooklyn Metal Stamping Corp., mfr.; Kresge Co., S. S., retailer)	BMS /BMS;OHC (called "Pandora" by mfr., earlier)	$1.00	1929	C
Boldo Radio & Elec. Co., (address unknown)	Crystal /XXX;XXX	$1.50	1923	O
	Crystal /XXX;XXX	$6.00	1923	O
	Boldo /XXX;XXX	$2.00	1926	O
Borchet Manegold Engineering & Mfg. Co., Milwaukee WI	Crystal /XXX;XXX	X	E	V
Bowman, (address unknown), mfr. (See also Radio Equip. Co.; Telephone Maint. Co., distributors)	Airophone /UWI;OHC	$20.00	1922	RCO
Bowman & Co., A. W., Cambridge MA (see Sears, Roebuck & Co., retailer)	Bowman Type 1A-120 / O--;OHC (called "Beginner's Unit" by S., R. & Co.)	$21.39	1921	CV
Branston, Chas. A., Buffalo NY	Crystal /XXX;XXX	X	1924	O
Briggs & Stratton Co., Milwaukee WI	BASCO 70300 /UWI;EHC	$12.00	1922	RQCO

MANUFACTURER	SET NAME/CASE; DETECTOR	PRICE	YEAR	REF.
(unknown mfr.)	Bronswick /BWS;OHC	X	E	O
Bronx Radio Equipment Co., NY NY (BRECO)	BRECO /XXX;XXX	X	1925	O
Brooklyn Metal Stamping Corp., Brooklyn NY (see also Kresge, S. S., retailer)	Pandora (2 types) /BMS;OHC (called "BMS" by a retailer in 1929)	$2.50	1922	RCVO
Brownie Mfg. Co., San Francisco CA (see also Rivero & Co., distributor)	Brownie Airphone /TMS;OHC	$7.50	1924	RVO
Brush Pottery Co., Zanesville OH	The Bug /MPS;EHM or EHC	$7.50	1926	RO
	The Rolling Pin /MWS;EHC	$7.50	1926	RO
	Wall Pocket /MPS;EHC	$7.50	1926	RO
Burstein-Applebee Co., Kansas City MO, retailer (see Melodian Labs., mfr.)	Melody King No. 2 /TFM;OHC	$0.59	1937	CO
Burwell Co., Lee D., Minneapolis MN	Crystal /XXX;XXX	X	1925	O
Buscher Co., C. A., Kansas City MO, distributor (mfr. of BUSCO sets, Kansas City Radio Mfg. Co.; mfr. of Fullatone, Star Mfg. Co.)	Atchison /UWI;OHC	$7.50	1923	C
	BUSCO Jr. /UWI;OHC	$5.00	1923	CV
	BUSCO Special /UWI;OHC	$9.00	1923	CV
	Fullatone /CWI;OHC	$5.00	1923	C
	Kleer-Tone Model B /CWI;EHC	$5.00	1923	CV
	Pathfinder /CWI;EHC	$10.00	1923	C
(See Duck Co., Wm. B.)	C /LWI;OHC	$31.50	1922	RO
(See Elec. City Novelty & Mfg. Co.)	C /XXX;XXX	$7.50	1923	O
(See Edgcomb-Pyle Wireless Mfg. Co.)	C-1 /UWI;OHC	$35.00	1914	C
	C-2 /UWI;OHC	$45.00	1914	C
	C-3 /UWI;OHC & OHP	$60.00	1914	C
(See Kodel Mfg. Co.)	C-1 /UWS;OHC	$5.00	1925	RO
(unknown mfr.; Ben Cooper Sales Dept., San Francisco CA, distributor)	California Diamond Jubilee /MFM;OHC	X	1925	V
California Radio Co., Los Angeles CA	Angeleno /XXX;XXX	X	E	O
C. & R. Radio Shop, Los Angeles CA	Rogwell Receiver, The Eagle UWI;OHC	X	1922	VO
Cannon Co., C. F., Springwater NY	Cannon Ball Baby Grand /BMS;OVC	$1.50* (*$0.88 in 1934)	L	CV

MANUFACTURER	SET NAME/CASE; DETECTOR	PRICE	YEAR	REF.
Carter Mfg. Co., Cleveland OH	CARCO /MWS;OHC	$7.50	1925	RV
Carter Radio Co., Chicago IL (see also Cunningham)	Carter /LWL;EHC (same as Cunningham)	X	E	V
Case Toy & Sales Co., Pittsburgh PA	Crystal /XXX;XXX	X	1925	O
(See Little Giant Radio Co.)	Cathedral Pocket Radio /MPM;EHM	$2.99	1937	M
(See Morvillo & Sons)	Cat's Meow /XXX;XXX	X	1925	O
(See Crain Bros. Radio Shoppe)	C-B (also was called "Aeriola X" earlier in 1925, "Crain Craft Jr." by 1926) /LWS;OHC	$12.00	1925	RVO
(See Man-Day Radio Corp.)	CDR-1 /UWI;OHC	X	1923	O
(See Tanner Radio Co., C. D.)	C. D. T. /UWI;OHC	X	1924	O
Chambers & Co., F. B., Philadelphia PA	No. F.B. 084 /LWL;OHC	$44.00	1914	C
	No. F.B. 086 /UWL;OHC	$38.00	1914	C
	No. F.B. 088 /BWL;OHC	$22.00	1914	C
	No. 760 /CWI;OHC	$5.00	1914	C
Champion Radio Products Co., Chicago IL	Champion /UWS;OHC	X	1924	VO
(unknown mfr.)	Chatterbox /UMS;EHM	X	E	O
Cheever Co., Wm. E., Providence RI	Wecco Gem /BWS;OHC	X	1926	VO
Cherington Radio Labs., Waukegan IL	Cherington /MMS;OHC	$2.50	1925	VO
(See Orator Radio Corp.)	Chico /XXX;XXX	X	1926	O
Clapp-Eastham Co., Cambridge, MA (see also Sears, Roebuck & Co., retailer in 1916 of Cruiser)	Clapp-Eastham /BWI;OHC & OHC	X	1913	O
	Blitzen /UWI;OHC	$33.00	1913	CO
	Cruiser /BWI;OHC	$19.25	1916	CO
	Type D /UWL;XXX	$180.00	1913	CO
	Type AF /UWL;XXX	$550.00	1914	O
	Lodge /XXX;XXX	X	1915	O
	Type ASA /UWL;XXX	X	1916	O
	Type S /UWL;XXX	X	1916	O
	Radion (A1377) /BWI;OHC	$21.25	1914	C
Clark Co., Thos. E. (TECLA Co.), Detroit MI	TECLA "Thirty" /LFI;OHC	X	E	VO

MANUFACTURER	SET NAME/CASE; DETECTOR	PRICE	YEAR	REF.
Cleartone Radio Electrical Supply Co., Providence RI	Perfect /UWI;EHM	$7.50	1925	RO
(unknown mfr.; see Radio Surplus Corp., retailer)	Clin-Tone-Ia /TPS;OHC	$2.00	1926	QV
Clinton Radiophone Co., Chicago IL	Clinton /LWI;OHC	X	1922	R
Cloverleaf Mfg. Co., Chicago IL	Wow /CWI;OHC	X	1925	O
Coast-To-Coast Radio Corp., NY NY, retailer (see Philmore Mfg. Co., mfr.)	Coast-To-Coast, Jr. (brand name for Philmore Blackbird) /UMI;EVM	$1.77	1932	C
(See Racon Elec. Co.)	Coherer Receiver Set /XXX;XXX	$2.00	1923	VO
(unknown mfr.)	Colonnade Jr. /UFS;XXX	X	E	O
Columbia Radio Corp., Chicago IL	Gem of the Air Model A /LWI;OHC	X	E	V
(See Aerial Elec. Co.)	Combination King-1 /TPI;OVC	$10.95	1926	RO
	Combination King-3 /XXX;XXX	$19.95	1926	RO
Commerce Radiophone Co., Detroit MI	Commerce Radiophone /LWI;OHC	$25.00	1919	RVO
(See Duck Co., Wm. B.)	Commercial Type D /LWL;OHC	$95.00	1914	C
(See Electro Importing Co.)	Commercial /LWL;OVC & OVE	$75.00	1916	MC
(See RCA; Wireless Specialty Apparatus Co.)	Concert, Radiola:			
	AR-1375 /LMI;OHC	$40.00	1922	RQMCO
	AR-1382 /UWI;OHC	$70.00	1922	C
Concert Radio Phone Co., Cleveland OH	Concert Jr. /UWI;OHC	$3.50	1925	RO
Connecticut Tel. & Elec. Co., Meriden CT	Type JR-3 /LWI;EHM	$30.00	1922	CVO
Cooper Sales Dept., Ben, San Francisco CA, distributor (unknown mfr.)	California Diamond Jubilee /MFS;OHC	X	1925	V
Corcoran Lamp Co., Thos. J., Cincinnati OH	Crystal /XXX;XXX	X	1925	O
Corona Radio Corp., NY NY	Crystal /UWI;OHC	$4.75	1922	Q
Cosmos Elec. Co., NY NY	Cosmophone BWI;OHC	$7.50	1914	O
(See Radio Service Co.)	(Type) CR /UWI;EVC	X	E	V

MANUFACTURER	SET NAME/CASE; DETECTOR	PRICE	YEAR	REF.
Crain Bros. Radio Shoppe, Oakland CA	Aeriola X (also called "C-B" later in 1925, "Crain Craft Jr." by 1926) /LWS;OHC	$6.00	1925	RVO
(See Petite Radio Corp.)	Crista-Phone /MPM;EHM	X	E	V
Crosley Mfg. Co., Cincinnati, OH (see also Precision Equipment Co.)	Crosley /UWI;OHC	$7.00	1921	RQMO
	Harko /UWI;OHC	$9.00	1921	RQMO
	Crosley Jr. /MWS;OHC	X	E	V
	Model No. 1 /UWI;OHC	$25.00	1922	RQMCO
(See Sears, Roebuck & Co.; also Clapp-Eastham Co.)	Cruiser /BWI;OHC	$19.95	1916	C
(See Wilkins Radio Co.)	Crystal Artay /XXX;XXX	$5.00	1926	O
(See Bethlehem Radio Corp.)	Crystal-Dyne /MMI;EVC	X	E	V
(See Aerial Elec. Co.)	Crystal Mystery /LWS;OHC	$10.00	1925	RO
	Crystal King /XXX;XXX	$8.00	1926	RO
(See Radio Distributing Co./ RADISCO)	Crys-ton Type R-300-A /LWI;OHC	X	E	V
(See Ajax Elec. Specialty Co.)	CST Jr. /CWI;OHC	$3.00	1925	RO
(See Home Radio Corp.)	CT /LWI;EVC	X	E	O
Cunningham, Detroit MI (see also Carter Radio Co.)	Cunningham's Radio Outfit /LWL;EHC (same as Carter)	X	E	V
Curry & Coutellier Labs., Los Angeles CA	Radio Argentite /XXX;XXX	X	1923	O
Custer, Dayton OH	Air Bug / CWM;OHC	X	E	V
(See Duck Co., Wm. B.)	CV /LWI;OHC	$30.00	1922	RO
(See Clapp-Eastham Co.)	D /UWL;XXX	X	1913	O
(See United Wireless Co.)	D /XXX;XXX	X	1912	O
Dank & Co., M. Carlton, Brooklyn NY	Dank Kit Set /BWI;OHC	X	E	O
DeForest Radio Tel. & Tel. Co., NY NY (see also General Radio Co.; Liberty Elec. Co.)	BC-14A (also called "Type SCR-54", "Type SCR-54A") /LWL;EVC	X	1918	VO
	Everyman Radiophone (DT-600) /LWI;EHC	$25.00	1922	RQMCVO
Deitrickson Radio Co., (address unknown) (see also Simplex Electrical Lab.)	Simplex /XXX;XXX	$8.50	1923	O
	Deluxe /XXX;XXX	$15.00	1923	O
(See Waterbury Button Co.)	Detector /MWS;EHM	X	1927	R

MANUFACTURER	SET NAME/CASE; DETECTOR	PRICE	YEAR	REF.
Detroit Coil Co., Detroit MI	Crystal /BWI;OHC	X	E	O
Detroit Radio Co., Detroit MI	Maxitone No. 1 /UWS;OHC	$17.50	1922	R
	Maxitone No. 2 /UWI;OHC	$30.00	1922	R
	Airmuse /XXX;XXX	$6.00	1923	O
(See United Metal Stamping & Radio Co.)	Diamond /XXX;XXX	$4.00	1924	O
Doron Bros. Electrical Co., Hamilton OH	Portable Receiving Set:			
	No. 1 /LWI;OHC	$53.00	1920	C
	No. 2 /LWI;OHC	$60.00	1920	C
	Receiving Set No. 380 /BWI;OVC	$25.30	1920	C
	Students' No. 381 /UWI;OHC & OHP	$71.60	1920	C
(See Adams-Morgan Co.)	Double Crystal, The Arlington /UWI;OHC & OHC	X	E	O
Douglass Auto Appliance Co., Minneapolis MN	Douglass /BMM;OHC	X	E	V
Duck Co., J. J., Toledo OH	61 x 12 /O--;OVC	$1.85	1913	C
	61 x 13 /O--;OVC	$5.00	1913	C
	61 x 14 /O--;OVC	$6.25	1913	C
	61 x 15 /O--;OVE	$7.40	1913	C
	61 x 16 /O--;OVC	$11.40	1913	C
	61 x 17 /O--;OVM	$12.00	1913	C
	61 x 18 /O--;OVC	$13.90	1913	C
	61 x 19 /O--;OVC	$19.95	1913	C
	61 x 20 /O--;OVC	$24.00	1913	C
	61 x 21 /O--;OVC	$32.00	1913	C
Duck Co., Wm. B., Toledo OH	A505 /BWI;OVC	$47.50	1914	C
	A1002 /BWI;OVC	$8.80	1914	C
	Arlington (R40) /BWI;OHC	$17.50	1916	C
	Type C /LWI;OHC	$31.50	1922	O
	Type CV /LWI;OHC	$30.00	1922	RO
	Commercial Type D /LWL;OHC	$95.00	1914	C
	Jr. Receiver Set:			
	No. A580 /BWI;OVC	$4.00	1914	C
	No. A585 /BWI;OVC	$6.00	1914	C
	R41 /BWL;OHC or OVC	$32.00	1916	C
	R42 /BWL;OHC	$47.00	1916	C
	(also same 11 sets listed above for Duck, J. J. Co.)	(from $1.85 to $29.00)	1914	C
Duncan, Donald F., Chicago IL	Duncan Simple /LWS;OHC	X	E	VO
Durban Elec. Co., Chicago IL	Crystal /XXX;XXX	X	1925	O

MANUFACTURER	SET NAME/CASE; DETECTOR	PRICE	YEAR	REF.
(See Palmer Elec. & Mfg. Co.)	DU-WA (Dual Wave) /UMS;OHC	X	1924	VO
(See Electrical Products Mfg. Co.)	Dymac /XXX;XXX	X	1925	O
(See United Wireless Co.)	E /MWL;OHC	X	1912	M
(See C. & R. Radio Shop)	(Rogwell Receiver) The Eagle /UWI;OHC	X	1923	VO
(The) Eastern Specialty Co., Philadelphia PA	TESCO Type A /BWM;OHC	$1.25	1923	MO
	TESCO Type B /LWI;OHC	$5.00	1923	MO
	Crystal /LWI;OHC	X	E	O
(See Snellenburgs)	Eastern Type /LWS;OHC	$10.00	1923	O
Eclipse Mfg. Co., Los Angeles CA	Eclipse /BWS;OHC	X	E	O
Edelman, P. E., NY NY	Aladdin Multitone (or Edelman's Multitone):			
	Large Set /UWI;OHC	$22.00	1922	R
	Smaller Set /UWS;OHC	$15.00	1922	R
Edgcomb-Pyle Wireless Mfg. Co., Pittsburgh PA	Type A-1 /BWI;OHC	$5.00	1914	C
	Type A-2 /BWI;OHC	$11.50	1914	C
	Type A-3 /BWI;OHC	$14.50	1914	C
	Type A-4 /BWI;OHC	$13.50	1914	C
	Type A-5 /BWI;OHC	$17.50	1914	C
	Type A-6 /BWI;OHC	$21.00	1914	C
	Type A-7 /UWI;OHC	$25.00	1914	C
	Type C-1 /UWI;OHC	$35.00	1914	C
	Type C-2 /UWI;OHC	$45.00	1914	C
	Type C-3 /UWI;OHC & OHP	$60.00	1914	C
Edgecomb, A. J./Wireless Shop, The, Los Angeles CA	Wireless Shop /LWI;OHC	X	E	O
Electrad Corp. of Am., NY NY	Crystal /XXX;XXX	$10.00	1923	O
Electrical Products Mfg. Co., Providence RI	Dymac /XXX;XXX	X	1925	O
Elec. City Novelty & Mfg. Co., Schenectady NY	C /XXX;XXX	$7.50	1923	O
	Radjo /XXX;XXX	$3.50	1923	O
Elec. Machine Corp., Indianapolis IN	ELMCO /XXX;XXX	$3.00	1922	M
Electro Importing Co. (EICO), NY NY	Commercial /LWL;OVC & OVE	$75.00	1913	MC
	Inter-ocean /BWI;OVC	$8.00	1916	C
	Interstate /CPS;OVC	$3.75	1913	MC
	Transcontinental /BWI;OVC	$24.00	1913	MC
	Trans-Pacific /BWI;OHC	$10.00	1916	C

MANUFACTURER	SET NAME/CASE; DETECTOR	PRICE	YEAR	REF.
Electro Importing Co. (*continued*)	Wireless Experimenter (Kit) /MMM;OHC	$1.25	1916	C
(See Westwyre Co.)	Elf /XXX;XXX	X	1925	O
Ell Ess Radio Exchange, (address unknown)	Journal /XXX;XXX	$9.00	1924	O
(See Elec. Machine Corp.)	ELMCO /XXX;XXX	$3.00	1922	M
El Paige Radio Co., (address unknown)	Crystal /XXX;XXX	$5.00	1926	O
Empire Radio Equipment Co., NY NY	Empire /UWI;OHC	$16.00	1921	RMO
(See New York Coil Co.; Pioneer Wireless Mfrs.)	Entertain-A-Phone Model 1 /BWI;OHC	$15.00	1922	CV
(See General Elec. Co.; RCA)	ER 753 /UMI;OHP	$18.00	1922	RQMCVO
	ER 753A (Radiola I) /LWI;OHP	$25.00	1922	RQMCVO
Ernest Elec. Co., St. Louis MO	Crystal (Kit) /UWI;EHC	$4.35	1923	R
Essex Mfg. Co., Newark NJ (see Paige Radio Co.)	Pall Mall /XXX;XXX	$14.75	1923	O
Essex Specialty Co., Berkeley Heights NJ	Aerophone /XXX;OHC	X	E	V
(See Radio Apparatus Co.)	Etherphonette /MMM;XXC	$12.50	1923	O
(See Pacific Screw Co.)	Eureka /XXX;XXX	X	1925	O
Everett Radio Co., Chicago IL (see Pacific Screw Co.)	Acme /XXX;XXX	$3.50	1925	O
Everyhome Radio Co., Ft. Wayne IN	Everyhome /LFS;OHC	X	E	VO
(See DeForest Radio Tel. & Tel. Co.)	Everyman Radiophone (DT-600) /LWI;EHC	$25.00	1922	RQMCVO
(See Sterling Radio Mfg. Co.)	Excello Galena /LWS;OHC (same as Greg-Sor)	X	E	VO
(See McIntosh Stereopticon Co.)	Excellophone /XXX;XXX	X	1924	O
Federal Telephone & Radio Corp., Newark NJ	MacKay Marine Radio Receiver Type 123 BX /O--;OHC	X	L	O
Federal Tel. & Tel. Co., Buffalo NY	Federal Jr. /UMI;OVC	$25.00	1922	RQMVO
Ferro Mfg. Co., Belvidere IL	Ferro /UWI;OHC	X	1924	VO

MANUFACTURER	SET NAME/CASE; DETECTOR	PRICE	YEAR	REF.
Firth & Co., John (FIRCO), NY NY	FIRCO Midget Type 35-A /O--;OHC	$15.00	1921	RO
	Type 35-B /O--;OHC	X	1922	O
Flash Radio Corp., Boston MA	Flash /XXX;XXX	X	1925	O
(See Lightrite Co.; Moore Mfg. Co.)	Flivver /TMM;OHC (same as Moore)	X	1925	VO
(See Hearwell Elec. Co.)	Forbes Lyric /O--;OHC	$5.00	1925	RO
	Forbes Selective /XXX;XXX	$3.50	1926	R
Ford Co., K. N., Los Angeles CA	Radiogrand /UFS;OHC	X	1924	O
(unknown mfr.)	Fordham /XXX;XXX	X	E	O
Forman & Co., NY NY	Crystal /XXX;XXX	$1.25	1923	O
Freed-Eisemann Radio Corp., NY NY (also called Marvel Radio Mfg. Co.; Radio Mfg. Co.; Radio Receptor Co.)	Marvel Model 101 /UWS;OHC	$8.00	1921	RQMCVO
(See Parker Radio Co.)	(Liberty Jr.) Freedom of the Air /MMS;OHC	X	E	VO
(See Buscher Co., C. A.; also see Star Mfg. Co.)	Fullatone /CWI;OHC (Star appears to be same set as Fullatone)	$5.00	1923	CV
Fuller Co., P. H., NY NY	Fuller's Sparta /XXX;XXX	X	1925	O
Furr Radio, Phoenix & Mesa AZ	The Mighty Atom /BWS;OHC	$1.50	1936	VO
Gardner Labs., Los Angeles CA	Gardner Broadcast Transformer /CFI;OHC	X	E	VO
Gardner-Rodman Corp., NY NY, mfr. (became Garod Corp. in 1923; see also Radio Distributing Co.)	Heliphone Pocket Radio Receiver /LWS;OHC	$5.00	1921	RVO
(See Gibbons-Dustin Radio Mfg. Co.)	G-D 1 /XXX;XXX	X	1922	O
Gehman & Weinert, NY NY	G. W. /BWS;OVC	X	E	V
(See Columbia Radio Corp.)	Gem of the Air Model A /LWI;OHC	X	E	V
(See Metro Electrical Co.; United Specialties Co./ U.S. Co.)	(Little) Gem /MMS;OVC	$6.50	1922	MCVO
(See Cheever Co., Wm. E.)	(Wecco) Gem /BWS;OHC	X	E	V

MANUFACTURER	SET NAME/CASE; DETECTOR	PRICE	YEAR	REF.
General Elec. Co., Schenectady NY (see also RCA)	AR-1300 /UMI;OHP	$50.00	1922	RQMCVO
	ER-753 /UMI;OHP	$18.00	1922	RQMCVO
	ER-753A (Radiola I) /LWI;OHP	$25.00	1922	RQMCVO
General Radio Co., Cambridge MA (see also DeForest Radio Tel. & Tel. Co.; Liberty Elec. Co.)	BC-14A (also called "Type SCR-54," "Type SCR-54A") /LWL;EVC	X* (*$4.95 in 1931 as Army Surplus item)	1918	CVO
(See Metro Electrical Co./ Metropolitan Radio Corp.)	(Little) Giant /LWS;OHC	$15.00	1922	RMO
(See Philmore Mfg. Co.)	(Little) Giant (or Super Tone/ Supertone/Super) /TMS,TPS; EVC	$1.50* (*price in 1931)	1926	CVO
(See Johnson Smith & Co.)	(Little) Giant Pocket Radio (also called Midget Pocket Radio) /MFM;EHM	$3.50	1936	CV
Gibbons-Dustin Radio Mfg. Co., Los Angeles CA	G-D 1 /XXX;XXX	X	1922	O
(See Standard Radio & Elec. Co.)	Giblin Radioear /LWI;EVC	X	1921	VO
Gilbert Co., A. C., New Haven CT	A. C. Gilbert No. 4016 /LWI;OHC	$10.00	1922	CVO
Gilfillan Bros., Los Angeles CA	R-475 /XXX;XXX	$5.00	1923	O
(See Voges Co., Glen F.)	Glen /XXX;XXX	X	1925	O
(See Lyons Co., G. E.)	Glen Radio Model K-12 /UWI;OHC	X	E	V
Globe Phone Mfg. Co., Reading MA	Globe /XXX;XXX	$6.50	1923	O
(See Kodel Mfg. Co.)	Gold Star S-14 /XXX;XXX	$6.00	1925	RO
Gordon Radio & Elec. Mfg. Co., Seattle WA	Gordon /XXX;XXX	X	1925	O
(See Ritter Radio Corp.)	Grand /CWS;OHC	$2.50	1923	MO
(See Sterling Radio Mfg. Co.)	Greg-Sor /LWS;OHC (same as Excello Galena)	X	E	O
Guarantee Sales Co., NY NY	Crystal /CXI;OHC	X* (*premium; not sold)	1923	RO
Gundlach-Manhattan Optical Co., Rochester NY (see also Liberty Mail Order House)	Gundlach /CWI;OHC (same as Liberty)	$5.00	1924	RCVO
	Korona /XXX;XXX	$6.00	1925	RO
(See Gehman & Weinert)	G. W. /BWS;OHC	X	E	V

MANUFACTURER	SET NAME/CASE; DETECTOR	PRICE	YEAR	REF.
Haller, W. B., Pittsburgh PA	Crystal /XXX;XXX	$5.00	1923	O
	Hallerio, Model:			
	III /CWS;OHC	$3.00	1924	RO
	IV /CWS;OHC	$4.00	1924	RO
	V /UWS;OHC	$5.00	1924	RO
	1½ /CWS;OHC	$1.50	1925	RO
	2½ /XXX;XXX	$2.50	1926	O
	3½ /UWS;OHC	$3.50	1925	RO
Hamburg Bros., Pittsburgh PA	Pennsylvania:			
	No. 1 /XXX;XXX	$4.50	1924	O
	No. 2 /XXX;XXX	$3.00	1924	O
	Jr., No. 1 /CWI;OHC	$2.50	1926	O
	Jr., No. 2 /XXX;XXX	$2.00	1926	O
Hammandson, New Castle PA	Hammandson /LWS;EHC	X	E	O
Handel Elec. Co., NY NY	Wireless Station /BWI;OHC	$10.10	1916	MO
Harcourt Radio Co., Toronto, Ontario CANADA (mfd. in Canada; sold in USA)	HARCO /TPS;EHM	$3.00	1925	RO
Hargraves, C. E. & H. T., Lakewood RI	Mel-O-tone /XXX;XXX	X	1924	O
(See Crosley Mfg. Co.; Precision Equipment Co.)	Harko (Crosley) /UWI;OHC	$9.00	1921	RQMO
Harrison Mfg. Co., Detroit MI	Crystal /XXX;XXX	$5.00	1922	R
Heard Co., St. Louis MO	Heard /LWL;OHC	X	E	V
Hearwell Elec. Co. (also called Hearwell Radio), Boston MA	Crystal /XXX;XXX	X	1924	O
	Variometer Lyric /O--;OHC	$10.00	1925	RO
	Whole Wave /O--;OHC,OHC	$8.00	1925	RO
	Forbes Lyric /O--;OHC	$5.00	1925	R
	Forbes Selective /XXX;XXX	$3.50	1926	R
	Hearwell /XXX;XXX	$3.75	1926	O
Hecco Elec. Co., Menominee MI	Arlington /BWI;OHC	$7.50	E	O
Heinemann Elec. Co., Philadelphia PA	Sensory /XXX;XXX	$10.00	1923	O
Heintz-Kohlmoos, San Francisco CA	Crystal /XXX;XXX	X	1925	O
(See Gardner-Rodman Co.; Radio Distributing Co.)	Heliphone Pocket Radio Receiver /LWS;OHC	$5.00	1921	RVO
Herke Radio Product Co., (address unknown)	Crystal 1 /XXX;XXX	$2.75	1924	O
	Crystal 2 /XXX;XXX	$10.00	1924	O

MANUFACTURER	SET NAME/CASE; DETECTOR	PRICE	YEAR	REF.
(See Radio Receptor Co.)	Home-o-fone No. 2 /LWI:EHC	$24.00	1922	RO
Home Radio Corp., NY NY	Type CT /LWI;EHC	X	E	O
(See McCorkle Co., D. H.)	Home Radio /XXX;XXX	X	E	O
Home Supply Co., NY NY	Crystal /BWS;OHC	X* (*premium; not sold)	1922	R
Howe Auto Products Co., Chicago IL	Howe /UMS;OHC	$1.75	1925	RVO
	No. 1 /XXX;XXX	$5.00	1926	O
	No. 2 /XXX;XXX	$7.00	1926	O
Hunt & McCree, NY NY	No. 797 /BWS;OVE	$1.75	1912	CVO
Hyman & Co., Henry, Chicago IL	Bestone /UWI;EHC	$22.50	1923	RO
(See Insuline Corp. of Am.)	ICA /BMS;OVC	X	E	O
	ICA Top Notch /TMS;EVC	$1.50	1939	C
Inman Specialty Store, Holyoke MA	Crystal /XXX;XXX	$5.00	1925	O
Insuline Corp. of Am., NY NY	ICA /BMS;OVC	X	E	O
	Pied Piper /MPS;EVC	$4.50	1934	CO
	ICA Top Notch /TMS;EVC	$1.50	1939	C
(See Electro Importing Co.)	Inter-Ocean /BWI;OVC	$8.00	1916	C
	Interstate /CPS;OVC	$3.75	1913	MC
(See Wireless Specialty Apparatus Co.)	IP-76 (later, called IP-111) /UWL;OHP	X	1907	O
	IP-77 /UWL;XXX	X	1919	O
	IP-500 /UWL;XXX	$425.00	1918	O
Jaynxon Labs., (address unknown)	Crystal /XXX;XXX	X	1923	O
(See J. K. Corp.)	Jewel /UWI;OHC	X	E	O
(See Patterson Elec. Co.)	Jewel /UWI;OHC	$12.00	1922	MO
(See Manhattan Electrical Supply Co.)	Jewellers' Time /UWL;OHC	$66.00	1914	C
Jewett Mfg. Corp., NY NY (see also Wireless Equip. Co., name of mfr. in 1920)	ABC /UWI;OHC	$40.00	1922	R
J. K. Corp., (unknown address)	Jewel /UWI;OHC	X	E	O
	Victor /UWI;OHC	X	E	O
Johnson Smith & Co., Detroit MI, retailer (see Little Giant Radio Co., mfr.)	Little Giant Pocket Radio (also called "Midget Pocket Radio") /MFM;EHM	$2.99	1936	MCV

MANUFACTURER	SET NAME/CASE; DETECTOR	PRICE	YEAR	REF.
(See Ell Ess Radio Exchange)	Journal /XXX;XXX	$9.00	1924	O
(See Conn. Tel. & Elec. Co.)	(Type) JR-3 /LWI;EHM	$30.00	1922	CVO
Jubilee Mfg. Co., Omaha NE	Jubilee /BMS;OHC	X	1926	VO
(See Duck Co., Wm. B.)	Junior:			
	No. A580 /BWI;OVC	$4.00	1914	C
	No. A585 /BWI;OVC	$6.00	1914	C
(See Remler Radio Mfg. Co.)	(Remler) Junior No. 1 /BWS;OHC	$5.00	1921	CO
(unknown mfr.)	Juvenola /XXX;XXX	X	E	O
Kansas City Radio Mfg. Co., Kansas City MO (see also Buscher Co., C. A.)	Radioceptor /UWI;OHC (BUSCO Special appears to be same set)	$18.00	1922	RQCVO
Kenrad Radio Corp., NY NY	Last Word, Model B-12 /UWI;OHC	X	E	V
Kilbourne & Clark Mfg. Co., Seattle WA	K & C /XXX;XXX	$13.50	1922	RO
	K-C Junior No. 1 (kit set) /XXX;XXX	$20.00	1920	C
	K-C Junior No. 2 (kit set) /XXX;XXX	$35.00	1920	C
Killark Elec. Mfg. Co., St. Louis MO	Crystal /XXX;XXX	$4.50	1923	O
Kingbridge Labs., NY NY (mfd. in France; sold in USA)	The Ondophone /MMS;EHM	$7.25	1914	O
(See Aerex Radiophone Co.)	King of the Air /LWI;OHC	X	1922	O
(See Baughman Radio Engineering Co.)	King Tut /XXX;XXX	X	1924	O
(See Radio Tel. & Tel. Co.)	Kismet /UMI;OHC	X	E	V
(See Buscher Co., C. A.)	Kleer-Tone Model B /CWI;EHC	$5.00	1923	CV
Kodel Mfg. Co. (also called Kodel Radio Corp.), Cincinnati OH	S-1 /UWS;OHC	$5.00	1924	RMVO
	C-1 /UWS;OHC	$5.00	1925	RO
	Gold Star S-14 /XXX;XXX	$6.00	1925	RO
	Quality /XXX;XXX	X	1925	O
(See Gundlach-Manhattan Optical Co.)	Korona /XXX;XXX	$6.00	1925	RO
Kresge Co., S. S., Detroit MI, retailer (see Brooklyn Metal Stamping Co., mfr. of BMS; see Pal Radio Co., mfr. of Pal)	BMS /BMS;OHC (called "Pandora" by mfr. earlier)	$1.00	1929	C
	Pal /BMX;OHC (same as Metro Jr.)	$0.89* (*price in 1929)	1926	RCO

MANUFACTURER	SET NAME/CASE; DETECTOR	PRICE	YEAR	REF.
(unknown mfr.), Attleboro MA	LaBelle /UWI;OHC	X	E	O
LaBidite Radio Products Co., NY NY	Crystal /XXX;XXX	X	1926	O
Lafayette Radio, NY NY (see Addendum—different sets in 1949 were called "Tee-nie")	Tee-nie:			
	Midget /UWS;XXM	$0.88	1939	C
	DeLuxe /UWS;XXM	$2.35	1939	C
Lalley Elec. Co., (address unknown)	Lalley /XXX;XXX	X	E	O
Lamb Co., F. Jos., Detroit MI	Lamb /BWI;OHC	$10.00	1922	R
Lambert Radio Co., Leon, Wichita KS	Crystal /MFM;OHC	X	E	VO
	Leon Lambert Long Distance /UWS;OHC	X	1926 (1925?)	VO
	Lambert Long Distance /MFS;OHC	X	1926	VO
(See Kenrad Radio Corp.)	Last Word, Model B-12 /UWI;OHC	X	E	V
Laurence Radiolectric Co., Cincinnati OH	Portable /XXX;XXX	$15.00	1922	R
Lawsam Elec. Co., NY NY	Lawsam Baby /UWS;OHC	$17.50	1922	RQV
(See United Metal Stamping Co.)	L. D. R. /XXX;XXX	$4.90	1924	O
(unknown mfr.)	Lectro /UWI;EVC	X	E	O
Lee Elec. & Mfg. Co. (LEMCO), San Francisco CA	LEMCO No. 340 /LWS;OHC	$7.50	1923	RVO
	LEMCO No. 340-A /LWS;OHC	$6.00	1924	RVO
	LEMCO No. 340-B /LWS;OHC	$6.00	1924	RVO
	LEMCO Portable /LFI;EHC	$9.00	1923	VO
Le Roy Elec. Co., Alhambra CA	Le Roy /UWI;EVM	X	1925	VO
(See Radio Shop)	Leroy /XXX;XXX	$15.00	1924	O
Liberty Elec. Corp., Port Chester NY (see also DeForest Radio Tel. & Tel. Co.; General Radio Co.)	BC-14A (also called "Type SCR-54," "Type SCR-54A") /LWL;EHC	X	1918	VO
(See Parker Radio Co.)	Liberty Jr., Freedom of the Air /MMS;OHC	X	E	VO
Liberty Mail Order House, NY NY, (see Gundlach-Manhattan Optical Co., mfr.)	Liberty /CWS;OHC (same as Gundlach)	$3.95	1924	C

MANUFACTURER	SET NAME/CASE; DETECTOR	PRICE	YEAR	REF.
Liberty Starter Co., (address unknown)	Crystal /XXX;XXX	X	E	O
Lightrite Co., Bloomfield NJ (see also Moore Mfg. Co.)	Soundrite /XXX;XXX	X	1925	O
	Flivver /TMM;OHC (same as Moore)	X	E	VO
Listen-in Publishing Co., Cambridge MA	Talking Book /MFS;EHM	$5.00	1926	RMO
(See Nu-Tone Radio Co.)	Little Freak /XXX;XXX	X	1926	O
(See Metro Electrical Co.; United Specialties Co./ U. S. Co.)	Little Gem /MMS;OVC	$6.50	1922	MCVO
(See Metro Electrical Co.; Metropolitan Radio Corp.)	Little Giant /LWS;OHC	$15.00	1922	RMO
(See Philmore Mfg. Co.)	Little Giant (or Super/ Super Tone/Supertone) /TMS,TPS;EVC	$1.50* (*price in 1931)	1926	CVO
(See S & M Radio Co.)	(Pocket Radio) Little Giant Radiophone /TFM;XXX	$2.00	1924	R
Little Giant Radio Co., Chicago IL (see also Johnson Smith & Co.)	Little Giant Pocket Radio (also called "Midget Pocket Radio") /MFM;EHM	$2.99	1936	MCV
	Cathedral Pocket Radio /MPM;EHM	$2.99	1937	M
(See Marinette Elec. Corp.; Telephone Maintenance Co.)	Little Tattler /CWS;OHC	$6.00	1923	CO
(See Philmore Mfg. Co.)	Little Wonder (later Sky Rover) /TMM,TPM;OHC	$1.00* (*price in 1931)	1926	CVO
(See Am. Electro Technical Appliance Co.; Radio Service & Mfg. Co.)	Little Wonder Portable Type S-8 /LWS;OHC	$6.00	1920	RMO
(See Clapp-Eastham Co.)	Lodge /XXX;XXX	X	1915	O
(See Lambert Radio Co., Leon)	(Leon Lambert) Long Distance /UWS;OHC	X	1926 (1925?)	VO
	(Lambert) Long Distance /MFS;OHC	X	1926	VO
(See Nichols Elec. Co.)	Long Distance /BWI;OVP	$6.75	1914	MCO
(See Steinite Labs.)	(Steinite) Long Distance /UWI;EHC	$6.00	1924	RMVO

MANUFACTURER	SET NAME/CASE; DETECTOR	PRICE	YEAR	REF.
(See Sears Roebuck & Co.)	Loose Coupler Crstal Outfit /BWI;OHC	$22.08	1923	C
	Loose Coupler Set Mounted with two detectors /BWI;OHC & OVC	$12.90	1917	C
Los Angeles Radio Supply Co., Los Angeles CA	Crystal /UWS;EHM	X	1925	O
Louisville Elec. Mfg. Co., Louisville KY	Crystal /UWI;OHC	X	E	O
(See Philmore Mfg. Co.)	Luxor /UFI;EVC (label variation of early Selective; same as Peerless)	$5.00* (*price of Selective)	1931	O
Lyons Co., G. E. St. Louis MO (see also Voges Co., Glen F.; both St. Louis mfrs. made a "Glen")	Glen Radio Model K-12 /UWI;OHC	X	E	V
(See Hearwell Elec. Co.)	Lyric, Forbes /O--;OHC	$5.00	1925	RO
	Lyric, Variometer /O--:OHC	$10.00	1925	RO
(See Federal Telephone & Radio Corp.)	MacKay Marine Radio Receiver Type 123 BX /O--;OHC	X	L	O
McCorkle Co., D. H., Oakland CA	Home Radio /XXX;XXX	X	E	O
McIntosh Stereopticon Co., (address unknown)	Excellophone /XXX;XXX	X	1924	O
McKay Instrument Co., Portland OR	McKay /UWS;OHC	$2.95	1928	RO
McMahon & St. John, South Norwalk CT	Type 5-A /BWI;OHC	$6.00	1919	R
(See Radio Specialties Co.)	M-2 Master /XXX;XXX	$15.00	1925	O
Magic, Portland OR	Magic /LWI;OHC	X	E	O
Magnus Elec. Co., NY NY (by 1925, called Magnus Elec. & Radio Mfg. Corp.)	4 /XXX;XXX	$12.00	1923	O
	55 /XXX;XXX	$12.00	1923	O
	77 /XXX;XXX	$16.00	1923	O
	872 /XXX;XXX	$10.00	1923	O
	873 /XXX;XXX	$12.00	1923	O
	876 /XXX;XXX	$10.00	1923	O
Man-Day Radio Corp., NY NY	Type CDR-1 /UWI;OHC	X	1923	O

MANUFACTURER	SET NAME/CASE; DETECTOR	PRICE	YEAR	REF.
Manhattan Electrical Supply Co. (MESCO), NY NY	Jewellers' Time /UWL;OHC	$66.00	1914	C
	MESCO Type B Receiving Set No. 458 /LWI;OVC	$7.80	1910	CO
Marinette Elec. Corp., Marinette WI (see also Telephone Maintenance Co.)	Little Tattler /CWS;OHC	$6.00	1923	CO
Marquette Radio Corp., Chicago IL (made for Quaker Oats Co.)	Quaker Oats /CFI;OHC or EHC	$1.00* (*plus 2 Quaker Oats box labels)	1921	VO
Martian Mfg. Co., Newark NJ (see also Blair Co., Mitchell; White Mfg. Co.)	Martian Big 4 /MMS;OHC (same as Blairco "4")	$7.50	1923	VO
	Martian Special /BMS;OHC	X	1924	VO
Marvel Radio Mfg. Co., NY NY (also called Freed-Eisemann Radio Corp.; Radio Mfg. Co.; Radio Receptor Co.)	Marvel Model 101 /UWS;OHC	$8.00	1921	RQMCVO
(See Radio Marvel Co.)	(Radio) Marvel /XXX;XXX	$2.00	1924	M
Master Elec. Co., Chicago IL (by 1925, Master Radio Co.)	Crystal /XXX;XXX	$6.00	1923	O
(See Detroit Radio Co.)	Maxitone No. 1 /UWS;OHC	$17.50	1922	R
	Maxitone No. 2 /UWI;OHC	$30.00	1922	R
(See North Ward Radio Co.)	Mayer's Wonder Set /BWI;EVC	X	1922	R
Meepon, (address unknown)	Meepon /LWI;EHC	X	1923	O
Melodian Labs., Independence MO (see Burstein-Applebee Co.)	Melody King /XXX;XXX	$4.00	1923	O
	Melody King No. 2 /TFM;OHC	$0.59	1937	CO
(See Hargraves, C. E. & H. T.)	Mel-O-tone /XXX;XXX	X	1924	O
Mengel Co., Jersey City NJ; Louisville KY; St. Louis MO (see Shamrock Radio Sales Corp.; United Metal Stamping & Radio Co.)	Mengel Type M.R. 101 /LWS;OHC (same as Diamond; Shamrock Radiophone A)	$4.00 (*price of Diamond crystal set)	1924	VO
(See Manhattan Electrical Supply Co.)	MESCO Type B Receiving Set No. 458 /LWI;OVC	$7.80	1910	CO
Meteor Radio Labs., Piqua OH	Meteor /UWI;OHC	X	E	O
Metro Electrical Co., Newark NJ (also called Metropolitan Radio Corp.) (see United Specialities Co./U.S. Co.; Pal Radio Co.)	Little Giant /LWS;OHC	$15.00	1922	RMO
	Little Gem /MMS;OVC	$6.50	1922	MCVO
	Metro Jr. /BMS;OHC (same as Pal)	$2.50	1924	RVO

MANUFACTURER	SET NAME/CASE; DETECTOR	PRICE	YEAR	REF.
(See Aldine Radio & Mfg. Co.)	Micr-o-phone /CWI;OHC	$10.00	1922	R
(See Audiola Radio Co.)	Midget /XXX;XXX	X	1925	O
(See Orator Radio Corp.)	Midget /XXX;XXX	$15.00	1925	O
(See United Metal Stamping & Radio Co.)	Midget /XXX;XXX	$2.00	1924	O
(See Firth & Co., John)	(FIRCO) Midget Type 35-A /O--;OHC	$15.00	1921	R
(See Johnson Smith & Co.; Little Giant Radio Co., mfr.)	Midget Pocket Radio (also called "Little Giant Pocket Radio") /MFM;EHM	$2.99	1936	MCV
Midget Radio Co., Kearney NE (earlier called Tinytone Radio Co.; later Pa-Kette Elec. Co.; Midway Co.; see also Western Mfg. Co., in Addendum)	Midget Pocket Radio /MPM;EHM	$2.99	1939	MVO
(See Radio Products Corp. of Am.)	(RPC) Midget Radio Pocket Receiver /LWM;OHC	$3.00	1923	RO
(See Siegal Elec. Supply Co.)	Midwest Jr. /BWI;XXX	$14.00	1922	R
(See Furr Radio)	(The) Mighty Atom /BWS;OHC	$1.50	1936	VO
Mignon Wireless Corp., Elmira NY (later called Universal Radio Corp.; Mignon Mfg. Corp.; Mignon Mfg. Export Corp.; Mignon System Mfg. Co.)	RC1 /UWS;XXX*	$10.00	1915	O
	RC2 /UWI;XXX*	X	1915	O
	RLC2 /UWI;XXX*	$30.00	1915	O
	RLC2 Special /UWI;XXX*	X	1915	O
	RLC3 Special /UWI;XXX* (*external detector required)	X	1915	O
Miller Radio Co., A. H., Detroit MI	Miller /XXX;XXX	X	1925	O
(See Uncle Al's Radio Shop)	Miracle /UWL;EVC	$12.75	1925	RVO
	Miracle Model 2 /UWL;EVC	$12.75	1925	RVO
	Miracle Jr. /XXX;XXX	X	E	V
Mississippi Valley Radio Co., St. Louis MO	M-V /XXX;XXX	$4.00	1924	O
Mohawk Battery & Radio Co., Los Angeles CA	Mohawk /UWI;OHC	X	1925	VO
	Mohawk /UWS;OHC	X	1925	VO
(See Philmore Mfg. Co.)	Monarch /UMS;EVC	X	E	O
(See Rippner Bros. Mfg. Co.)	Monarch /MMS;OHC	X	1926	RVO

MANUFACTURER	SET NAME/CASE; DETECTOR	PRICE	YEAR	REF.
Monroe Radio Mfg. Co., Monroe MI	Monrona /XXX;XXX	$8.00	1925	O
Montgomery Ward & Co., Chicago IL	Airline Jr. /UWI;EHC	$7.40	1923	C
	Airline /UWI;EHC or OHC	$3.95	1925	C
Mooney Radio Co., (address unknown)	Mooney /XXX;XXX	X	1925	O
Moore Mfg. Co., Bloomfield NJ; later, Nutley NJ (see Lightrite Co.)	Moore /TMM;OHC (same as Flivver)	X	E	O
Morse Mfg. Co., Newark NJ	Morse-O-Phone (Type M-7) /UWI;OVC	$22.50	1922	R
Morvillo & Sons, Providence RI	Cat's Meow /XXX;XXX	X	1925	O
Multiphone Co., Oakland CA	Multiphone /LWS;EHM	X	1924	RO
(See Edelman, P. E.)	(Aladdin) Multitone (or Edelman's Multitone):			
	Large set /UWS;OHC	$22.00	1922	R
	Smaller set /UWI;OHC	$15.00	1922	R
Murdock Co., Wm. J. Chelsea MA (sold by J. J. Duck Co.)	Murdock No. 505 /BWI;OVC	$50.00	1914	O
	Type 103 /UWL;OHC	X	E	O
Musio Radio Co., Pittsburgh PA	Musio /UWI;EVC	$12.50	1922	RV
(See Mississippi Valley Radio Co.)	M-V /XXX;XXX	$4.00	1924	O
Myers & Blackwell, Portland OR	Crystal /BWL;OHC	X	E	O
	Crystal /CWS;OHC	X	E	O
(See Berstan Radio Products Co.)	Mysto-C /XXX;XXX	$3.00	1924	O
(See National Motor Accessories Corp.)	NACO /UWI;OHC	$20.00	1922	M
National Airphone Corp., NY NY	National Airphone G /TPS;EVT	$12.50	1922	RMO
National Elec. Supply Co. (NESCO), (address unknown)	National Electric (NESCO) CN-113 /XXX;XXX	X	E	O
	NESCO Jr. /XXX;XXX	X	E	O
	NESCO CN-239 /UWL;XXX	$425.00	1917	O
(See Sears, Roebuck & Co.)	National Guard Field Receiving Set /LPI;OHC	$22.75	1917	C

MANUFACTURER	SET NAME/CASE; DETECTOR	PRICE	YEAR	REF.
National Industries, (address unknown)	Black Diamond /XXX;XXX	$2.50	1923	O
National Motor Accessories Corp., NY NY (NACO)	NACO /UWI;OHC	$20.00	1922	M
National Radiophone, (address unknown)	National Radiophone /UWI;OHC	X	1922	O
National Radio Products Corp., NY NY	Radiolean Jr. /UWI;XXX	$12.50	1923	RO
(See Sears, Roebuck & Co.)	National Wireless Receiving Set /BWI;OVC	$10.50	1916	C
New Era Wireless Corp., NY NY	NEWCO /LWL;EVC	X	E	O
New York Coil Co., NY NY (see also Pioneer Wireless Mfrs.)	Entertain-A-Phone Model 1 /BWI;OHC	$15.00	1922	CV
Nichols Elec. Co., NY NY	Long Distance /BWI;OVP	$6.75	1914	MCO
	Time Signal Receiving Station /BWI;OHC	$10.85	1915	C
(See Novelty Radio Mfg. Co.)	Nickelette /XXX;XXX	X	1924	O
Niehoff & Co., Paul G., Chicago IL (see also Radio Electric Service/R. E. S.)	Resodon /XXX;XXX	X	1922	R
Non-Skid Crystal Mfg. Co., Kansas City MO	Crystal /XXX;XXX	X	1925	O
Northern Elec. Sales Co., (address unknown)	Roycroft /XXX;XXX	$2.50	1923	O
	Crystal /XXX;XXX	$5.00	1923	O
North Ward Radio Co., Newark NJ	Mayer's Wonder Set /BWI;EVC	X	1922	R
Novelty Radio Mfg. Co., St. Louis MO	Nickelette /XXX;XXX	X	1924	O
Nu-Tone Radio Co., Salt Lake City UT	Little Freak /XXX;XXX	X	1926	O
(See Kingbridge Labs.)	Ondophone /MMS;EHM	$7.25	1914	O
Orator Radio Corp., NY NY	Chico /XXX;XXX	X	1926	O
	Midget /XXX;XXX	$15.00	1926	O
(unknown mfr.)	O-SO-EZ /BWS;OVC	X	E	V
Pacent Elec. Co., NY NY	Pacent /UWI;EVC	X	1921	MO
Pacific Screw Co., Portland OR (see also Everett Radio Co.)	Crystal /XXX;XXX	$3.00	1923	O
	Crystal /XXX;XXX	$5.00	1923	O
	PASCO /XXX;XXX	$3.00	1923	O

MANUFACTURER	SET NAME/CASE; DETECTOR	PRICE	YEAR	REF.
Pacific Screw Co. (*continued*)	Acme /XXX;XXX	$3.50	1925	O
	Eureka /XXX;XXX	X	1925	O
Paige Radio Co., (address unknown) (see Essex Mfg. Co.)	Crystal /XXX;XXX	$5.00	1924	O
	Pall Mall /XXX;XXX	$14.75	1923	O
Palmer Elec. & Mfg. Co. Toledo OH	Du-Wa (Dual Wave) /UMS;OHC	X	1924	VO
Pal Radio Co., Jersey City NJ (see also Metro Electrical Co.)	Pal /BMS;OHC (same as Metro Jr.)	$1.50	1926	RCVO
Pal Radio Corp., North Bergen NJ	Pal Radio /LFM;OHC	$1.00	1924	RO
(unknown mfr.)	Palset /UWI;OHC	X	E	O
(See Brooklyn Metal Stamping Co.)	Pandora (2 types) /BMS;OHC (called "BMS" by a retailer in 1929)	$2.50	1922	RCVO
Parker Metal Goods Co., Worcester MA	Crystal /XXX;XXX	X	1923	O
Parker Radio Co., Cincinnati OH	Liberty Jr., Freedom of the Air /MMS;OHC	X	E	VO
Parkin Mfg. Co., San Rafael CA	Parkin /UWI;EVC	X	1922	O
(See Pacific Screw Co.)	PASCO /XXX;XXX	$3.00	1923	O
(See Buscher Co., C. A.)	Pathfinder /CWI;EHC	$10.00	1923	C
(unknown mfr.)	Patrick Henry Jr. /XXX;XXX	X	E	V
Patterson Elec. Co. (later, Patterson Radio Co.), Los Angeles CA	Jewel /UWI;OHC	$12.00	1922	MO
(See Philmore Mfg. Co.)	Peerless /UFI;EVC (label variation of early Selective; same as Luxor)	$5.00* (*price of Selective)	1931	V
(unknown mfr.)	Peerless /UWI;OHC	X	E	O
Peerless Wireless Co., Detroit MI	Crystal /BWI;OHC	X	E	O
Penberthy Injector Co. (also called Penberthy Products), Detroit MI	Penberthy Model 4R /UWI;EVC	X	1925	VO

MANUFACTURER	SET NAME/CASE; DETECTOR	PRICE	YEAR	REF.
(See Hamburg Bros.)	Pennsylvania:			
	No. 1 /XXX;XXX	$4.50	1924	O
	No. 2 /XXX;XXX	$3.00	1924	O
	Jr., No. 1 /CWI;OHC	$2.50	1926	O
	Jr., No. 2 /XXX;XXX	$2.00	1926	O
Pentz Radio Factory, Minneapolis MN	Pentzlyne /XXX;XXX	$7.50	1925	O
	Pocket Radio /XXX;XXX	$2.00	1925	O
	Power /XXX;XXX	$25.00	1925	O
(See Cleartone Radio Supply Co.)	Perfect /UWI;EHM	$7.50	1925	RO
Perfection Radiofone Co., NY NY	Perfection No. 1 /UWI;OHC	$27.50	1922	R
Perry & Sons, Frank B., Providence RI	Radio Blinker /UMS;EVC	$12.00	1922	RO
Petite Radio Corp., Boston MA	Crista-Phone /MPM;EHM	X	E	V
Philmore Mfg. Co., NY NY (see also Addendum for other, more recent sets)	Ajax (early version of Supertone?) /XXX;XXX	X	1925	O
	Blackbird /UFI,UMI;EVC	$3.00	1931	CVO
	De Luxe /UPS;EHM	$1.95	1939	C
	Kompact /MOM;EVM	$1.95	1939	C
	Little Giant (also called "Super Tone," "Supertone," "Super") /TMS,TPS;EVC	$1.50* (*price in 1931)	1926	CVO
	Little Wonder (later, Sky Rover) /TMM,TPM;OHC	$1.00* (*price in 1931)	1926	CVO
	Luxor /UFI;EVC (label variation of early Selective; same as Peerless)	$5.00* (*price of Selective)	1931	O
	Monarch /UMS;EVC	X	E	O
	Peerless /UFI;EVC (label variation of early Selective; same as Luxor)	$5.00* (*price of Selective)	1931	V
	Pocket Radio /UFM;EHM	$1.00	1938	MV
	Selective /UFI,UMI;EVC (see Addendum for entirely different Selective/VC1000)	$5.00	1931	CVO
	Supertone (or Super Tone, Super, or Little Giant) /TMS,TPS;EVC	$1.50* (*price in 1931)	1926	CVO
(See Insuline Corp. of Am.)	Pied Piper /MPS;EVC	$4.50	1934	CO
Pinkerton Elec. Equipment Co., NY NY	Pink-a-tone /LWI;EHC	$25.00	1922	RQO

MANUFACTURER	SET NAME/CASE; DETECTOR	PRICE	YEAR	REF.
Pioneer Wireless Mfrs., NY NY (see also New York Coil Co.)	Entertain-A-Phone Model 1 /BWI;OHC	$15.00	1922	CV
(unknown mfr.)	Pocketphone /TPM;OHC	X	E	O
Pocket Radio Corp., Des Moines IA	Vest-O-Fone /MPM;OHM	X	E	V
(See Am. Leader Products Co.)	Pocket Radio /UWS;XXM	X	L	O
(See Pentz Radio Factory)	Pocket Radio /XXX;XXX	$2.00	1925	O
(See Philmore Mfg. Co.)	Pocket Radio /UFM;EHM	X	L	V
(See S & M Radio Co.)	Pocket Radio Little Giant Radiophone /TFM;XXX	$2.00	1924	R
(See Johnson Smith & Co.; Little Giant Radio Co., mfr.)	(Little Giant) Pocket Radio (also called "Midget Pocket Radio") /MFM;EHM	$2.99	1936	MCV
(See Midget Radio Co.)	(Midget) Pocket Radio /MPM;EHM	X	1932	VO
(See Spencer Radio Labs.)	(Spencer) Pocket Radio /LFM;OHC	X	1929	V
(See Triangle Mfrs.)	(Vest) Pocket Radio /XXX;XXX	$1.50	1926	R
(See Tinytone Radio Co.)	(Tiny Tone) Pocket Radio /UFM;EHM	$2.99	1932	MVO
(See Gardner-Rodman Corp.; Radio Distributing Co.)	(Heliphone) Pocket Radio Receiver /LWS;OHC	$5.00	1922	RV
(See Radio Products Corp. of Am.)	(RPC Midget Radio) Pocket Receiver /LWM;OHC	$3.00	1923	RO
(See Doron Bros. Electrical Co.)	Portable No. 1 /LWI;OHC Portable No. 2 /LWI;OHC	$53.00 $60.00	1920 1920	C C
(See Laurence Radiolectric Co.)	Portable /XXX;XXX	$15.00	1922	R
(See Lee Elec. & Mfg. Co.)	(LEMCO) Portable/LFI;EHC	$9.00	1923	VO
(See Sears, Roebuck & Co.)	Portable Jr. /CWI;OVC	$3.60	1916	C
(See Turney, E. T. Labs.)	(Turney) Portable /LWL;EVT	$87.50	1915	C
(See Winn Radio & Elec. Mfg. Co.)	(Winn) Portable /LWI;EHC	$50.00	1920	RQ
Ports Mfg. Co., Fresno CA	PRMCO (Kit Set) /XXX;XXX	X	1924	O
(See Pentz Radio Factory)	Power /XXX;XXX	$25.00	1925	O

MANUFACTURER	SET NAME/CASE; DETECTOR	PRICE	YEAR	REF.
Powertone Radio Products Co., Minneapolis MN	Powertone /XXX;XXX	$9.00	E	O
Power-Tone, Detroit MI	Power-Tone /BWL;OHC	X	E	O
Precision Equipment Co., Cincinnati OH (see also Crosley Mfg. Co.)	Harko /UWI;OHC	$9.00	1921	RQMO
	Crystal /XXX;XXX	$23.50	1921	R
Premier Dental Mfg. Co. Philadelphia PA	PRAMCO Unit (Kit Set) /XXX;OHC	$2.00	1922	R
(unknown mfr.)	Premier Jr. /LWL;OHC	X	E	V
Progressive Specialty Co., (address unknown)	Crystal /XXX;XXX	X	1924	O
Pyramid Products Co., Chicago IL	Crystal /XXX;XXX	X	1925	O
Quaker Oats Co., Chicago IL (distributed as premium; Marquette Radio Corp., mfr.)	Quaker Oats Crystal Set /CFI;OHC or EHC	$1.00* (*plus 2 Quaker Oats box labels)	1921	VO
(See Duck Co., Wm. B.)	R40 (Arlington) /BWI;OHC	$17.50	1916	C
	R41 /BWL;OHC or OVC	$32.00	1916	C
	R42 /BWL;OHC	$47.00	1916	C
(See Gilfillan Bros.)	R-475 /XXX;XXX	$5.00	1923	O
(See Adams-Morgan Co.)	RA /BWS;OVC	$4.45	1913	C
Racon Elec. Co., NY NY	Racon Coherer Receiver Set /XXX;XXX	$2.00	1923	VO
Radio Apparatus Co., Detroit MI	Etherphonette /MMM;XXC	$12.50	1923	O
Radio Apparatus Co., Pittsburgh PA	Speer /LWI;EHC	$9.00	1922	R
(See Curry & Coutellier Labs.)	Radio Argentite /XXX;XXX	X	1923	R
(See Perry & Sons, Frank B.)	Radio Blinker /UMS;EVC	$12.00	1922	RO
Radio Cabinet Co. of Detroit, Detroit MI	Crystal /CWI;EHC	$3.95	1922	R
(See Kansas City Radio Mfg. Co.)	Radioceptor /UWI;OHC	$18.00	1922	RQCVO
(See Taylor Elec. Co.)	Radioclear /XXX;XXX	$2.90	1924	O
Radio Corp. of America (See RCA)				

MANUFACTURER	SET NAME/CASE; DETECTOR	PRICE	YEAR	REF.
Radio Distributing Co. (RADISCO), Newark NJ (see also Gardner-Rodman Corp.)	Crys-ton Type R-300A /LWI;OHC	X	E	V
	Heliphone Pocket Radio Receiver /LWS;OHC	$5.00	1921	RVO
(See Standard Radio & Elec. Co.)	(Giblin) Radioear /LWI;EVC	X	1921	VO
Radio Elec. Service (R. E. S.), Chicago IL (see also Niehoff, Paul G. & Co.)	Resodon /XXX;XXX	X	1922	R
Radio Engineering Labs, L. I. City NY	Radio Eng. Labs /O--;EHP	X	E	O
Radio Equipment Co. (RADECO), Boston MA (see also Bowman; Telephone Maintenance Co.)	Airophone /UWI;OHC	$20.00	1922	RCO
Radioette Mfg. Co., Los Angeles CA	Radio-ette /LFM;OHC	$12.50	1922	MO
	Radioette /UFS;EHC	X	1922?	O
Radio Exchange & Supply Co., (address unknown)	Airscout /XXX;XXX	X	1923	O
Radiofone Corp., Detroit MI	Radiofone Type No. 1-B /UWS;OHC	$3.10	E	V
Radiogem Corp., NY NY	Radiogem /CFS;OHC	$1.00	1922	RMO
(See Ford Co., K. N.)	Radiogrand /UFS;OHC	X	1924	O
(See General Elec. Co.; RCA)	Radiola I (ER 753A) /LWI;OHP	$25.00	1922	RQMCVO
(See Wireless Specialty Apparatus Co.; RCA)	Radiola AR 1375 (Radio Concert) /LWI;OHC	$40.00	1922	RMCO
	Radiola AR 1382 (Radio Concert) /UWI;OHC	$70.00	1922	C
(See National Radio Products Corp.)	Radiolean Jr. /UWI;XXX	$12.50	1923	RO
(See Aeolus Corp.; Benson Engineering Co.)	Radiolette /MMS;EHM	$4.00	1924	RO
Radio Mail Order Co., (address unknown)	Crystal /XXX;XXX	$0.60	1924	O
Radio Mfg. Co., NY NY (also called Freed-Eisemann Radio Corp.; Marvel Radio Mfg. Co.; Radio Receptor Co.)	Marvel Model 101 /UWS;OHC	$8.00	1921	RMCVO
Radio Marine Corp. of Am., NY NY (subsidiary of RCA)	Crystal Type B /XXX;XXX	X	L	O

MANUFACTURER	SET NAME/CASE; DETECTOR	PRICE	YEAR	REF.
Radio Marvel Co., Chicago IL	Radio Marvel /XXX;XXX	X	1924	M
(See Clapp-Eastham Co.)	Radion (A1377) /BWI;OHC	$21.25	1914	C
(unknown mfr.)	Radionette /XXX;XXX	X	E	V
Radionola International Corp., (address unknown)	Receptor Radionola Pocket Radio /MFM;OHC	X	E	V
(See Commerce Radiophone Co.)	(Commerce) Radiophone /LWI;OHC	$25.00	1919	RVO
(See National Radiophone)	(National) Radiophone /UWI;OHC	X	1922	O
Radio Products Corp. of Am., NY NY (RPC)	RPC Midget Radio Pocket Receiver /LWM;OHC	$3.00	1923	RO
Radio Products Mfg. Co., (address unknown)	RPM /XXX;XXX	$6.00	1923	O
Radio Products Sales Co., San Francisco CA	Supreme /XXX;XXX	X	1926	O
Radio Receptor Co., NY NY (also called Freed-Eisemann Radio Corp.; Marvel Radio Mfg. Co.; Radio Mfg. Co.)	Marvel Model 101 /UWS;OHC	$8.00	1921	RQMCVO
	Home-o-fone No. 2 /LWI;EHC	$24.00	1922	RO
Radio Service Co., Los Angeles CA	Type CR /UWI;EVC	X	E	V
Radio Service & Mfg. Co., NY NY (in 1920, Am. Electro Technical Appliance Co.)	Little Wonder Portable Type S-8 /LWS;OHC	$6.00	1920	RMO
	Type S-100 /UWI;OHC	$25.00	1922	R
Radio Shop, Chicago IL, Long Beach & Sunnyvale CA	Leroy /XXX;XXX	$15.00	1924	O
Radio Specialties Co., San Francisco CA	M-2 Master /XXX;XXX	$15.00	1925	O
Radio Supply Co. of Calif., Los Angeles CA (RAD-SCO)	RAD-SCO No. 101 /UWI;EVC or OHC	X	E	O
Radio Surplus Corp., Boston MA, retailer; mfr. unknown	Clin-Tone-Ia /TPS;OHC	$2.00	1926	QV
(See Tuska Co., C. D.)	(No. 4007—with:) Radiotector /UWS;OHC	$25.00	1920	C
Radio Tel. & Tel. Co., NY NY	Kismet /UMI;OHC	X	E	V
Radiowunder Specialty Co., Boston MA	Radiowunder /MFM;OHC	X	E	V

MANUFACTURER	SET NAME/CASE; DETECTOR	PRICE	YEAR	REF.
(See Elec. City Novelty & Mfg. Co.)	Radjo /XXX;XXX	$3.50	1923	O
R & O Mfg. Co., San Francisco CA (ROMCO)	ROMCO /TMS;OHC	X	E	V
(unknown mfr.)	Rapid Radio /XXX;XXX	X	E	O
Ray-Di-Co Organization, Chicago IL	Crystal /O--;OHC	$8.50	1921	Q
	Baby Wonder /LWI;OHC	$35.40	1922	R
Rayphone Radio, NY NY	Rayphone /UWS;OHC	X	E	V
(See Baldor Elec. Co.)	Ray Tuner /LWS;OHC	X	E	O
(See Adams-Morgan Co.)	RB /BWI;OVC	$6.60	1913	C
	RC /BWI;OVC	$8.80	1913	C
(See Mignon Wireless Corp.)	RC1 /UWS;XXX*	$10.00	1915	O
	RC2 /UWI;XXX* (*external detector required)	X	1915	O
RCA, NY NY (see also General Elec. Co.; Westinghouse Elec. & Mfg. Co.; Wireless Specialty Apparatus Co.)	AR 1300 /UMI;OHP	$50.00	1922	RQMCVO
	AR 1375 (Concert, Radiola) /LMI;OHC	$40.00	1922	RMCO
	AR 1382 (Concert, Radiola) /UWI;OHC	$70.00	1922	C
	ER 753 /UMI;OHP	$18.00	1922	RQMCV
	ER 753A (Radiola I) /LWI;OHP	$25.00	1922	RQMCVO
	RE (Aeriola Jr.) /LWI;OHP	$25.00	1921	RQMCVO
(See Reynolds Radio Specialty Co.)	RCR-30 /UWI;EVC	$25.00	1921	RQ
(See Adams-Morgan Co.)	RD /UWL;OHC	$20.00	1913	C
	RE /UWL;OHC	$45.00	1913	C
(See Tavel Radio Co.)	Real Radio /BWS;EHC	$4.90	1922	RM
(unknown mfr.)	Recep-ton, Type R 100A /XXX;OHC	X	E	V
(See Radionola International Corp.)	Receptor Radionola Pocket Radio /MFM;OHC	X	E	V
(See Am. Specialty Mfg. Co.)	Regal Jr. /UWI;OHC	$20.00	1922	RO
Remler Radio Mfg. Co., Chicago IL & San Francisco CA	Remler Junior No. 1 /BWS;OHC	$5.00	1921	CO
(See Niehoff & Co., Paul G.; Radio Elec. Service/ R. E. S.)	Resodon /XXX;XXX	X	1922	R
Reynolds Radio Specialty Co., Colorado Springs CO	RCR-30 /UWI;EVC	$25.00	1921	RQ

MANUFACTURER	SET NAME/CASE; DETECTOR	PRICE	YEAR	REF.
Rippner Bros. Mfg. Co., Cleveland OH	Monarch /MMS;OHC	X	1926	RVO
Rittenhouse Co., A. E., (address unknown)	Crystal /XXX;XXX	$5.00	1923	O
Ritter Radio Corp., NY NY	Ritter /BWS;OHC	$5.00	1922	MO
	Grand /CWS;OHC	$2.50	1923	MO
Rivero & Co., San Francisco CA, distributor (see also Brownie Co., mfr.)	Brownie Airphone /TMS;OHC	$7.50	1924	RVO
(See Mignon Wireless Corp.)	RLC2 /UWI;XXX*	$30.00	1915	O
	RLC2 Special /UWI;XXX*	X	1915	O
	RLC3 Special /UWI;XXX* (*external detector required)	X	1915	O
(See Brush Pottery Co.)	Rolling Pin /MPS;EHC	$7.50	1926	RO
Roll-O Radio Co. (or Roll-O Radio Corp.), Cincinnati OH	Roll-O Super Set (5 in 1 detector) /TPS;OHC	$6.00	1924	VO
(See R & O Mfg. Co.)	ROMCO /TMS;OHC	X	E	V
(See Northern Elec. Sales Co.)	Roycroft /XXX;XXX	$2.50	1923	O
(See Radio Products Corp. of Am.)	RPC Midget Radio Pocket Receiver /LWM;OHC	$3.00	1923	RO
R-W Mfg. Co., Chicago IL	R-W /UWS;EHC	$12.50	1924	RO
(See Clapp-Eastham Co.)	S /UWL;XXX	X	1916	O
(See Kodel Mfg. Co.)	S-1 /UWS;OHC	$5.00	1924	RMVO
	S-14 (Gold Star) /XXX;XXX	$6.00	1925	RO
(See Am. Technical Appliance Co.; Radio Service & Mfg. Co.)	S-8 (Little Wonder) /LWS;OHC	$6.00	1920	RMO
	S-100 /UWI;OHC	$25.00	1922	RMO
St. Marks Radio Co., NY NY	Crystal /XXX;XXX	X	1925	O
S & M Radio Co., Chicago IL	Pocket Radio Little Giant Radiophone /TFM;XXX	$2.00	1924	R
Schenectady Radio Corp., Schenectady NY	Scyrad SR-12 /XXX;XXX	$7.00	1922	O
	Crystal /XXX;XXX	$5.00	1923	O
(unknown mfr.)	Scientific American /XXX;XXX	X	E	O
Sears, Roebuck & Co., Chicago IL (see also Bowman & Co., A. W., mfr. of the Beginner's Unit; see Clapp-Eastham Co., mfr. of Cruiser)	Acme Mounted:			
	Single-slide Tuner /BWI;OHC & OHP	$5.50	1917	C
	Double-slide Tuner /BWI;OHC & OHP	$7.00	1917	C

MANUFACTURER	SET NAME/CASE; DETECTOR	PRICE	YEAR	REF.
Sears, Roebuck & Co. (*continued*)	Beginner's Unit /O--;OHC (called "Bowman Type 1A-120" by mfr.)	$7.95	1921	C
	Beginners' Sets:			
	Type 1 /UWI;OHC	$8.57	1923	C
	Type 2 /UWI;OHC	$6.55	1923	C
	2-Panel /BWI;OHC	$21.39	1923	C
	3-Panel /BWI;OHC	$26.81	1923	C
	Loose Coupler Crystal Outfit /BWI;OHC	$22.08	1923	C
	Tuning Coil Crystal Outfit /BWI;OHC	$18.95	1923	C
	Cruiser /BWI;OHC	$19.95	1916	C
	Double-slide Tuner Set Mounted with Two Detectors /BWI;OHC & OVC	$7.75	1917	C
	Loose Coupler Set Mounted with Two Detectors /BWI;OHC & OVC	$12.90	1917	C
	National Guard Field Receiving Set /LPI;OHC	$22.75	1917	C
	National Wireless Receiving Set /BWI;OVC	$10.50	1916	C
	Portable Jr. /CWS;OVC	$3.60	1916	C
	Superior /UWS;OHP	$10.75	1916	C
	University /UWI;OHC	$30.00	1917	C
(See Spielman Elec. Co.)	SECO /XXX;XXX	$15.00	1923	O
(See Philmore Mfg. Co.)	Selective /UFI,UMI;EVC (early Selective same as Luxor, Peerless)	$5.00	1931	CVO
Selectrol Radio Corp., NY NY	Selectrol /XXX;XXX	X	1925	O
(See Heinemann Elec. Co.)	Sensory /XXX;XXX	$10.00	1923	O
Shamrock Radiophone Sales Corp., NY NY (see Mengel Co., mfr.)	Shamrock Radiophone Crystal Model A /LWS;OHC (Same set as Mengel Type M.R. No. 101, also Diamond)	$4.00* (*price of Diamond crystal set)	1924	VO
Sidbenel Radio Mfg. Co., NY NY	No. 1205 /UWS;OHC	$7.50	1922	R
	No. 1206 /LWS;OHC	$21.50	1922	R
Siegal Elec. Supply Co., Chicago IL	Midwest Jr. /BWI;XXX	$14.00	1922	R
Signal Elec. Mfg. Co., Menominee MI	Crystal /CWI;OHC,EHC	X	E	VO
	Type 110 /UWI;OHC	X	1922	RQ

MANUFACTURER	SET NAME/CASE; DETECTOR	PRICE	YEAR	REF.
(See ARJO Radio Products Co.)	Simple-X /CWI;OVC	$4.75	E	VO
Simplex Electrical Lab., Brooklyn NY (Basub label also has Simplex Radio Co.; see also Deitrickson Radio Co.)	Simplex /XXX;XXX Deluxe /XXX;XXX Basub /UWI;OHC	$8.50 $15.00 X	1923 1923 E	O O VO
Simplex Radio Co., Philadelphia PA (see also Simplex Electrical Lab.)	Basub /UWI;OHC Simplex Jr. /O--;OHC	X $9.50	E 1921	VO R
(See American Can Co.)	Skeezix /MMS;OHC	X	E	O
Snellenburgs, Philadelphia PA, retailer; mfr. unknown	Eastern Type /LWS;OHC	$10.00	1923	O
(See Lightrite Co.)	Soundrite /XXX;XXX	X	1924	O
(See Fuller Co., P. H.)	(Fuller's) Sparta /XXX;XXX	X	1925	O
(See Steel Products Corp. of California)	SPCO /UMS;OHC	X	E	V
(unknown mfr.), San Francisco CA	Speed-X /XXX;XXX	X	E	O
(See Radio Apparatus Co.)	Speer /LWI;EHC	$9.00	1922	R
Spencer Radio Labs., Akron OH	Spencer Pocket Radio /LFM;OHC	X	1929	V
Spielman Elec. Co., NY NY	SECO /XXX;XXX	$15.00	1923	O
Splendid Radio Co., Chicago IL (see Audiola Radio Co.)	Splendid /UWS;OVC (same as Audiola)	$10.00* (*price of Audiola)	1922	VO
Splitdorf Radio Corp., Newark NJ	Splitdorf /CWS;XXC	X	E	O
Stafford Radio Co., Medford Hillside MA	Crystal /XXX;XXX Crystal /XXX;XXX Crystal /XXX;XXX	$2.00 $5.00 $6.50	1925 1926 1926	O O O
Standard Radio & Elec. Co., Pawtucket RI	Giblin Radioear /LWI;EVC	X	1921	VO
Stanford Elec. Co., (address unknown)	Standford /MPS;OHP or EHC	X	E	O
Star Mfg. Co., Turtle Creek PA (see also Buscher Co., C. A.)	Star /CWI;OHC (Fullatone appears to be same set as Star)	$5.00	1923	RCV
States Radio Corp., (address unknown)	Crystal /UWI;OHC	X* (*premium; not sold)	1923	M
Steel Products Corp. of California, San Francisco CA	SPCO /UMS;OHC	X	E	V

MANUFACTURER	SET NAME/CASE; DETECTOR	PRICE	YEAR	REF.
Steinite Labs., Atchison KS	Steinite Long Distance /UWI;EHC	$6.00	1924	RMVO
	Steinite De Luxe (has Steinite Rectifier for detector) /UWI;EVM	$10.00	1925	O
	Steinite "New" /UWI;EHC	X	1926	O
Steinmetz Wireless Mfg. Co., Pittsburgh PA	Superior /UWS;OHC	$4.75	1922	RQMO
Sterling Radio Mfg. Co., Berkeley CA	Excello Galena /LWS;OHC	X	E	VO
	Greg-sor /LWS;OHC (both sets have same appearance)	X	E	VO
(See Doron Bros. Electrical Co.)	Students' Receiving Set No. 381 /UWI;OHC & OHP	$71.60	1920	C
(See Philmore Mfg. Co.)	Super (or Super Tone, Supertone, or Little Giant) /TMS,TPS;EVC	$1.50* (*price in 1931)	1926	CVO
(See Roll-O Radio Co.; later called Rollo-O Radio Corp.)	(Roll-O) Super Set (5 in 1 detector) /TPS;OHC	$6.00	1924	VO
(See Sears, Roebuck & Co.)	Superior /UWS;OHP	$10.75	1916	C
(See Steinmetz Wireless Mfg. Co.)	Superior /UWS;OHC	$4.75	1922	RQMO
Superior Mfg., Cleveland OH	Superior /BWS;OVC	X	E	O
(See Radio Products Sales Co.)	Supreme /XXX;XXX	X	1926	O
(See Ward & Bonehill)	Sure-set Radio-de-tectaphone /LWS;OHC	X	1923	V
(See Listen-in Publishing Co.)	Talking Book /MFS;EHM	$5.00	1926	RMO
Tanner Radio Co., C. D., Los Angeles CA	C. D. T. /UWI;OHC	X	1924	O
Tavel Radio Co., Detroit MI	Real Radio /BWS;EHC	$4.90	1922	RM
Taylor Elec. Co., Providence RI	Radioclear /XXX;XXX	$2.90	1924	O
(See Clark Co., Thos. E.)	TECLA "Thirty" /LFI;OHC	X	E	VO
(See Lafayette Radio; see Addendum—different sets in 1949 were called "Tee-nie")	Tee-nie:			
	Midget /UWS;XXM	$0.88	1939	C
	DeLuxe /UWS;XXM	$2.35	1939	C
Telephone Maintenance Co., Chicago IL (TELMACO) (see also Bowman; Radio Equipment Co.; Marinette Elec. Co.)	TELMACO /CWI;XXX	$12.00	1921	Q
	Airophone /UWI;OHC	$20.00	1922	RCO
	Little Tattler /CWS;OHC	$6.00	1923	CO

MANUFACTURER	SET NAME/CASE; DETECTOR	PRICE	YEAR	REF.
Teleradio Co., Newark NJ	Teleradio No. 1 /LWI;EHC	X	E	V
Teleradio Engineering Corp., NY NY	Crystal /XXX;XXX	X	1922	R
Teletone Corp. of Am., NY NY	Model C Teletone /LFS;OHC	X	1925	O
Tel-Radion Co., NY NY	Tel-Radion /BWI;OHC	$10.45	1917	M
(See [The] Eastern Specialty Co.)	TESCO Type A /BWM;OHC	$1.25	1923	MO
	TESCO Type B /LWI;OHC	$5.00	1923	MO
T. E. S. Radio Co., Kansas City MO	Wetota /BWS;OHC	X	E	V
Thoma Radio Co., J. H., Pittsburgh PA	Crystal /XXX;XXX	X	1925	O
(See Nichols Elec. Co.)	Time Signal Receiving Station /BWI;OHC	$10.85	1915	C
Tinytone Radio Co., Kearney NE (later called Midget Radio Co.; Pa-Kette Elec. Co.; Midway Co.; see Western Mfg. Co., in Addendum)	Tiny Tone Pocket Radio /UFM;EHM	$2.99	1932	MVO
	Tinytone /UFM;EHM (2 later versions of "Tinytone" in Addendum)	$2.99	1937	M
Tippins & Sprengle Radio Mfg. Co., Pittsburgh PA	Tippins & Sprengle /LWL;OHC	X	E	VO
(See Insuline Corp. of Am.)	(ICA) Top Notch / TMS;EVC	$1.50	1939	C
Towner Radio Mfg. Co., Kansas City MO	Crystal (detector called "Radetec") /XXX;EHM	X	1923	R
(See Electro Importing Co.)	Transcontinental /BWI;OVC	$24.00	1913	MC
	Trans-Pacific /BWI;OHC	$10.00	1916	C
(See United Metal Stamping & Radio Co.)	Travellers /XXX;XXX	$3.00	1924	O
(See Tri-City Radio Elec. Supply Co. / TRESCO)	Trescola /XXX;XXX	X	1925	O
Triangle Mfrs., Chicago IL	Vest Pocket Radio /XXX;XXX	$1.50	1926	R
Tri-City Radio Elec. Supply Co./TRESCO, Davenport IA	Trescola /XXX;XXX	X	1925	O
Trio Radio Lab., Oakland CA	Trio /XXX;XXX	X	E	O
Trippe Reflector Corp., Chicago IL	Trippe /XXX;XXX	X	1925	O
(See Wolverine Radio Co.)	Trix /XXX;XXX	X	1925	O

MANUFACTURER	SET NAME/CASE; DETECTOR	PRICE	YEAR	REF.
True Tone Radio Mfg. Co., Chicago IL	Crystal /XXX;XXX	$8.00	1923	O
(See Sears, Roebuck & Co.)	Tuning Coil Crystal Outfit /BWI;OHC	$18.95	1923	C
Turney Labs., Eugene T., Newark NJ	Turney Portable /LWL;EVT	$87.50	1915	C
	Voxola /XXX;XXX	X	1922	R
Tuska Co., C. D., Hartford CT	No. 4007 (with Radiotector) /UWS;OHC	$25.00	1920	C
	Tuska /UWI;OHC	X	E	V
Tustin Radio & Elec. Co., San Francisco CA	Tustin /XXX;XXX	X	1925	O
U.C. Battery & Elec. Co., Berkeley CA	Crystal /LWS;OHC	X	1923	V
Uncle Al's Radio Shop, Oakland CA	Miracle /UWL;EVC	$12.75	1925	RVO
	Miracle Model 2 /UWL;EVC	$12.75	1925	RVO
	Miracle Jr. /XXX;XXX	X	E	V
United Metal Stamping & Radio Co., Cincinnati OH (see Mengel Co., mfr. of the Diamond crystal set)	Diamond /LWS;OHC (same set as Mengel M.R. 101)	$4.00	1924	O
	L. D. R. /XXX;XXX	$4.90	1924	O
	Midget /XXX;XXX	$2.00	1924	O
	Travellers /XXX;XXX	$3.00	1924	O
	United Cabinet Crystal Unit /LWS;OHC	X	E	O
United Specialties Co./ U.S. Co., Newark NJ (see Metro Elec. Co., also called Metropolitan Radio Corp.)	Little Gem /MMS;OVC	$6.50	1922	MCVO
United States Elec. Corp., Springfield MA	USECO /XXX;XXX	X	E	V
United Wireless Co., NY NY	Type D /XXX;XXX	X	1912	O
	Type E /MWL;OHC	X	1912	M
United Wireless Tel. Co., (address unknown)	Crystal /BWI;OVC	X	1919	M
Universal Wireless Co., NY NY	Crystal /BWI;OVC	$3.45	1915	MO
(See Sears, Roebuck & Co.)	University /UWI;OHC	$30.00	1917	C
(See United States Elec. Corp.)	USECO /XXX;XXX	X	E	O
Van Valkenberg Co., L. D., Holyoke MA	Van Fixed /TWS;EHM	$3.50	1925	RVO

MANUFACTURER	SET NAME/CASE; DETECTOR	PRICE	YEAR	REF.
(See Hearwell Elec. Co.)	Variometer Lyric /O--;OHC	$10.00	1925	RO
(See Whitehurst, B. W.)	VB Radio /UWS;OHC	$1.25	1923	M
(See Pocket Radio Corp.)	Vest-O-Fone /MPM;OHM	X	E	V
(See Triangle Mfrs.)	Vest Pocket Radio /XXX;XXX	$1.50	1926	R
(See Beaver Machine & Tool Co.)	(Baby Grand) Vest Pocket / TWM or TPM;OHC	$7.50	1923	CMVO
(See J. K. Corp.)	Victor /UWI;OHC	X	E	O
Voges Co., Glen F., St. Louis MO (see also Lyons Co., G. E.; both St. Louis mfrs. made a "Glen" crystal set)	Glen /XXX;XXX	X	1925	O
Volta Engineering Co., Brooklyn NY	Volta /UWI;OVC	$7.50	1922	RV
(See Turney Labs., Eugene T.)	Voxola /XXX;XXX	X	1922	R
(See Ajax Elec. Specialty Co.)	VS&L /UWS;OHC	$2.50	1926	RMVO
Wade-Twichell Co., Chicago IL	Wade /UWI;OVC	X	E	O
(See Brush Pottery Co.)	Wall Pocket /MPS;EHC	$7.50	1926	RO
Ward & Bonehill, Rochester NY	Sure-set Radio-de-tectaphone /LWS;OHC	X	1923	V
Waterbury Button Co., Waterbury CT	Detector /MWS;EHM	X	1926	RO
(unknown mfr.)	WEB /CPS;EHC	X	E	VO
(See Cheever Co., Wm. E.)	Wecco Gem /BWS;OHC	X	1926	VO
Western Coil Co., Racine WI	Type A /LWI;OHC	X	E	VO
Westinghouse Elec. & Mfg. Co., Pittsburgh PA (see RCA)	Aeriola Jr. (Model RE) /LWI;OHP	$25.00	1921	RQMCVO
Westwyre Co., Westfield MA	Elf /XXX;XXX	X	1925	O
(See T. E. S. Radio Co.)	Wetota /BWS;OHC	X	E	V
Whitehurst, B. W., Newark NJ	VB Radio /UWS;OHC	$1.25	1923	M
White Mfg. Co., Newark NJ (see Blair Co., Mitchell; Martian Mfg. Co.)	Martian Big 4 /MMS;OHC (same as Blairco "4")	$7.50	1923	RVO
(See Hearwell Elec. Co.; also called Hearwell Radio)	Whole Wave /O--;OHC,OHC	$8.00	1925	RO

MANUFACTURER	SET NAME/CASE; DETECTOR	PRICE	YEAR	REF.
Wilkins Radio Co., Los Angeles CA	Crystal Artay /XXX;XXX	$5.00	1926	O
Willemin, E. A., Providence RI	Willemin 1100 /BW&MS;OHC	X	E	O
Wilson, K. K., Buffalo NY	Wilson /UMS;OHC	X	E	O
Winchester Repeating Arms Co., New Haven CT	Winchester /BWS;OHC	X	E	V
Winn Radio & Elec. Mfg. Co., Chicago IL	Winn Portable /LWI;EHC	$50.00	1920	RQ
Wireless Equipment Co., NY NY (see also Jewett Mfg. Co., mfr. of this set in 1922)	ABC /UWI;OHC	$24.50	1920	R
(See Electro Importing Co.)	Wireless Experimenter (kit) /MMM;OHC	$1.25	1916	C
Wireless Improvement Co., NY NY	178 /XXX;XXX	X	1922	O
Wireless Shop, The / A. J. Edgecomb, Los Angeles CA	Wireless Shop /LWI;OHC	X	E	O
Wireless Specialty Apparatus Co., Boston MA (before 1912, NY NY) (see also RCA)	IP-76 (later called IP-111) /UWL;OHP	X	1907	O
	IP-77 /UWL;XXX	X	1919	O
	IP-500 /UWL;XXX	$425.00	1918	O
	Model AR-1375, Radio Concert, Radiola /LMI;OHC	$40.00	1922	RMCO
	Model AR-1382, Radio Concert, Radiola /UWI;OHC	$70.00	1922	C
(unknown mfr.)	Wizard /XXX;XXX	X	E	V
Wolverine Radio Co., Detroit MI	Trix /XXX;XXX	X	1925	O
(See Bethlehem Radio Corp.)	World /TMS;EVC	X	1929	VO
(See Cloverleaf Mfg. Co.)	Wow /CWI;OHC	X	1925	O
X-L Radio Co., Los Angeles CA	X-L 3-Way Radio /UWI;OHC	$8.75	1924	RO
Young & McCombs, Rock Island IL	YM-9 /UWI;OVE	$24.50	1920	R

ADDENDUM: SOME OF THE MORE RECENT CRYSTAL SETS (1940–1955)

MANUFACTURER	SET NAME/CASE; DETECTOR	PRICE	YEAR	REF.
(See Electronic Age Mfrs.)	Air-Champ (kit set) /BFS;OHC	X	R	VO

MANUFACTURER	SET NAME/CASE; DETECTOR	PRICE	YEAR	REF.
Allen, Alva F., Clinton MO (all kit sets)	Distite No. 1 /BWM;OHC	$1.50* (*$0.73 in 1952)	R*	CO
	Melodian No. A4 /BWS;OHC	$3.85	1952	C
	Melody King No. 2 /BWS;OHC	$1.60	1952	C
	Silver Bell No. A5 /BWS;OHC	$5.00	1952	C
American Merchandising Co., Montgomery AL (see Modernair Corp.)	Radaradio /MPM;EHM	$7.98	1947	RO
(See Lafayette Radio)	Beginner's Crystal (kit set) /BFI;OHC	$2.45	1949	C
Carron Mfg. Co., Chicago IL (see also Lafayette Radio)	Tee-nie Model 100 /UPS;OVC	$1.62	1941	C
	Tee-nie Model 250 /UMS;OVC	$3.92	1949	CV
(See Midway Co.)	CD (Civil Defense Pocket Size Radio) /MPM;EHM	$4.99	1955	M
(See REMCO Sales)	Comet (kit set) /BWS;OHC	X	R	V
(See National Supply Service)	Cub Scout Radio Kit (kit set) /BFS;OHC	X	R	V
DA-MYCO Products Co., NY NY	Easy Built (Kit No. 1) /BPS;OHC	X	1946	MO
(See Johnson Smith & Co.)	Dick Tracy Two-Way Wrist Radio /RPM;EHM	$3.98	1950	C
(See Allen, Alva F.)	Distite No. 1 (kit set) /BWM;OHC	$1.50* (*$0.73 in 1952)	R*	CO
(See WEZCO)	Dol Model 010 (kit set) /UWS;OHC	X	R	V
(See Lafayette Radio)	Eagle (kit set) /BFS;OHC	$1.15	1955	C
(See DA-MYCO Products Co.)	Easy Built (Kit No. 1) /BPS;OHC	X	1946	MO
Electronic Age Mfrs., Brooklyn NY	Air-Champ (kit set) /BFS;OHC	X	R	VO
Fine Co., Bernard, NY NY	Rada-Phone /MPM;EHM	$2.98	1947	M
(See Lafayette Radio)	Germanium Diode Radio (kit set) /BWS;OHM	$3.29	1955	C
(See Superex)	Germanium Diode Loop-stick Radio (kit set) /XMS;EHM	$3.87	R	C
Heath Co., Benton Harbor MI	Heathkit Model CR-1 /UPS;EHM	$7.95 (*price in 1958)	R*	VO
Hobby Specialities Co., Milwaukee WI	Midget Crystal Radio (kit set) /MFM;OHC	$2.25	1946	M

MANUFACTURER	SET NAME/CASE; DETECTOR	PRICE	YEAR	REF.
Johnson Smith & Co., Detroit MI (see also Midway Co., mfr.—called this "Midget" crystal set "Pakette")	Dick Tracy Two-Way Wrist Radio /RPM;EHM	$3.98	1950	C
	Midget Pocket Size Radio /MPM;EHM	$3.99	1950	C
Kitcraft Products Co., Los Angeles CA	Kitcraft No. 102 (kit set) /BPM;OHC	$2.95	1946	MV
	Kitcraft Ready Built, Fixed Germanium Crystal, No. 130 /BPM;EHM	X	R	V
Lafayette Radio, NY NY (see also Carron Mfg. Co.) (see Lafayette Radio in main section—different sets called Tee-nie in 1939)	Beginner's Crystal (kit set) /BFI;OHC	$2.45	1949	C
	Eagle (kit set) /BFS;OHC	$1.15	1955	C
	Germanium Diode Radio (kit set) /BWS;OHM	$3.29	1955	C
	Tee-nie Model 100 /UPS;EHM	$1.62	1941	C
	Tee-nie Model 250 /UMS;EHM	$3.92	1949	CV
Lombard Co., E. T., Oakland CA	Lombard Multiphone /UFS;EHC	X	R	V
(See Superex)	Loopstick Crystal Radio Kit (kit set) /XMS;EHM	$2.60	R	C
	Loopstick Crystal Radio /TPS;EHM	$3.48	R	C
Magna Kit Co., Los Angeles CA	Magna (kit set) /UWI;OHC	X	R	V
(See Hobby Specialties Co.)	Midget Crystal Radio (kit set) /MFM;OHC	$2.25	1946	M
(See National/NAFICO)	Midgette Radio /MPM;EHM	$4.95	1955	M
(See Johnson Smith & Co.; Midway Co., mfr., who called this set "Pakette")	Midget Pocket Size Radio /MPM;EHM	$3.99	1950	C
Midway Co., Kearney NE (earlier, called Tinytone Radio Co.; Midget Radio Co.; Pa-Kette Elec. Co.; see also Western Mfg. Co., in Addendum; also see Johnson Smith & Co., in Addendum)	CD (Civil Defense Pocket Size Radio) /MPM;EHM	$4.99	1955	M
	Mitey Pocket Radio /MPM;EHM	$3.99	1948	MO
	Pa-Kette Radio / MPM;EHM	$3.99	1946	RMVO
	Pakette Radio /MPM;EHM	$3.99	1947	RMCVO
	Pee Wee Pocket Radio /MPM;FHM	$6.99	1956	MVO
	Ti-Nee Pocket Radio /MPM;EHM	$6.99	1957	MO
	Tinymite Radio /MPM;EHM	$3.99	1948	RMO
	Tinytone Radio /MPM;EHM	$4.99	1951	M
	Tinytone Pocket Radio /MPM;EHM	$6.99	1959	MVO
(See Allen, Alva F.)	Melodian No. A4 (kit set) /BWS;OHC	$3.85	1952	C
	Melody King No. 2 (kit set) /BWS;OHC	$1.60	1952	C

MANUFACTURER	SET NAME/CASE; DETECTOR	PRICE	YEAR	REF.
(See Midway Co.)	Mitey Pocket Radio /MPM;EHM	$3.99	1948	MO
Modernair Corp., Los Angeles CA (see Am. Merchandising Co.)	Radaradio /MPM;EHM	$7.98	1947	RO
Modern Radio Labs., Garden Groves CA	MRL (various kit sets) /XXX;XXX	X	R	CO
(See Lombard Co., E. T.)	(Lombard) Multiphone /UFS;EHC	X	R	V
National/NAFICO, Toledo OH	Midgette Radio /MPM;EHM	$4.95	1955	M
National Supply Service, NY NY	Cub Scout Radio Kit (kit set) /BFS;OHC	X	R	V
Pa-Kette Elec. Co., Kearney NE (see Midway Co.; Johnson Smith & Co.)	Pa-Kette Radio /MPM;EHM	$3.99	1946	RVO
	Pakette Radio /MPM;EHM	$3.99	1947	RMCVO
Patrick Labs., Indianapolis IN	Patrick Personal Radio /MPM;EHM	$5.00	1949	M
(See Midway Co.)	Pee Wee Pocket Radio /MPM;EHM	$6.99	1956	MVO
(See Shees Mfg. Co.)	Personal Radio /MFM;EHM	$4.95	1947	M
Philmore Mfg. Co., NY NY (see also main section above for earlier sets)	Panel Set /O--;OHC	X	R	V
	Selective Model No. VC-1000 (kit) /BWI;OHC Selective in main section)	$5.00	1955	CVO
	Space Patrol /TFS;OHC	X	R	O
(See Scientific Products of Indianapolis)	Pocket Size Radio /MPM;EHM	$7.95	1954	M
(See Fine Co., Bernard)	Rada-Phone /MPM;EHM	$2.98	1947	M
(See Am. Merchandising Co.; Modernair Corp.)	Radaradio /MPM;EHM	$7.98	1947	RO
Radio Marine Corp. of Am., NY NY (subsidiary of RCA)	Crystal Type D /O--;OHC	X	1944	O
REMCO Sales, Berkeley CA	Comet (kit set) /BWS;OHC	X	R	V
Scientific Products of Indianapolis, Indianapolis IN	Pocket Size Radio /MPM;EHM	$7.95	1954	M
(See Philmore Mfg. Co.)	Selective Model No. VC-1000 (kit) /BWI;OHC	$5.00	1955	CVO
Shees Mfg. Co., Lake Geneva WI	Personal Radio /MFM;EHM	$4.95	1947	M

MANUFACTURER	SET NAME/CASE; DETECTOR	PRICE	YEAR	REF.
(See Allen, Alva F.)	Silver Bell No. A5 (kit set) /BWS;OHC	$5.00	1952	C
(See Philmore Mfg. Co.)	Space Patrol /TFS;OHC	X	R	O
Superex, Buffalo, NY	Germanium Diode Loopstick Radio (kit set) /XMS;EHM	$3.87	R	C
	Loopstick Crystal Radio Kit (kit set) /XMS;EHM	$2.60	R	C
	Loopstick Crystal Radio /TPS;EHM	$3.48	R	C
(See Carron Mfg. Co.; also see Lafayette Radio)	Tee-nie Model 100 /UPS;OVC	$1.62	1941	C
	Tee-nie Model 250 /UMS;OVC	$3.92	1949	CV
(See Midway Co.)	Ti-Nee Pocket Radio /MPM;EHM	$6.99	1957	MO
	Tinymite Radio /MPM;EHM	$3.99	1948	RMO
	Tinytone Radio /MPM;EHM	$4.99	1951	M
	Tinytone Pocket Radio /MPM;EHM (2 earlier versions of "Tinytone" in main section)	$6.99	1959	MVO
Western Mfg. Co., Kearney NE (parent company for Tinytone Radio Co., Midget Radio Co., Pa-Kette Elec. Co., Midway Co.—see these companies in both main section and Addendum)				
WEZCO, Seattle WA	Dol Model 010 (kit set) /UWS;OHC	X	R	VO

Crystal Sets Advertisements

ICA Top Notch Crystal Set

No. J320
List $1.50

Your Cost
88c

A super-sensitive crystal mounted under a dust-proof glass cover—brings in stations with clarity and volume. Has beautifully engraved calibrated dial and is housed in a handsome black crystaline metal case. Selector coil wound on solenoid form assuring clear and perfect reception.

Burstein-Applebee Co., *Radio* Catalog No. 55 (1939), p. 107

Make your own Radio Set

Parts consist of: Inclosed Detector, Mounted Galena, Formica Panel (holes bored), Tuning Coil, Switch, Switch Points, Binding Posts, Solderall, Wire, Tubing and Screws.

Sent postpaid for only $3.45

Cabinet 90c extra

Try to match at three times the cost

ERNEST ELECTRIC CO., *Radiophone Manufacturers*
4849 Easton Avenue St. Louis, Mo.

Radio News (Jan. 1923), p. 1408

AMERICAN RADIO AND RESEARCH CORPORATION

15 Park Row
New York

Address all Communications
to New York Office

Factory and Laboratory
Medford Hillside, Mass.

You Can't Go Wrong

in getting this new complete receiving equipment. For reasons of good taste we rarely indulge in superlatives in our advertising, but after very thorough trial we unhesitatingly suggest that the new **Amrad Receiving Set, illustrated above, is the most compact, convenient and efficient apparatus ever offered.**

Anyone can operate it with excellent results. Anyone able to use a few simple tools can in an hour erect a complete station and actually hear radio telephone and telegraph messages. This wonderful little set is not an "electrical toy"; it is a high-grade commercial product —Amrad made—yet available at a popular price. Rated range 100 miles. Can receive up to 500 miles under favorable conditions.

Nothing to get out of order. Can be carried in your suit case—size only 5" x 5" x 7". No batteries to charge or replace. Elaborate aerial not necessary.

Three wavelength ranges—180 meters minimum, 750 meters maximum with 60 foot aerial.

Need never be discarded. May be combined with Amrad Units, illustrated below, as user elaborates his Set.

Really, you'll have to inspect this complete equipment at your dealer's to appreciate it. If he can't help you, write us.

PRICES

Amrad Crystal Receiver, as illustrated, $23.50.
Double 2000 ohm Murdock telephone set, $4.50 extra.
100' Antenna and Ground Equipment, $12.00 extra.

Radio News (Jan. 1921), p. 432

This ad shows the AMRAD Crystal Receiver TypeE.

TRADE NAME: Cleartone Perfect Crystal Set.
TYPE: Fixed crystal detector. Cleartone circuit.
CONTROLS: One.
AERIAL: Outside.
PRICE: $7.50 without accessories.
MANUFACTURER'S NAME: Cleartone Radio Supply Co.

TRADE NAME: Concert Junior.
TYPE: Crystal set.
CONTROLS: One.
AERIAL: Outdoor.
PRICE: $3.50 without accessories.
MANUFACTURER'S NAME: The Concert Radio Phone Co.

Radio News (Mar. 1925), p. 1652

JUST PERFECTED

Wireless Receiving and Sending Stations

$10.10

Sends 12 to 20 miles. Receive hundreds of miles. Your money refunded if it fails to do this.
Unassembled parts of this station. $9.05

Send stamp for our large catalog "E" of remarkable values

THE HANDEL ELECTRIC CO.
66 Vesey St., N. Y.

Everyday Mechanics (Dec. 1916), p. XXV

A Complete "MARVEL" Receiving Set for $15.00

You need not spend a large sum for a COMPLETE RADIOPHONE RECEIVING OUTFIT. Why spend your money for parts with which to assemble your own Receiving Set? You can buy a MARVEL RADIOPHONE RECEIVING OUTFIT for only $15.00. This OUTFIT will receive radiophone, speech, and music within a radius of fifty miles of the large broadcasting stations. You can also receive wireless telegraph messages from large stations hundreds of miles away.

The OUTFIT comes complete in every detail. No additional parts are required—as simple to operate as a phonograph. Can be set up in twenty minutes.. Full instructions with every OUTFIT.

The "MARVEL"—the most talked about OUTFIT in Radio

The "MARVEL" RADIOPHONE RECEIVING OUTFIT is the lowest priced, complete, guaranteed, crystal-receiving OUTFIT on the market, with a wave length range of 180 to 2600 meters. The OUTFIT is thoroughly practical and dependable. It can be quickly and easily set up for use. No source of power—no batteries—no license—no special knowledge necessary for setting up and operating.

The "MARVEL" RADIOPHONE RECEIVING OUTFIT, model 105 (patents applied for), consists of a "MARVEL" RADIO RECEIVER, model 101 (patents applied for)—(a highly efficient radio receiver, completely enclosed in a handsomely finished mission oak cabinet)—150 feet copper antenna wire, 25 feet insulated wire, 5 porcelain insulators, 1000 ohm telephone with leather-covered head band and telephone cord, one extra "MARVEL" tested galena Radio-crystal, antenna switch, ground clamp, code chart, abbreviation chart and complete instructions for setting up and operating. Nothing else required! Results guaranteed! $15.00

Model 110, same as above but with a 2,000 ohm telephone double headset $18.00

Buy from your Radio, Electrical or Sporting Goods Dealer, or Department Store. If they have not as yet received their supply, we will ship to you direct, on receipt of Parcel Post or Express Money Order.

Absolutely guaranteed to equal or surpass in performance any outfit selling at double the price—a statement our engineers can prove by actual test.

FREED-EISEMANN RADIO CORPORATION
FORMERLY RADIO MANUFACTURING COMPANY
Offices and Sales Department, 255 FOURTH AVE., NEW YORK CITY, U. S. A.

Radio News (Feb. 1922), p. 763

Here's the Radiolean Jr.

Everybody wants this popular price Radio Set.

The "Radiolean Jr." PRICE $12.50 without Headset

$15.00 with Headset

Looks like a Detector Tube Set and can be changed into one at small cost. The Biggest Value ever offered by any Radio manufacturer.

This wonderful crystal receiving set is being sold for $12.50 retail price which includes all equipment except phones. Head phones from $3.00 upward. Get our prices on any radio goods you need and compare same with other dealers.

SPECIAL XMAS OFFER

WESTERN ELECTRIC HEAD-SETS (Army type)	$7.50
U. S. NAVY TUBES (Amplifying, Detector and Transmitting)..	$4.25
PORCELAIN RHEOSTATS Each	35c

Order the Radiolean Jr. at any Radio Dealer, Department Store, etc., or Direct from

NATIONAL RADIO PRODUCTS CORP.
509 Fifth Avenue
New York

Radio News (Jan. 1923), p. 1364

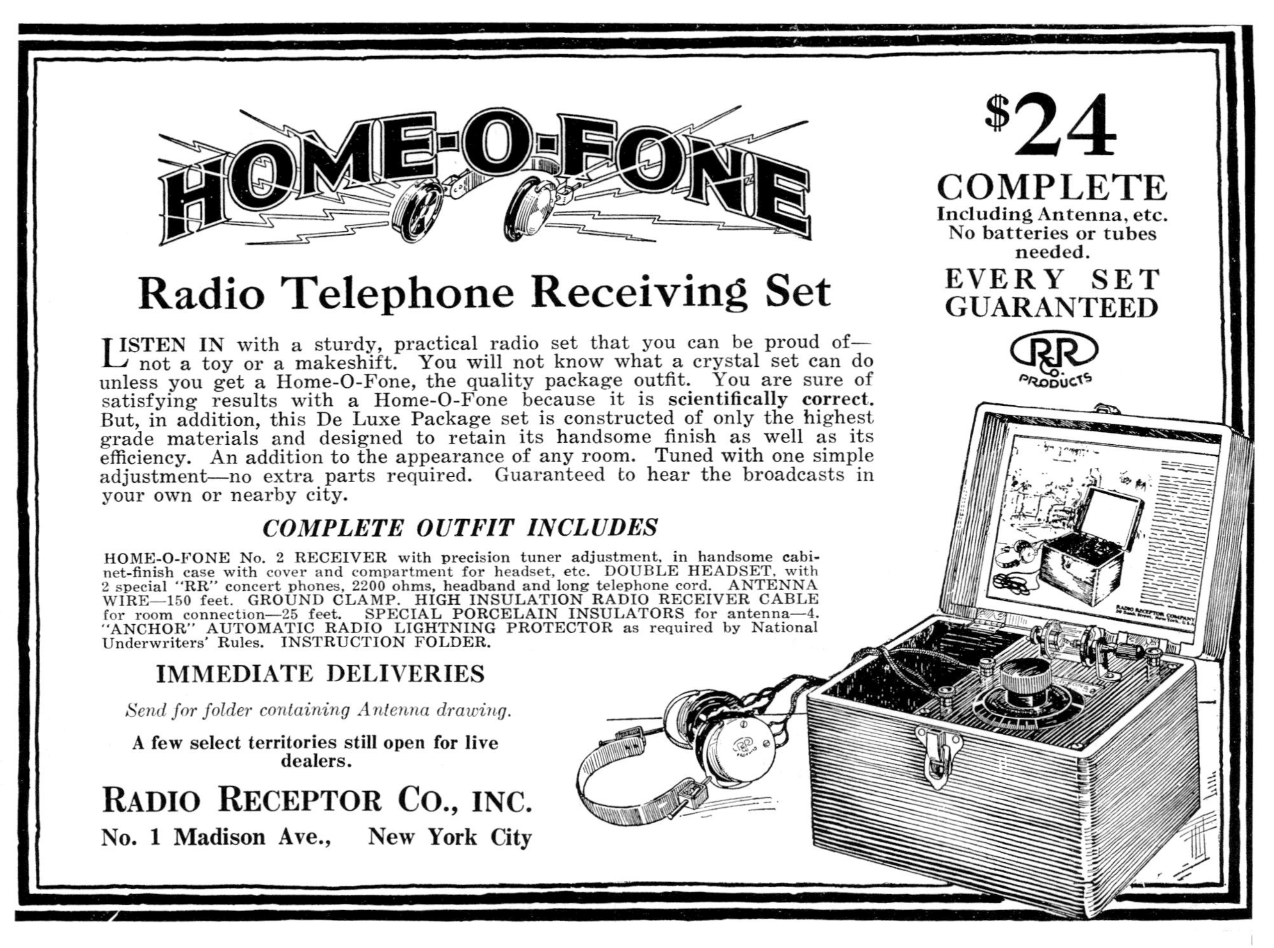

Radio News (July 1922), p. 157

National Airphone was the first *crystal set* awarded the Radio News Laboratory Certificate of Merit ("Approval")—No. 7. The preceding six certificates were for radio components and accessories.

The Wild Waves Are Saying

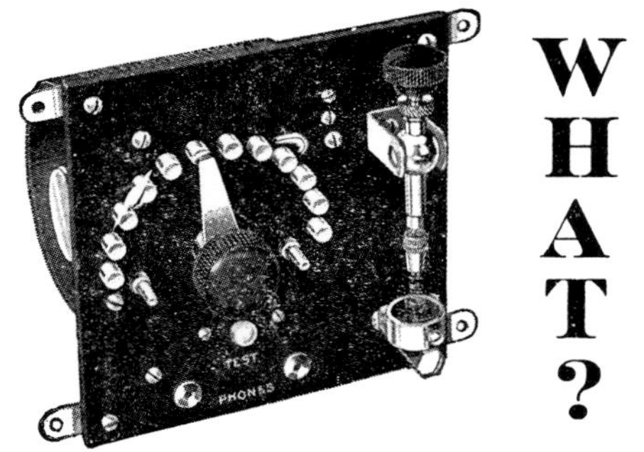

A SURPRISE awaits you. GET STARTED in this fascinating science.

A COMPLETE receiving set composed of a LITZ WOUND TUNER, Fixed Condenser IMPROVED DETECTOR making all points of crystal accessible, CONSTANT spring TENSION, LATEST CORD TIP JACKS, all MOUNTED on a $\frac{3}{16}$" GRAINED BAKELITE panel 5" x 5" equipped with lugs so more panels can be added.

Receiving set as described $8.50. With test buzzer, battery case and push button $10.50.

Post Paid in the U. S. A.

2000 ohm receivers $5.00. Tested crystals $0.25

Antenna Wire, 7 strands #22 copper 85c per ft. Postage extra.

MOTOR GENERATORS, DYNAMOTORS and C.W. EQUIPMENT. COMPLETE RECEIVING AND TRANSMITTING SETS.

THE RAY-DI-CO ORGANIZATION
1547D N. Wells St., Chicago, Ill.

QST (Dec. 1921), p. 114

NATIONAL AIRPHONE

The crystal receiver, shown in the illustration is manufactured by the National Airphone Corporation, 16 Hudson Street, New York City.

This apparatus is neat and well designed. A single circuit is employed with fixed inductances and a variable tuning condenser. The inductances are wound on a tube 3" long and 1" in diameter with

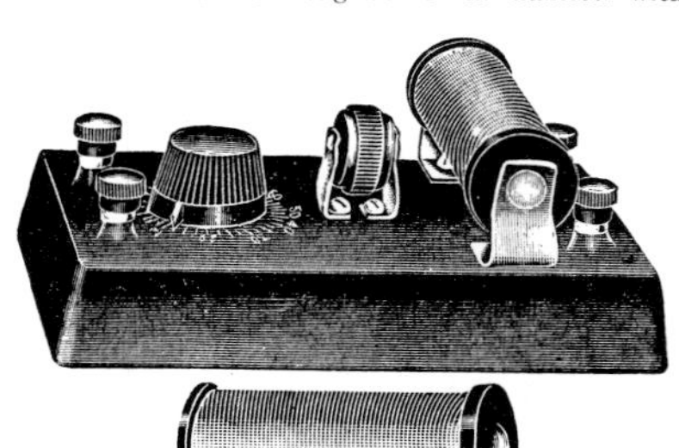

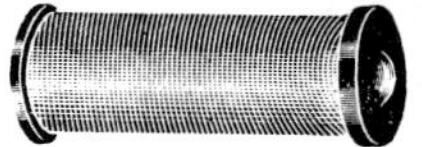

enameled wire. Tested on a 100' aerial, single wire, one inductance when placed in the clips, gave a range of wave lengths of 200 to 380 meters. The other coils allowed tuning over the band 580-875 meters. The variable condenser has mica dielectric and is moulded into the base.

A dustproof detector of novel design is provided and gave excellent results under actual receiving conditions. Mechanically, the set is of good quality.

The set was carefully packed in a double corrugated cardboard box.

AWARDED THE RADIO NEWS LABORATORIES CERTIFICATE OF MERIT No. 7.

Radio News (Sep. 1922), p. 445

Radio News (June 1924), p. 1832

PHILMORE CRYSTAL SETS

THE BLACKBIRD

The "Blackbird" is a **first quality crystal set** presented at the lowest possible price. It requires no tubes or batteries and delivers programs loud and clear without distortion. Housed in crystalline finish metal cabinet with attractive front panel. It employs a dust-proof glass-enclosed supersensitive crystal detector making possible the best of results. **A fast seller everywhere.** Accessories of good quality are listed below. Shpg. wt., 3 lbs.
No. E5940. List, $3.00.
YOUR PRICE **$1.58**

LITTLE GIANT

A rigidly constructed crystal set which will give unlimited service. Gives the utmost in entertainment on local stations and with the extensive chain programs now in effect reception from stations all over the country is available. A sensitive crystal enclosed in a dust proof glass casing helps **bring stations in loud and clear.** Shpg. wt., 2 lbs.
No. E5933. List, $1.50.
YOUR PRICE **69c**

THE LITTLE WONDER

Compact in size but big in results. This set will receive broadcasting within 25 miles of a station. The open type detector permits adjustments to be made to the finest degree. Set includes a supersensitive crystal assuring quick results when looking for a station because the entire surface of the crystal is sensitive. Shpg. wt., 2 lbs.
No. E5931. List, $1.00.
YOUR PRICE **58c**

PHILMORE SELECTIVE

Designed to give most efficient results and possessing unusual selectivity. Built with a tapped coil matched with a **.00035 Mfd. variable condenser.** The reception on this set is **loud and clear** and will bring in one station after another, tuning out all other stations. It has double connection posts making possible the **use of two pairs of headphones** at one time. The next thing to a tube set at the lowest possible price. Contained in a crackle finish cabinet with a panel of modernistic design. Shpg. wt., 3 lbs.
No. E5932. List, $5.00.
YOUR PRICE **$2.45**

Allied Radio Corp., *Radio* Catalog (1933), p. 22

LITTLE TATTLER RECEIVING SETS

This durable Crystal receiving set is built according to the United States Bureau of Standards, Circular No. 120, and incorporates a tuner and detector unit, headset, 100 ft. copper-clad aerial wire, two formica insulators and full detailed instructions; all packed securely in a truly attractive box.

No. D118. Little Tattler Receiving Set, complete. List Price $6.00

Wernes & Patch, *Radio* Catalog (1923), p. 5

Standard Long Distance Receiving Set

This set has a range of from about 1,500 to 2,000 miles. It will do the work of many commercial sets. By means of the large loose coupler, accuracy in tuning out unwanted stations is accomplished. Static and other disturbances are almost entirely eliminated with this set, and it is tunable within very wide limits. This outfit is made up of the following items:

- 1 3,000-Meter Professional Receiving Transformer, No. 57A7169.
- 1 Standard Rotary Variable Condenser, capacity .0004 microfarad, No. 57A7166.
- 1 Tubular Fixed Condenser, No. 57A7200.
- 1 Peerless Detector Stand, No. 57A7107.
- 1 2,000-Ohm Head Set, Murdock No. 55 Receivers, No. 57A7114.
- 1 Electrose Wall Insulator, No. 57A7241.
- 1 Aerial Connector Block, No. 57A7172.
- 1 Ground Clamp, No. 57A7213.
- 275 feet Copper Aerial Wire, No. 57A7289, enough to make a four-wire aerial, 60 feet long with four leads.
- 25 feet No. 14-Gauge Rubber Covered Wire, No. 57A7955, for connecting the set.
- 2 Electrose Strain Insulators, 10½ inches long, No. 57A7239.
- 10 Pony Strain Insulators, No. 57A7173.
- 2 Aerial Suspension Pulleys, No. 57A7258.
- 100 feet Aerial Suspension Rope, No. 57A7259.
- 1 Copy of "Amateur's Wireless Handy Book," No. 57A7254.
- 1 Copy of our Instruction Book.

No. 57A7337 Standard Long Distance Receiving Set....... **$20.80**

Shipping weight, about 45 pounds.

Page Thirty SEARS, ROEBUCK AND CO., CHICAGO, ILL.

Sears, Roebuck & Co., *Wireless Apparatus* Catalog (1916), p. 30

NACO
Radio Receiving Sets
$20.00

If you wish to hear the human voice clearly and distinctly get a NACO SET. The wonderful satisfaction obtained surely is pleasing. A sick person, lying in bed, can listen to the music and be happy, making the hours fly by. Send your order at once with check.

DEALERS WANTED EVERYWHERE

Knock down sets, radio parts and phones at reasonable prices

"Our motto is to please always"

National Motor Accessories Corporation
1446-1448 Woolworth Building
New York City

Scientific American (June 1922), p. 425

'MIRACLE'

SUPER-POWERFUL AND SELECTIVE CRYSTAL SET

Gives you choice of stations without interference.
You Buy Direct from Mfgr. & Originator Factory Price
Good material, workmanship, handsome cabi net and engraved panel. $12.75 for set. $18.5(with $6.00 phones and aerial outfit.
Sold Under a 10-Day Trial. If Not Satisfactory Return and Full Amount Paid Will Be Refunded

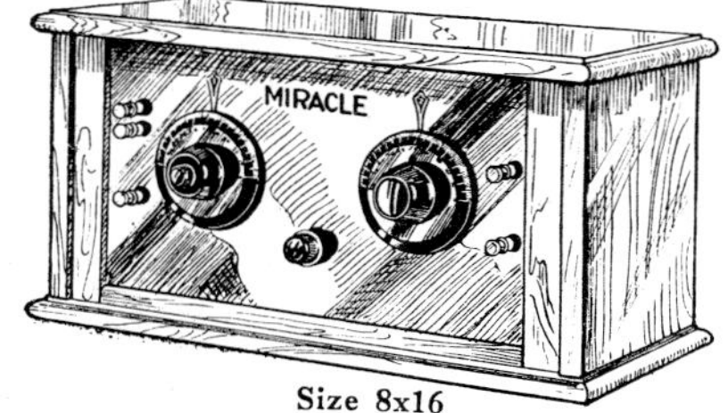

Size 8x16

No tubes or batteries. Always ready for use Simple to operate. Nothing to get out of order No howls or other disagreeable noises.

Sent Express Collect

UNCLE AL'S RADIO SHOP
3015 Dakota Street, Oakland, Calif

Radio News (May 1925), p. 2187

CROSLEY CRYSTAL RECEIVER

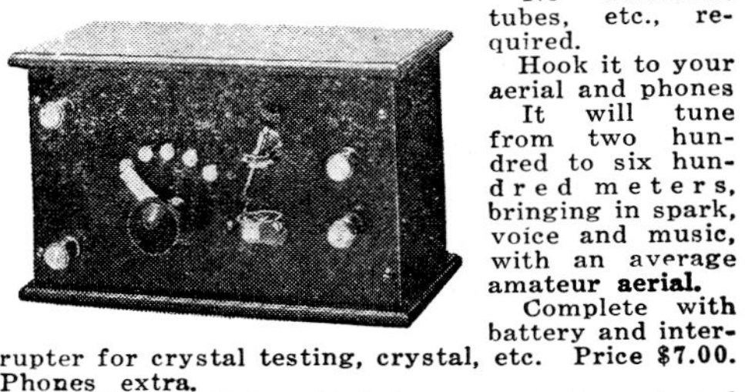

No batteries, tubes, etc., required.
Hook it to your aerial and phones. It will tune from two hundred to six hundred meters, bringing in spark, voice and music, with an average amateur **aerial.**
Complete with battery and interrupter for crystal testing, crystal, etc. Price $7.00. Phones extra.
DEALERS: This will help you get 'em started

CROSLEY MANUFACTURING COMPANY

CROSLEY MFG. CO.
Radio Dept. Q-5 Cincinnati, O.

Every article bearing the name "CROSLEY" is GUARANTEED to give absolute satisfaction or money will be refunded.
We shall be pleased to send literature describing the above mentioned and other radio apparatus to any one free of charge upon request. Get your name on our mailing list to receive latest Bulletins of other new Crosley products. If your dealer does not handle our goods, order direct and send us his name.

Radio Dept. Q-4B. Cincinnati, Ohio

QST (Oct. 1921), p. 60

The first "Crosley" crystal receiver (Oct. 1921).

HARKO RADIO RECEIVER

No batteries, tubes, etc., required.
Hook it to your aerial and phones. It will tune from two hundred to six hundred meters, bringing in spark, voice and music, with an average amateur **aerial.**
Complete with battery and interrupter for crystal testing, crystal, etc. Price $7.00. Phones extra.
DEALERS: This will help you get 'em started

QST (Nov. 1921), p. 77

The same set was then called "Harko" (Nov. 1921).

CROSLEY
Better --- Cost Less

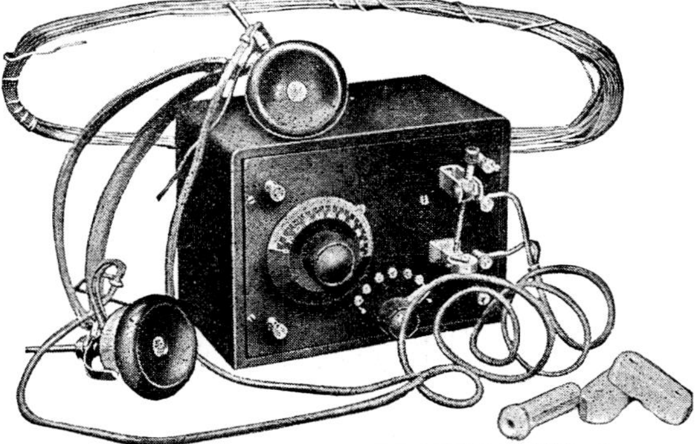

CROSLEY CRYSTAL RECEIVER NO. 1
Beginners in Radio will find this a very efficient unit. With a range from 200 to 600 meters, this set will receive broadcasting stations up to 25 or 30 miles, depending upon conditions and their power.
Complete with head phones, antenna wire insulators ready to install without any additional equipment.—$25.00.
Crosley Crystal Receiver No. 1 is made so that the Crosley Audion Detector Unit, Crosley Radio Frequency Tuned Amplifier and Crosley Two-Stage Audio Frequency Amplifier may be added if desired, to increase the range and volume of sound.

QST (Sep. 1922), p. 114

By September, 1922, the set was called "Crosley Crystal Receiver No. 1." A tuning dial was added to the tap-switch on the panel.

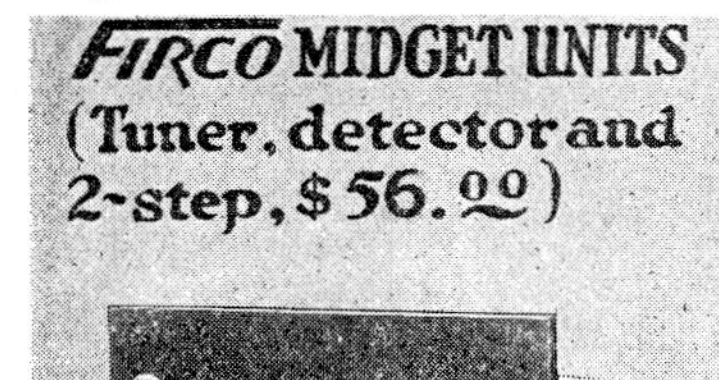

Firco Midget Units
Quality equal to Standard Apparatus, but greatly simplified. Set of three units (Tuner, detector and 2-step), $56.00; Tuner, $15.00; Detector, $11.00 and other units at equally reasonable prices.

Radio News (Sep. 1921), p. 181

The Recognized Symbol of Superior Performance

AMRAD CRYSTAL RECEIVING SET $27.00
A Beginner's Set De Luxe—One You Never Need Discard. Anyone Can Operate. Price includes Phones and 125 ft. aerial wire. Range 15-25 miles.

AMRAD COMBINATION B-I
Range for Radio telephone, 50 miles
Range for radio telegraph, 500 miles
Amrad Crystal Receiver, 2575 .... $21.50
VT Detector only, 2771 .... 12.50
AMRAD Load Coil only, 2962 .... 3.85
(The above is one of 15 Combinations)

AMRAD VT 2 Stage Amplifier, 2776 .... $38.00
Can be added to either of above Sets to triple range and audibility.

Do YOU Know What's NEW?

SUPPOSE some friend asked you to compare the merits of the several receiving sets designed for the beginner. Would YOU KNOW about the automatic wave-change switch, the plug-in crystal detector, the rear panel connections, the all formica insulation, the basketball variometer, the load coil terminals, the solid mahogany cabinet—all features found only in the new de luxe Amrad Crystal Receiver. You should SEE and EXAMINE this instrument. If your dealer does not stock, write for Bulletin M.

Do you know the Amrad Crystal Receiver, with detector removed, is a highly selective single circuit tuner when combined with an Amrad VT Detector also NEW? This combination, with Amrad Load Coil, will bring in Arlington, 500 miles away. And where can you find a modern VT control panel in a mahogany case for only $12.50? New Bulletin F describes it.

Do you know a VT 2-stage Amplifier can be used on a CRYSTAL with quite as good results as on a VT detector? This is being done. Amrad Crystal Receiver and VT 2-stage Amplifier in combination can be used with a loud speaker for local radiophone. Here is a set which your friend, just shaking hands with Radio, can operate. No critical adjustments. Tell him about it. Also described in Bulletin F.

Do you know all about the improved Amrad Unit System? Does your friend? Does he know he can buy the Amrad Crystal Receiver and expand his equipment without losing a cent—as he progresses in his knowledge of radio?

Do you know there are at least fifteen NEW Amrad instruments actually ON SALE this month? Seven Individual Receiving Units, a new Wavemeter, Send-Receive Switch, Electrolytic 38 mfd. Condenser, and Safety Devices for the antenna are among the more important. Better send five two-cent stamps for the latest Amrad Catalog. It will pay you to keep posted.

Radio News (Mar. 1922), p. 869

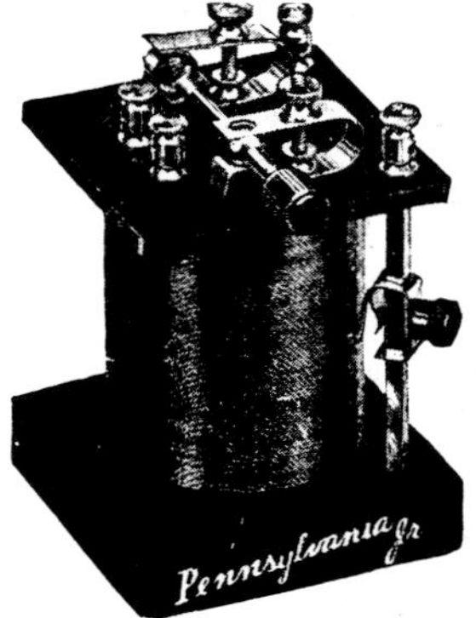

K-1 Crystal Set
List, **$2.50**

HAMBURG BROTHERS CO.
450-452 Seventh Avenue
PITTSBURGH, PA.

McGraw-Hill Radio Trade Catalog (Aug. 1926), p. 134

RADIOLETTE
CRYSTAL RECEIVING SET
COMPACT AND COMPLETE

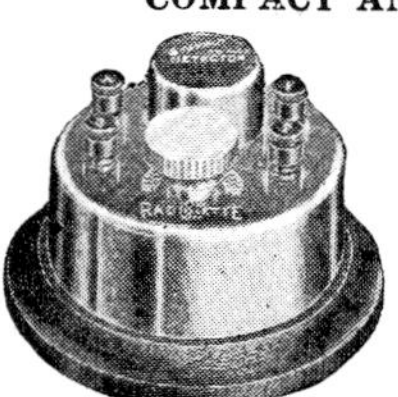

RADIOLETTES are beautiful in appearance, made of metal and are heavily plated in Nickel, and mounted on a polished base.
RADIOLETTES have the range and efficiency of any Crystal Set regardless of size or price.
They are equipped with our patented "Stay-put" detector, containing a supersensitive Crystal.
Packed in individual boxes, twenty-four to the carton. All sets are unconditionally guaranteed.
PRICE $4.00 EACH
BENSON MELODY CO., ***Distributors***
2125 N. Halsted Street, Chicago, Ill.
Dealers: Write for discounts.

Radio News (Mar. 1924), p. 1369

This Set Is Practical In Every Way

For the Junior Operator

A high grade, practical and instructive toy wireless receiving set, which has proved to be very efficient. The beginner should have no trouble in erecting and operating this set and in a short time should be able to receive messages from stations miles away.

Set consists of two receiving coils, primary and secondary, having a fixed coupling. Secondary is mounted inside of primary coil. Primary coil winding is variable by means of a nickel plated slider. A galena detector of the cat whisker type is also mounted on the base. A fixed condenser is included and is properly incorporated in the set. The head set consists of a receiver and cord, complete with headband.

The aerial-ground unit consists of bare copper wire, porcelain insulators, a ground clamp lead-in tube and connecting rod. Complete instructions for installing and operating the set are furnished, together with a pad of radiogram blanks, wall card, code charts, pliers, screwdriver and box of tested galena. Shipping weight, about 14 pounds.

6E9548—Receiving Set. Price, each.......................................**$15.25**

Sears, Roebuck & Co., *Radio* Catalog (1921), p. 50

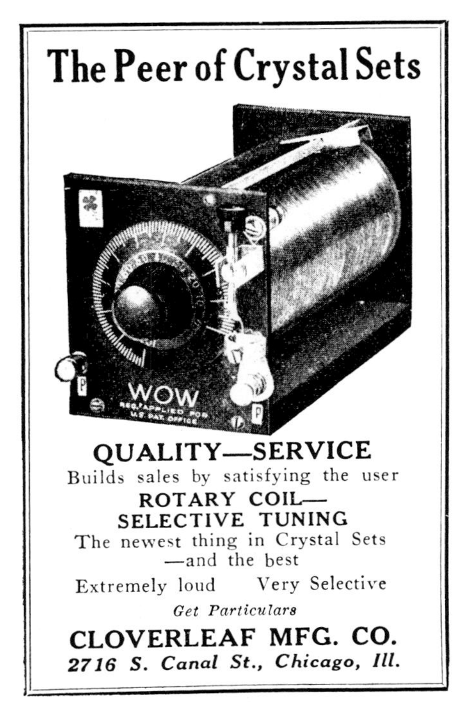

The Radio Trade Directory (Aug. 1925), p. 116

Popular Radio (Apr. 1923), p. 38

the INTRODUCING— Crystal Receiver

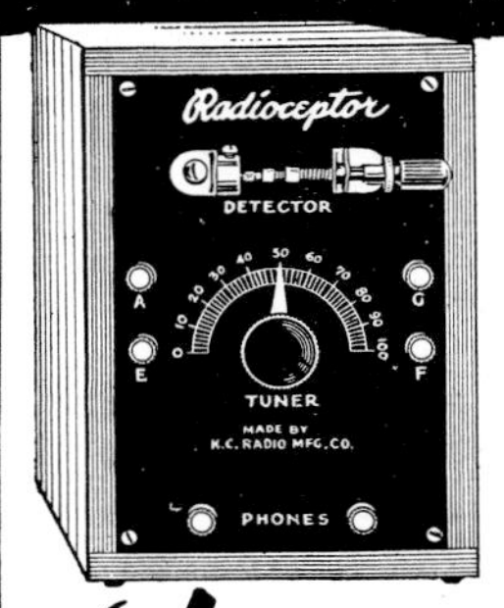

Order by Mail.
We Will Ship At Once!

If your dealer or jobber cannot supply you with a Crystal Receiver, write direct. List Price, $18.00. Murdock or Frost headset, $5.00. Antenna Completion Package, $5.00. Variometer, unmounted, $5.00. Send Post Office Money Order with your order.

A Radioceptor UNIT

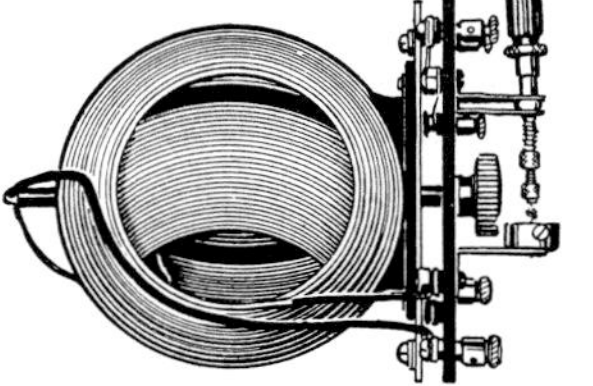

In this—the first of Radioceptor Units—you find a receiver which combines improved electrical efficiency with beautiful mechanical appearance.

A very unusual and exclusive design of variometer forms the tuner—giving greater range of wave length than variometers heretofore designed. Only one adjustment. Continuously variable. No overlapping.

Details of finish include dark mahogany box, highly polished and engraved bakelite panels, buffed nickel parts.

The Crystal Receiver, altho complete in itself, combines with other Radioceptor Units to form a high grade long distance set.

KANSAS CITY RADIO MFG. CO., 607 Grand Ave. Kansas City, Mo.

JOBBERS AND DEALERS—NOTICE!
RADIOCEPTOR UNITS ARE SOLD ONLY THROUGH REGULAR TRADE CHANNELS. *Write today for our proposition. Stock Radioceptor Units NOW and "cash in" on the popular demand.*

Radio News (July 1922), p. 181

RADIO SERVICE PRODUCTS

Radio Service Type S100
Complete Radio Receiving Set

A wonderfully designed and constructed Radio Instrument. Made to receive wireless telephone speeches and music from amateur and ship stations, farm and market reports, time signals from Government station at Arlington, etc. Contains a ***Variable Condenser*** for tuning.

Price complete $32.50
Receiving set only 25.00

Receiving Set Only, Type S8 $7.50

The Little Wonder Portable Radio Set

It's the wireless set everybody's talking about—built right and compact—perfected by four years' development and manufacture. Simple to operate.

Other Radio Service Products

Single VT Sockets Type S2	$1.50
Single VT Sockets Type S10	1.00
Double VT Sockets Type S3	2.25
Triple VT Sockets Type S4	3.25
Grid Condensers .0005 mf. Type S15	.35
Phone Condensers .002 mf. Type S16	.35
Grid Leak Condensers Type S30	.50
Crystal Detector Stand Type S20	1.00
Variable Grid Leak Type S40, ½ to 3 megohms	.75

We carry numerous other types of interest to Radio dealers and wireless men. Radio Service Products are for sale by all reliable dealers.

Type S3

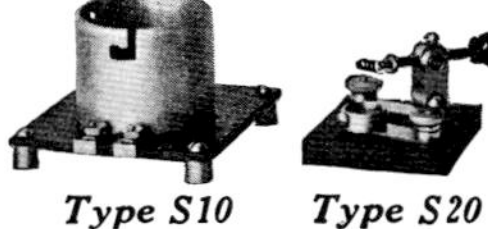

Type S10 *Type S20*

RADIO SERVICE & MFG. CO.

Factory: LYNBROOK Long Island

Sales Office: 110 West 40th St. New York City

Radio News (Apr.-May 1922), p. 1047

Throughout the length and breadth of the land, thousands of families are listening to music, concerts, news of the day, lectures, weather reports, church services, etc., right in their own homes with a $15.00 "Marvel" Radio Telephone Receiving Outfit. No longer is it necessary to have expensive apparatus or any knowledge or experience in wireless telegraphy.

Bring YOUR Home in Touch With The Outside World

Get a "MARVEL" Radio Telephone Receiving Outfit. Music, concerts, lectures, current news, sermons, Government reports, etc., will come to you daily. The cost is small—only $15.00 complete! Nothing else is needed. Receives speech and music, etc., within a 50 mile radius of the large radio telephone broadcasting stations, and wireless telegraph messages within many hundreds of miles.

The "MARVEL" RADIO TELEPHONE RECEIVING OUTFIT, is the only COMPLETE crystal receiving outfit on the market with a wave-length range of 180 to 2,600 meters. The OUTFIT is thoroughly practical, dependable and complete. It can be quickly and easily set up for use. No source of power, no batteries, no license and no special knowledge is necessary for installation.

THE COMPLETE "MARVEL" RADIO TELEPHONE RECEIVING OUTFIT, Model 105, consists of a "MARVEL" Radio Receiver (a completely enclosed, highly efficient, crystal receiver, mounted in a handsomely finished mission oak cabinet), 150 feet of solid copper wire (for the antenna), 5 porcelain insulators, a 1,000 ohm telephone with leather covered headband and telephone cord, antenna switch, ground clamp, code chart, abbreviation chart, and COMPLETE instructions for installation and operation. Nothing else is needed........ **$15.00**

Model 110, same OUTFIT, but with a 2000 ohm double headset **$18.00**

Quality Apparatus **Designed by Engineers** **Popularly Priced**

"MARVEL" RADIO TELEPHONE RECEIVING OUTFITS are for sale at all progressive Radio and Electrical Dealers. If your dealer has not yet received his stock of "MARVEL" OUTFITS, we will send you one prepaid on receipt of Postal or Express Money Order, or via Parcels Post C. O. D. ORDER NOW—DON'T DELAY.

Look for the Registered Trade Mark on every genuine "MARVEL" RADIO TELEPHONE RECEIVING OUTFIT (Patents applied for).

Bulletin R-101 on request.

DEALERS: "MARVEL" RADIO TELEPHONE RECEIVING OUTFITS are nationally advertised; they are the most popular of crystal receiving OUTFITS. Get your share of the growing demand, by selling them. Liberal discounts and selling co-operation are part of our sales campaign. ORDER NOW!

TRADE MARK REG.

MARVEL RADIO MANUFACTURING COMPANY, Inc.

OFFICES and SALES DEPARTMENT: **170 Fifth Avenue, New York City**

Radio News (Jan. 1922), p. 646

CRYSTAL DETECTORS TABLE

GUIDE TO ABBREVIATIONS USED IN TABLE FOR CRYSTAL DETECTORS:

BASE:

SHAPE
- S Square or Rectangular
- R Round or Oval
- M Miscellaneous other
- O No Base
- X Unknown

MATERIAL
- W Wood
- M Marble
- P Plastic, Bakelite, Hard Rubber, or Celluloid
- F Fiberboard
- O Other (includes metal, slate)
- X Unknown

SIZE*
- M Miniature
- S Small
- I Intermediate
- L Large
- X Unknown

*Size Baselines:
Miniature less than 1½ inches (length or diameter)
Small 1½ to 2¹⁵⁄₁₆ inches
Intermediate 3 to 4½ inches
Large greater than 4½ inches

DETECTOR:

STATE OF ENCLOSURE
- O Open
- E Enclosed
- X Unknown (or Other)

ORIENTATION*
- H Horizontal
- V Vertical
- X Unknown (or Other)

CONTACT TYPE
- C Catwhisker or Other Metal Point
- P Perikon (Mineral-to-mineral)
- E Electrolytic Type
- T Thumbwheel Type
- M Miscellaneous (Fixed, Cylinder or Cartridge Type)
- X Unknown

*Orientation of detector refers to whether the rod/arm is parallel ("horizontal") or at right angle ("vertical") to the underlying detector base.

PRICE: Earliest retail price found; X unknown.

YEAR:*
- E Early (Before or by 1929)
- L Later (1930–1939)

ADDENDUM TABLE ONLY:
- R Recent (1940–1955)

*If year is unknown or uncertain, one of the above letters is used for an approximation, based on available evidence; when year is listed, it is for the earliest references found.

REFERENCES:
- R *Radio News*
- Q *QST*
- M Other Magazine(s)
- C Catalogs
- V Viewed by Author
- O Other Source(s)

MANUFACTURER	DET. NAME/BASE; DETECTOR	PRICE	YEAR	REF.
(See Duck Co., Wm. B.)	A /SPI;OHC	$7.50	1914	C
(See Grant, Harry Jr.)	A-1 /XXX;XXC	X	1924	R
(See Lindell Elec. Shop)	A-1 Pacific Type /SPI;OVC (2 styles)	$4.00 $4.50	1914 1914	O O
(See Sears, Roebuck & Co.)	Acme Galena /SPI;OHC	$1.00	1916	C
Adams-Morgan Co. (AMCO), Upper Montclair NJ	AMCO /SPI;OHC	$2.50	1912	CO
	AMCO /SPI;OVC	$3.00	1913	C
	AMCO /SMI;OVC	$5.00	1913	C
	AMCO Carborundum (United Wireless type) /SML;OVP	$6.00	1913	C
	AMCO Electrolytic /SML;OVE	$2.50	1913	C
	Paragon No. 45 /SPS;OHC	$1.75	1920	RQCV
(See American Sales Co.; Wireless Specialty Apparatus Co. / WSA)	Adjustable /SPI;OHC (same as "Cleartone" detector of WSA)	$0.29 (*marketed by WSA in 1919)	1932*	C
(See Manhattan Electrical Supply Co.)	(MESCO) Adjustable /SPI;OVC	$4.00	1920	C
(See United Metal Stamping & Radio Co.)	Adjustable /SMI;OHC	X	1926	R
Advance Elec. Co., Los Angeles CA	Advance Fixed /RPM;EXM	$1.50	1924	RMCO
(See Essex Specialty Co.)	Aerophone /SPI;OHC	X	E	O
Aerovox Wireless Corp., NY NY	Aerovox /XXX;XXX	X	1925	O
(See American Electro Technical Appliance Co.)	AETACO /SPS;EHC	$2.00	1922	Q
Airex Co., NY NY	Airex /XXX;XXX	X	1926	O
(See Bernard's Radio Co.)	Airader /SOM;EXM	X	1925	RM
(See Philmore Mfg. Co.)	Ajax (Philmore) Glass Enclosed No. 309 (later clear plastic, No. 7008) /ROM;EVC	$0.40	1925	CVO
	Ajax (Philmore) Open No. 310 (later No. 7003) /SPS,SFM;OHC	$0.40	1925	CVO
	Ajax (Philmore) Unmounted No. 308, 101 (later No. 7010) /O--;OHC	$0.15	1925	CVO
Ajax Elec. Co., Cambridge MA	Detector /XXX;XXX	X	1925	O
Ajax Elec. Specialty Co., St. Louis MO	Detector /XXX;XXX	X	1925	O

MANUFACTURER	DET. NAME/BASE; DETECTOR	PRICE	YEAR	REF.
(See Maguire & Shotton)	Albany Combination /SPL;OHP or OHC	$2.00	1914	O
Allied Radio Corp., Chicago IL, retailer	Open Type /SFM;OHC	$0.12	1933	C
	MLO Fixed /RPM;EXM	$1.00* (*price in 1934)	E	C
All-Point Mfg. Co., Detroit MI	Detector /XXX;XXX	X	1925	O
(See Atlantic Radio Co.)	A. M. /XXX;XXX	$1.75	1921	R
(See Edgcomb-Pyle Wireless Mfg. Co.)	AM5 /SWS;OHC	$0.50	1914	C
Ambrose Radio Co., Brooklyn NY	Ambrose Vernier /SPM;OHC	X	1924	RO
(See Adams-Morgan Co.)	AMCO /SPI;OHC	$2.50	1912	CO
	AMCO /SPI;OVC	$3.00	1913	C
	AMCO /SMI;OVC	$5.00	1913	C
	AMCO Carborundum (United Wireless type) /SML;OVP	$6.00	1913	C
	AMCO Electrolytic /SML;OVE	$2.50	1913	C
American Electro Technical Appliance Co., NY NY	AETACO /SPS;EHC	$2.00	1922	Q
American Radio & Research Corp., Medford Hillside MA & NY NY (AMRAD)	AMRAD Duplex Type C /SPI; EHC & EHC	$4.50	1919	RQC
	AMRAD Single Type C-1 /SPS;EHC	$2.50	1919	RQC
American Sales Co., NY NY (mfr., Wireless Specialty Apparatus Co. / WSA)	Adjustable /SPI;OHC (same as "Cleartone" detector of WSA)	$0.29	1932* (*marketed by WSA in 1919)	C
American Specialty Mfg. Co., Bridgeport CT	Regal No. 174 /SPI;OHC	$1.25	1922	RC
Andrae Co., Julius & Sons, Milwaukee WI (see also Signal Elec. Mfg. Co.)	Standard Galena No. 41 /SPI;OHC	$1.50	1921	RQCVO
Andrea, Frank A. D., NY NY (see Barawik Co.)	FADA Type 101-A /SPS;EVC	$3.00	1921	RQCVO
	FADA Type X101 /SPS;EHC	$2.00	1922	RQC
(See Aylsworth & Robinson)	A-R Semi-fixed /RPM;EHC	X	1925	M
Argentite Radio Corp., Los Angeles CA	Argentite /SPM;OHC	X	1925	RO
(See Riley-Klotz Mfg. Co.)	Arkay /SPS;OHC	$1.00	1922	R

MANUFACTURER	DET NAME/BASE; DETECTOR	PRICE	YEAR	REF.
Atlantic Radio Co., Boston MA	A. M. /XXX;XXX	$1.75	1921	R
Atwater Kent Mfg. Co., Philadelphia PA	Model 2A Twin Unit (Part No. 4200) /RPI;OHC & OHC	X	1923	O
Aylsworth & Robinson, Los Angeles CA	A-R Semi-fixed /RPM;EHC	X	1925	M
(See Duck Co., Wm. B.)	B (Universal) /SPI;OHC	$5.00	1914	C
(See Electro Importing Co.)	(E. I.) Baby (or Miniature) /SPM;OHC	$0.25	1915	CVO
(See Radio Specialty Co.)	(RASCO) Baby /SPM;OHC	$0.50	1921	RQMV
(See Doron Bros. Electrical Co.)	(Doron) Ball and Socket /SPI;OVC	$4.00	1920	C
Barawik Co., Chicago IL, retailer (see Andrea, Frank A. D., mfr. of FADA 101-A)	Enclosed (FADA 101-A, without logo) /SPS;EVC	$2.25	1921	RMV
	Galena /SPS;OHC	$1.39	1922	RM
(See Electro Importing Co.)	Bare Point Electro-lytic /SPI;OVE	$1.50	1909	MCV
Barr Mercury-Cup Detector, Washington DC	Barr Mercury-Cup /SPI;OVE	$4.50	1919	RQC
(See Briggs & Stratton Co.)	BASCO 70429 /O--;EHC	$1.00	1922	C
(See Radio Service Co.)	BD /XXX;XXX	$1.50	1923	Q
(See Central Elec. Co.)	Beacon /RPM;EVC	$1.50	1924	C
Bernard's Radio Co., Providence RI	Airader /SOM;EXM	X	1925	RM
(See Hyman & Co., Henry)	Bestone /SPS;EHC	X	1925	VO
Bierman-Everett Foundry Co., Irvington NJ	Bierman-Everett /XXX;XXX	X	1923	R
Bisby Mfg. Co., NY NY	Supertone /XXX;XXX	X	1925	O
B-Metal Refining Co., Detroit MI	B-Metal Reflex, Type C Ever Ready Tube Detector (Fixed) /SPM;EVC	$1.50	1923	RM
	B-Metal Adjustable Type D /SPM;EXM	$1.00	1924	RC
(See Miller Radio Co., A. H.)	(Miller) B-Metal /XXX;XXX	X	1926	M
(See Brooklyn Metal Stamping Corp.)	BMS Fixed /RPM;EHM	$1.00	1925	CO

MANUFACTURER	DET NAME/BASE; DETECTOR	PRICE	YEAR	REF.
Bowman & Co., A. W. Cambridge MA (see also Sears, Roebuck & Co.)	Bowman /SPL;OHC (called "Marine Mineral Detector" by S,R & Co.)	$1.50	1920	RQCV
Brach Mfg. Co., L. S., Newark NJ (see Radio Distributing Co.; Telephone Maintenance Co.)	Brach Fixed /SPM;EHM	$0.35	1925	CVO
	Pur-a-tone /SPI;EHC	$2.75	1922	RCO
Branch Tool Co., Providence RI	Branch /XXX;XXX	$2.25	1923	RO
Briggs & Stratton Co., Milwaukee WI (BASCO)	BASCO 70429 /O--;EHC	$1.00	1922	C
Bronx Radio Equipment Co., NY NY	BRECO /XXX;XXX	X	1925	O
Brooklyn Metal Stamping Corp., Brooklyn NY	BMS Fixed /RPM;EHM	$1.00	1925	CO
Brownlie & Co., Roland, Medford MA	Brownlie Vernier /O--;EHM	$2.00	1924	RMCVO
	Brownlie /SPM;OHC	X	1925	RO
(See Hearwell Elec. Co.)	Bug /XXX;XXX	X	1925	O
Bunnell & Co., J. H., NY NY	Ghegan /SPI;OHC	X	E	V
	Jove /SPI;OHC	$2.00	1916	RQMCVO
	Jumbo-Jove /SML;OHC	$10.00	1916	C
	Molybdenite /SPI;OVC	$6.00	1916	C
Burke Mfg. Co., Newark NJ	Burke /RPS;OVC (looks like a vacuum tube; has plug-in pins)	$2.00	1924	R
Buscher Co., C. A., Kansas City MO, distributor	SpeeD /RPS;EVC	$0.85	1923	C
	UMCO /SPI;OHC	$0.85	1923	C
B. W. Battery & Radio Corp., NY NY	Detector /XXX;XXX	X	1925	O
(See Lindell Elec. Shop)	C-1 Catwhisker Type /SPS;OHC	$1.50	1914	O
(See Edgcomb-Pyle Wireless Mfg. Co.)	C6 /SWL;OHC,OHP,OHC & OHP	$25.00	1914	C
California Radio Products Co., Los Angeles CA	Calrad /XXX;XXX	X	1925	O
Carborundum Co., Niagara Falls NY	Carborundum Detector Unit Fixed /O--;EHM	$1.50	1925	RMCVO
	Carborundum Detector Unit Adjustable /O--;EHM	X	E	VO

MANUFACTURER	DET NAME/BASE; DETECTOR	PRICE	YEAR	REF.
Carborundum Co. (*continued*)	Carborundum Stabilizing Detector Unit /SPI;EHM	$3.50	1926	RMCVO
(See Adams-Morgan Co.)	(AMCO) Carborundum Detector, United Wireless Type /SML;OVP	$6.00	1913	C
(See Duck Co., J. J.)	Carborundum Detector, United Wireless Type /SPI;OHP	$3.00	1913	CO
(See Electro-Set Co.)	Cat-whisker /SPI;OHC	$1.50	1916	C
(See Nichols Elec. Co.)	Cat-whisker /SPS;OHC	$0.50	1915	C
Celerundum Radio Products Co., Boston MA	Celerundum DE-TEX-IT /O--;EHM	$1.25	1924	RMO
Central Elec. Co., Chicago IL	Beacon /RPM;EVC	$1.50	1924	C
Chambers & Co., F. B., Philadelphia PA	Our Favorite /RPI;OHC	$1.75	1914	C
	Our Latest /SPI;OHC	$2.75	1914	C
	Our Own /SPI;OHP	$2.50	1914	C
	Special Pick /SPS;OHC	$1.50	1914	C
Champion Radio Products Co., Chicago IL	Champion /SPS;OHC	$0.75	1923	CVO
Cheever Co., Wm. E., Providence RI	Wecco /XXX;XXX	X	1926	O
Chelten Elec. Co., Philadelphia PA	Chelten /SPS;EH*C (*semi-horizontal)	X	1925	VO
Clapp-Eastham Co., Cambridge MA (see also Duck Co., J. J.; Duck Co., Wm. B.)	Clapp-Eastham (Universal):			
	Ferron /SML;OVC	$3.25	1913	CVO
	Ferron /SML;OHC or OHP	$3.25	1914	CVO
Clearco Crystal Co., Milwaukee WI	Clearco /SFS;OHC	X	L	V
(See Wireless Specialty Apparatus Co., mfr.; Crosley Mfg. Co.; Manhattan Electrical Supply Co.; Precision Equipment Co.; Ship Owners Radio Service, Inc.; Westmore-Savage Co.)	Cleartone /SPI;OHC (contract companies listed at left used their own labels)	$1.60	1919	CVO
(See RCA; Wireless Specialty Apparatus Co.)	Cleartone Radio Concert Model UD 1432 /SPI;OHC	X	E	V

MANUFACTURER	DET NAME/BASE; DETECTOR	PRICE	YEAR	REF.
(See Edgcomb-Pyle Mfg. Co.)	CM4 /SPL;OHC & OHP	$7.50	1914	C
(See Wireless Specialty Apparatus Co.)	Combined Silicon & Galena (Type IP202) /SPL;OHP & OHC	$25.00	1919	C
(See Electro Importing Co.)	Commercial /SMI;OVC	$3.75	1912	CV
Connecticut Telephone & Elec. Co., Meriden CT (see also Turney Co., Eugene T., mfr. of Crystaloi until 1920)	Crystaloi:			
	Type A* /RPS;EVM	$3.50	1916*	RCVO
	Type AA /RPI;EVM	$6.00	1916*	QMCO
	Type O* /RPS;EVM	$3.50	1916*	RQMCV
	(*A and O appear same)	(*mfr. changed in 1920)		
	Connecticut Vacuum Crystal Detector Type J112 /O--;EHM	$2.50	1922	C
Continental Elec. Co., NY NY	No. 1200 /SPI;EHC	$2.45	1922	Q
	No. 1201 /SPS;EVC	$1.50	1922	Q
Crosley Mfg. Co., Cincinnati OH (see Wireless Specialty Apparatus Co., mfr.; also Manhattan Electrical Supply Co.; Precision Equipment Co.; Ship Owners Radio Service, Inc.; Westmore-Savage Co.)	Crosley /SPI;OHC (label lists both mfr. and contract company)	$1.60	1922	QCV
(See Connecticut Telephone & Elec. Co.; also Turney Co., Eugene T., original mfr., until 1920)	Crystaloi Type:			
	A* /RPS;EVM	$3.50	1916*	RCVO
	AA /RPI;EVM	$6.00	1916*	QMCO
	O* /RPS;EVM	$3.50	1916*	RQMCV
	(*Type A very similar to or same as Type O)	(*mfr. changed in 1920)		
(See Star Crystal Co.)	Crystalstat /XXX;XXX	X	1925	O
(Welty & Co., Wm. A.)	Crystector /MPS;EVC	$4.00	1925	RO
(See Radio Specialty Co./ RASCO)	Crystodyne (Zincite) /SPS;OHC	$1.75	1924	RM
Davidson Radio Corp., NY NY	Rasla Fixed /O--;EHM	$1.25	1923	CO
(See Westinghouse Elec. & Mfg. Co.)	DB /SPL;OVC & OVC	X	1921?	C
	DB /SPL;OHC & OHP	$5.00	1921	RQMCVO
(See Pacific Radio Specialty Co.)	Death Valley Permatect /O--;OHC (cartridge type)	$1.25	1925	O

MANUFACTURER	DET NAME/BASE; DETECTOR	PRICE	YEAR	REF.
DeForest Radio Tel. & Tel. Co., NY NY (later, Jersey City NJ)	D-101 /SPS;EVC	$2.25	1919	RQMCVO
	D-101-b /SPI;EHC	$2.75	1921	CV
	Detector, Cam Adjustable (Panel Mount) /SPM;OHC	$0.59* (*price in 1924)	1922	RVO
	UD-100 /O--;EVC	$4.00	1919	C
DeJur Products Co., NY NY	DeJur /XXX;XXX	X	1925	O
(See Wernes & Patch)	(No. 325) Deluxe Adjustable /RMM;EVC	$1.25	1923	C
(See Pyramid Products Co.)	De-Tec-Tone:			
	Type B (Panel Mount) /O--;EHM	$1.60	1924	RO
	Type C (Clip Mount) /O--;EHM	$1.50	1924	RO
(See Fitzpatrick, G.)	Detex /SXM;OHC	$1.00	1926	R
(See Celerundum Radio Products Co.)	(Celerundum) DE-TEX-IT /O--;EHM	$1.25	1924	RMO
Detroit Radio Co., Detroit MI	Maxitona /O--;OHC	$0.35	1922	R
Dixie Distance Crystal Co., Covington KY	Dixie Distance /XXX;XXX	X	1926	O
Doron Bros. Electrical Co., Hamilton OH	Doron:			
	Ball & Socket /SPI;OVC	$4.00	1920	C
	Galena /SPI;OHC	$4.30	1920	C
	Mineral No.115 /SPI;OHC	$3.20	1920	C
	Silicon No.110 /SPI;OVC	$4.20	1916	C
	Universal /SPI;OHC,OHP	$4.70	1916	RCO
(See Freshman Co., Chas. F.)	(Freshman) Double Adjustable /MPS;EVC	$1.50	1924	RQMVO
(See Wireless Improvement Co.)	Double Detector Type WI-101-A /SPI;OHC,OHC	X	1917	VO
Dubilier Condenser & Radio Corp., NY NY	Dubilier /SPI;EHC	X	E	V
Duck Co., J. J., Toledo OH (see Clapp-Eastham Co., re: Ferron detector)	Carborundum, United Wireless Type /SPI;OHP	$3.00	1913	CO
	Electrolytic /SPI;OVE	$1.25	1913	CO
	Ferron /SML;OVC	$5.00	1913	CO
	Mineral, No. 60 x 12 /SWS;OVC	$1.00	1913	CO

MANUFACTURER	DET NAME/BASE; DETECTOR	PRICE	YEAR	REF.
Duck Co., J. J. (*continued*)	Peroxide of Lead /SPI;OVM	$2.50	1913	CO
	Standard /SPI;OVC	$1.50	1913	CO
Duck Co., Wm. B., Toledo OH (see Clapp-Eastham Co., re: Type E Ferron detector)	Type A /SPI;OHC	$7.50	1914	C
	Type B (Universal) /SPI;OHC	$5.00	1914	C
	Type E (Universal) Ferron /SML;OHC or OHP	$5.00	1914	C
	Electrolytic /SPI;OVE	$1.25	1914	C
	Ever-Ready /SPI;OHC	$2.25	1914	C
	Type G /SPI;OHC	$5.00	1914	CV
	Junior:			
	Horizontal /RWS;OHC	$0.75	1914	C
	Vertical /RWS;OVC	$0.95	1914	C
	No. 60 x 12 /SWS;OVC	$0.75	1914	C
	No. A676 /SWI;OVC	$1.00	1914	C
	No. 1915 /SPI;OHC	$2.00	1914	C
	Turret Cup /SPI;OHC	$2.00	1914	C
	Standard /SPI;OVC	$1.50	1914	C
(See American Radio & Research Corp./AMRAD)	(AMRAD) Duplex Type C /SPI;EHC,EHC	$4.50	1919	RQC
(See Nichols Elec. Co.)	Duplex Special /SPI;OHC	$2.50	1915	C
(See Tel-Radion Co.)	Duplex Tel-Radion /SWI;XXX	$5.00	1916	M
(See Erisman Labs.)	Duratek /SPM;EHM	$2.00	1923	R
(See Duck Co., Wm. B.)	E (Universal) Ferron /SML;OHC or OHP	$5.00	1914	C
Edgcomb-Pyle Wireless Mfg. Co., Pittsburgh PA	Type:			
	AM5 /SWS;OHC	$0.50	1914	C
	C6 /SWL;OHC,OHP,OHC, & OHP	$25.00	1914	C
	CM4 /SPL;OHC & OHP	$7.50	1914	C
	R1 /SPS;OHC	$1.50	1914	C
	R2 /SPS;OHC	$2.00	1914	C
	R3 /SPI;OHP	$2.50	1914	C
	R4 /SPI;OHP	$3.00	1914	C
	R5 /SPI;OHP	$3.50	1914	C
	R6 /SPI;OHP	$3.00	1914	C
	R7 /SPI;OHC & OHP	$4.50	1914	C
	R12 /SPL;OHC & OHP	$10.00	1914	C

MANUFACTURER	DET NAME/BASE; DETECTOR	PRICE	YEAR	REF.
Electrical Research Labs., Chicago IL (see Scientific Research Labs.—ERLA and Scientific Fixed, same?)	ERLA Fixed /MFM, MPM; EHM	$1.00	1923	RMCVO
	ERLA Semi-fixed /O--;EHP	$1.00	1924	RMC
Electrical Specialty Co., Columbus OH	ESCO /SPS;EVC	$2.80	1920	RQ
Electric City Novelty & Mfg. Co., Schenectady NY	Radjo /SPS;OHC	X	1924	RO
Electric Mfg. & Sales Co., Newark NJ (EMSCO)	EMSCO Turret-Top /SPI;OHC	X	1922	RQ
Electro-Chemical Labs., Jersey City NY	Detector /XXX;XXX	X	1925	O
Electro Importing Co., NY NY (EICO)	E. I. Baby (or Miniature) /SPM;OHC	$0.25	1915	CVO
	Bare Point Electro-lytic /SPI;OVE	$1.50	1909	MCV
	Commercial /SMI;OVC	$3.75	1912	CV
	EICO (milk-glass base) /SOI;OHC	X	E	O
	Galena /SPI;OVC	$1.00	1914	CO
	Galena-Gas /SML;OVC	$4.00	1914	C
	E. I. Miniature (or Baby) /SPM;OHC	$0.25	1915	CVO
	Peroxide of Lead Dry Electrolytic /SPI;OVM	$1.25	1913	CO
	Radiocite /SPI;OHC	$3.50	1918	CV
	Radioson /SPI;EVE	$4.50	1914	RMCO
	Universal /SPI;OHC	$1.50	1913	CV
(See Adams-Morgan Co.)	Electrolytic /SML;OVE	$2.50	1913	C
(See Duck Co., J. J.; also Duck Co., Wm. B.—1914)	Electrolytic /SPI;OVE	$1.25	1913	CO
Electro-Set Co., Cleveland OH	Cat-whisker /SPI;OHC	$1.50	1916	C
	Detector /SPI;OHC	$0.45	1916	C
	Electrolytic /SPI;OVE	$1.50	1916	C
	Permanon /SPL;EHM	$2.50	1916	C
	Quick Change /SPI;OHC	$0.60	1916	C
	Senior /SPI;OHC	$1.50	1916	C
	Silicon /SPL;OVC	$4.50	1916	C
(See Elec. Mfg. & Sales Co.)	EMSCO Turret-Top /SPI;OHC	X	1922	RQ
(unknown mfr.)	ERCO /SPS;OHC	X	E	VO
Erisman Labs., NY NY	Duratek /SPM;EHM	$2.00	1923	R
	Pyratek Fixed /SPM;EHM	$1.25	1924	RCVO

MANUFACTURER	DET NAME/BASE; DETECTOR	PRICE	YEAR	REF.
(See Electrical Research Labs.; Scientific Research Labs.—ERLA and Scientific Fixed, same?)	ERLA Fixed /MFM, MPM;EHM ERLA Semi-fixed /O--;EHP	$1.00 $1.00	1923 1924	RMCVO RMC
(See Electrical Specialty Co.)	ESCO /SPS;EVC	$2.80	1920	RQ
Essex Specialty Co., Berkeley Heights NJ	Aerophone /SPI;OHC	X	E	O
(See Duck Co., Wm. B.)	Ever-Ready /SPI;OHC	$2.25	1914	C
Everset Lab., Providence RI	Everset /SFM;EHM	X	1925	R
(See Andrea, Frank A. D.)	FADA 101-A /SPS;EVC FADA X101 /SPS;EHC	$3.00 $2.00	1921 1922	RQCVO RQC
Federal Radio Products Co., Schenectady NY (see also Schwartz & Son, Maurice)	Whiz Bang! /SPS;EHP (appears same as Schwartz Adjustable)	$0.75	1925	RV
Federal Telegraph & Telephone Co., Buffalo NY	Federal No. 17 /SPI;OVC	X	1922	VO
(See Clapp-Eastham Co.)	(Universal): Ferron /SML;OVC Ferron /SML;OHC or OHP	 $3.25 $3.25	 1913 1914	 CVO CVO
(See Duck Co., J. J.)	Ferron /SML;OVC	$5.00	1913	CO
(See Duck Co., Wm. B.)	Ferron (Universal) Type E /SML;OHC or OHP	$5.00	1914	C
Firth & Co., John A., NY NY (FIRCO)	Firth (FIRCO) /SPI;OHC 3-Mineral /SPI;OHP	$2.50 X	1922 E	RQVO VO
Fitzpatrick, G., Chicago IL	Detex /SXM;OHC	$1.00	1926	R
(See Roll-O Radio Corp.)	5-in-1 /XXX;XXX	X	1924	O
(See Foote Mineral Co./ Foote Radio Corp.)	Fixedetector /O--;EHM	$1.25	1924	RVO
(See Hargraves, C. E. & H. T.)	Fixed Krystal /MFM;EHC (similar to ERLA Fixed)	X	1925	R
Flash Radio Corp., Boston MA	Detector /XXX;XXX	X	1925	O
Foote Mineral Co. (by 1925, Foote Radio Corp.), Philadelphia PA	Fixedetector /O--;EHM Variotector /O--;EHM Foote-Tector /XXX;XXX	$1.25 $1.50 X	1924 1924 1925	RVO RVO O
Ford Co., K. N., Los Angeles CA	Radiogrand /XXX;XXX	X	1925	O
Forman & Co., NY NY	UN-X-LD /XXX;XXX	X	1923	RO

MANUFACTURER	DET NAME/BASE; DETECTOR	PRICE	YEAR	REF.
Freed-Eisemann Radio Corp., NY NY	Freed-Eisemann /SPI;OHC	$0.70	1923	M
Freshman Co., Chas., NY NY	Freshman Double Adjustable /MPS;EVC	$1.50	1924	RQMVO
(See Duck Co., Wm. B.)	G /SPI;OHC	$5.00	1914	CV
(See Barawik Co.)	Galena /SPS;OHC	$1.39	1922	RM
(See Doron Bros. Electrical Co.)	(Doron) Galena /SPI;OHC	$4.30	1920	C
(See Electro Importing Co.)	Galena /SPI;OVC	$1.00	1914	CO
	Galena-Gas /SML;OVC	$4.00	1914	C
(See Wireless Specialty Apparatus Co.)	Galena Type IP201 /SPL;OHC	$10.00	1919	C
G & S Radio Research Lab., Los Angeles CA	G & S Silkcore /O--;EHC	$1.50	1925	RO
Gehman & Weinert, Newark NJ	G-W /SPI;OHC	$1.25	1922	R
(See Bunnell, J. H. & Co.)	Ghegan /SPI;OHC	X	E	V
Gilbert Co., A. C., New Haven CT (see also Tuska Co., C. D.)	Radiotector /SWL;OHC	X	1920	CVO
Gilfillan Bros., Los Angeles CA	Gilfillan /XXX;XXX	X	1925	O
(See National Airphone Corp.)	Gold Grain /SPS;EVM	$2.50	1922	RM
Goucher & Co., Newark NJ	Detector /XXX;XXX	X	1925	O
Grant, Harry Jr., Burlingame CA	A-1 /XXX;XXC	X	1924	R
(See Sears, Roebuck & Co.)	Great Lakes /SPI;OHC (appears same as Welch Scientific Co. 2615ff)	$0.95	1917	RC
Great Lakes Radio Co., Chicago IL	N-20 /SPI;EHC	$1.40	1922	R
	N-30 /SPI;OHC	$0.88	1922	R
Grebe & Co., A. H., Richmond Hills NY	Grebe /SPI;OHC	$2.00	1914	MO
	RPDA /SPI;EVC & EVC	$18.00	1919	RQC
	RPDB /SPI;EVC	$5.50	1919	RQCV
Grewol Mfg. Co., Newark NJ (see also Randel Wireless Co., sole distributor)	Grewol Fixed /RMS;EVC	$2.00	1922	RMCVO
(See Gehman & Weinert)	G-W /SPI;OHC	$1.25	1922	R

MANUFACTURER	DET NAME/BASE; DETECTOR	PRICE	YEAR	REF.
Haller-Cunningham Elec. Co., San Francisco CA (successor in 1919: Meyberg Co., Leo J., San Francisco CA)	Halcun /SPI;OHC	$2.00	1914	O
Haller, W. B., Pittsburgh PA	Hallerio /XXX;XXX	X	1925	O
(See Liberty Mail Order House —distributor; mfr. unknown; also see Harris Lab.)	Harco Permanent /SPS;EHP (appears same as Radio/Radjo Permanent)	$1.35	1925	C
Hargraves, C. E. & H. T., Lakewood RI	Fixed Krystal /MFM;EHC (similar to ERLA Fixed)	X	1925	RO
	Silver Dome /XXX;XXX	X	1925	RO
Harkness Radio Corp., Newark NJ (see Shamrock Mfg. Co.)	Shamrock-Harkness (panel mount) /O--;EHC	$2.00	1924	RMVO
Harris Lab., NY NY (see Liberty Mail Order House)	Radio/Radjo Permanent /SPS;EHP (appears same as Harco)	X	1923	RV
Harvey & Walter Mfg. Co., Cincinnati OH	Semi-permanent /SXS;EXC	X	1924	R
Haydon & Haydon, NY NY	Detector /XXX;XXX	X	1925	O
Hearwell Elec. Co., Boston MA	Bug /XXX;XXX	X	1925	O
Hedden Place Machine Co., E. Orange NJ (see also Radio Service Co.)	HPMCO /XXX;XXX	$1.50	1923	QO
(See Telephone Maintenance Co./TELMACO)	Hot Spot Semi-permanent /O--;EHC	$1.50	1923	C
(See Hedden Place Machine Co.; Radio Service Co.)	HPMCO /XXX;XXX	$1.50	1923	QO
Hyman & Co., Henry, Chicago IL	Bestone /SPS;EHC	X	E	V
Imperial Elec. & Mfg. Co., Chicago IL	Imperial /SFI;OHC	$2.00	1914	O
(See Star-King Co.)	James Blue Bullet Fixed /O--;EHM	$1.00	1925	R
J. K. Corp., (address unknown)	J. K. Corp. /SPI;OHC	X	E	O
Jones Radio Co., Brooklyn NY	Jones /SOI;OHC (slate base)	$2.00	1917	MO
(See Bunnell & Co., J. H.)	Jove /SPI;OHC	$2.00	1916	RQMCVO
	Jumbo-Jove /SML;OHC	$10.00	1916	C
(See Duck Co., Wm. B.)	Junior:			
	Horizontal /RWS;OHC	$0.75	1914	C
	Vertical /RWS;OVC	$0.95	1914	C

MANUFACTURER	DET NAME/BASE; DETECTOR	PRICE	YEAR	REF.
(See Tel-Radion Co.)	(Tel-Radion) Junior /SWI;XXX	$2.00	1916	M
(unknown mfr.)	K-3 /SOS;OHC	X	L	V
Kennedy Co., Colin B., St. Louis MO & San Francisco CA	Kennedy /SPS;OHC	X	E	VO
Keystone Products Co., Royal Oak MI	Nion Fixed /O--;EHM	$1.00	1924	RCO
Kilbourne & Clark Mfg. Co., Seattle WA (K & C)	Kilbourne & Clark (K & C) /SPI;OHC	$1.50	1922	CV
Klein Radio & Elec. Supply Co., NY NY	Klein /SPI;OHC	$1.50	1922	R
(See Radio Distributing Co./ RADISCO)	(RADISCO) Knock-down /O--;OHC	$0.75	1922	R
(See Radio Specialty Co./ RASCO)	Knocked-down /O--;OHC	$0.35	1923	R
Kolster Radio Corp., NY NY	Kolster /SPI;OVC	X	E	O
Krause, D. B., Schenectady NY	Z & T /SPS;EHP	X	E	V
Kresge Co., S. S., Detroit MI, retailer	Tellurium & Zincite /SPS;EHP	$0.25	1925	C
Kress & Co., S. H., NY NY	Kress Fixed /O--;EHC	X	E	O
Lamb Co., Jos. F., Detroit MI	Detector /XWX;OXC Detector /XFX;EXC Detector /O--;EHM	$1.00 $1.25 $0.75	1922 1922 1922	Q Q Q
(See Palmer & Palmer)	Lebnite Multipoint Fixt /O--;EHM	$0.75	1926	R
Lee Elec. & Mfg. Co., San Francisco CA (LEMCO)	LEMCO No. 300 Fixed /XXX;XXX	$0.60	1924	R
Lego Corp., NY NY	Lego Wonder Fixed /O--;EHM	$1.00	1924	RM
Lenzite Crystal Corp., Pasadena CA	Lenzite /SPS;EHC	$5.00	1917	MO
Liberty Mail Order House, NY NY, distributor; mfr. unknown (see Harris Lab.)	Harco Permanent /SPS;EHP (appears same as Radio Permanent)	$1.35	1925	C
Liberty Radio Co., NY NY	Liberty /XXX;XXX	$1.25	1923	R
Lincoln Mfg. Co., Los Angeles CA	Lincoln /RPM;EVC	$2.00	1924	RQMVO
Lindell Elec. Shop, Seattle WA	A-1 Pacific Type /SPI;OVC (2 styles) C-1 Catwhisker Type /SPS;OHC	$4.00 $4.50 $1.50	1914 1914 1914	O O O

MANUFACTURER	DET NAME/BASE; DETECTOR	PRICE	YEAR	REF.
Lip-Tone Detector Co., Revere MA	Lip-Tone /XXX;XXX	X	1925	O
Lowenstein Radio Co., Brooklyn NY (see also Wireless Specialty Apparatus Co. /WSA)	Triple Detector Type SE183-A (also called "IP203" by WSA) /SPL;OHC,OHP & OHC	$35.00* (*price in 1919 by WSA)	1917	CVO
(See Powertone Radio Products Co.)	(Powertone) Lockset /XXX;XXX	$0.50	E	O
McCarty Wireless Telephone Co., San Francisco CA	McCarty Micrometer /SPL;OHC	X	E	V
McCloy Elec. Co., Pittsburgh PA	Sta-Just /SPS;OHC	$0.65	1922	R
McCreary-Moore Co., Kansas City MO	Detector /SPI;OHC	$1.75	1914	O
McIntosh Elec. Corp., Chicago IL	McIntosh /SML;OHC	X	E	V
(See Multi-Audio-Fone)	M. A. F. /RPS;EVC	$4.25	1917	MO
Magnus Elec. Co., NY NY	Magnus /SPI;OHC	X	E	V
Maguire & Shotton, Albany NY	Albany Combination /SPL;OHP or OHC	$2.00	1914	O
Manhattan Electrical Supply Co. (MESCO), NY NY (for MESCO Galena detector, see Wireless Specialty Apparatus Co., mfr.; Crosley Mfg. Co.; Precision Equipment Co.; Ship Owners Radio Service, Inc.; Westmore-Savage Co.)	MESCO Perikon /SWL;OHP	X	E	V
	MESCO Adjustable /SPI;OVC	$4.00	1920	C
	MESCO No. 343 /SPI;OVC	$10.00	1914	C
	MESCO No. 454 /SWI;OVC	$1.80	1910	C
	MESCO Universal /SPI;OHC	$6.00	1916	RCMO
	MESCO Duplex /SPI;OHC & OHC	$16.00	1916	C
	MESCO Single /SPI;OHC	$6.00	1916	C
	No. 8496 /SPI;OVC	$4.50	1916	C
	MESCO Galena /SPI;OHC	$2.50	1922	C
(See Sears, Roebuck & Co.; also Bowman, A. W. & Co.)	Marine Mineral /SPL;OHC (same as Bowman)	$2.70	1921	C
Martin-Copeland Co., Providence RI	Mar-Co /SPS;EVT	X	1925	MVO
(See Detroit Radio Co.)	Maxitona /O--;OHC	$0.35	1922	R
Melodian Labs., Independence MO	Melody King /XXX;XXX	X	1925	O
(See Pacent Elec. Co.)	Merit /RPS;EVC (Same as Pacent No. 31)	X	1923	VO

MANUFACTURER	DET NAME/BASE; DETECTOR	PRICE	YEAR	REF.
(See Manhattan Electrical Co.)	MESCO /SWL;OHP	X	E	V
	MESCO Adjustable /SPI;OVC	$4.00	1920	C
	MESCO No. 343 /SPI;OHC	$10.00	1914	C
	MESCO No. 454 /SWI;OHC	$6.00	1914	C
	MESCO Universal /SPI;OHC	$3.00	1916	RCMO
	MESCO Duplex /SPI;OHC & OHC	$16.00	1916	C
	MESCO Single /SPI;OHC	$6.00	1916	C
	MESCO Galena /SPI;OHC	$2.50	1922	C
(See Radio Supply Co.)	Mexican Pyrites Tubular Fixed /O--;EHM	$0.60	1932	C
(See Sears, Roebuck & Co.)	Midget /SPS;OHC (Same as Murdock No. 324)	$0.45	1916	C
Miller Radio Co., A. H., Detroit MI	Miller B-Metal /XXX;XXX	X	1925	MO
(See M. P. M. Sales Co.)	(M. P. M.) Million Point Mineral Detector /XXX;XXX	X	E	M
(See Doron Bros. Electrical Co.)	(Doron) Mineral No. 115 /SPI;OHC	$3.20	1920	C
(See Duck Co., J. J. & Wm. B.)	Mineral, No. 60 x 12 /SWS;OVC	$1.00	1913	CO
(See Electro Importing Co.)	(E. I.) Miniature (or Baby) /SPM;OHC	$0.25	1915	CVO
(See Allied Radio Corp.)	MLO Fixed /RPM;EXM	$1.00* (*price in 1934)	E	C
(See Bunnell & Co., J. H.)	Molybdenite /SPI;OVC	$6.00	1916	C
Montgomery Ward & Co., Chicago IL (for Wizard/WMC, see Westfield Machine Co., mfr.)	Standard Galena /SPI;OVC	$1.43	1920	RQC
	Universal /SPI;OHC	$1.88	1920	RQ
	Wizard /SPS;OVC (called "WMC" by mfr.)	$0.89	1922	C
	Wizard /SPS;OHC (differs from the 1922 "Wizard")	$0.89	1923	C
Moore Mfg. Co., Bloomfield NJ	Moore /XXX;XXX	X	1925	O
M. P. M. Sales Co., Los Angeles CA	M. P. M. (Million Point Mineral) Detector /XXX;XXX	X	1921	M
Multi-Audio-Fone, Elizabeth NJ	M. A. F. /RPS;EVC	$4.25	1917	MO
(See Palmer & Palmer)	(Lebnite) Multipoint Fixt /O--;EHM	$0.75	1926	R

MANUFACTURER	DET NAME/BASE; DETECTOR	PRICE	YEAR	REF.
(See Noble, F. H. Co.)	Multi-Point Panel Detector /MMM;OHC	X	1926	R
Murdock Co., Wm. J., Chelsea MA (also see Sears, Roebuck & Co.)	Murdock No. 324 /SPS;OHC (also called "Chelsea," "Junior," or "Midget")	$1.00	1916	RQCV
	Silicon /SPL;OVC	$4.50	1913	CVO
(See Great Lakes Radio Co.)	N-20 /SPI;EHC	$1.40	1922	R
	N-30 /SPI;OHC	$0.88	1922	R
(See Radio Surplus Corp.)	NAECO No. 101 /SPS;EVC	X	E	VO
	NAECO Fixed /XXX;XXM	$0.50	1926	Q
National Airphone Corp., NY NY	Gold Grain /SPS;EVM	$2.50	1922	RM
New England Radio Co., Revere MA	Universal /SPI;OHC	$2.25	1921	Q
Newman-Stern Co., Teagle Radio Div., Cleveland OH	No. T-105 /SPI;OHC	$1.00	1922	Q
Niagara Sales Corp., NY NY	Niagara /SPI;EVC (white Bakelite base)	$2.25	1922	R
Nichols Elec. Co., NY NY	Cat-whisker /SPS;OHC	$0.50	1915	C
	Duplex, Special /SPI;OHC	$2.50	1915	C
	Universal /SPI;OHC	$0.75	1915	C
Nilsson, O. F., NY NY (see Radio Trading Co., retailer)	Puretone Adjustable /RPS;EHC	$0.98	1934	CV
(See Keystone Products Co.)	Nion Fixed /O--;EHM	$1.00	1924	RCO
Noble Co., F. H., Chicago IL	Multi-Point Panel Detector /MMM;OHC	X	1926	R
Novelty Radio Mfg. Co., St. Louis MO	Stayput /XXX;XXX	X	1925	O
(See Chambers & Co., F. B.)	Our Favorite /RPI;OHC	$1.75	1914	C
	Our Latest /SPI;OHC	$2.75	1914	C
	Our Own /SPI;OHP	$2.50	1914	C
Pacent Elec. Co., NY NY	Pacent No. 30 /RPS;EVC	$1.75	1922	RQCVO
	Pacent No. 31 /RPS;EVC	X	1923	VO
	Merit /RPS;EVC (same as Pacent No. 31)	X	1923	VO
Pacific Radio Specialty Co., Philadelphia PA	Death Valley Permatect /O--;EHM (cartridge type)	$1.25	1925	O
(See Lindell Elec. Shop)	(A-1) Pacific Type /SPI;OVC (2 styles)	$4.00	1914	O
		$4.50	1914	O

MANUFACTURER	DET NAME/BASE; DETECTOR	PRICE	YEAR	REF.
Palmer & Palmer, Buffalo NY	Lebnite Multipoint Fixt /O--;EHM	$0.75	1926	R
(See Adams-Morgan Co.)	Paragon No. 45 /SPS;OHC	$1.75	1920	RQCV
Parkin Mfg. Co., San Rafael CA	Parkin /SPS;OHC	$1.00	1925	CO
Parkston Co., Schenectady NY	Parkston /XXX;XXX	X	1926	O
(See Sears, Roebuck & Co.)	Peerless /SPS;OHC	$2.60	1916	C
(See S. A. M. Radio Co.)	(S. A. M.) Perfect /O--;EHM	X	1925	R
(See Wireless Specialty Apparatus Co.)	Perikon Type IP162A /SPI;OHP	X	E	V
	Perikon P520 /SPL;OHP	X	E	V
(See Electro-Set Co.)	Permanon /SPL;EHM	$2.50	1916	C
(See Pacific Radio Specialty Co.)	(Death Valley) Permatec /O--;EHM (cartridge type)	$1.25	1925	O
(See Duck Co., J. J.)	Peroxide of Lead /SPI;OVM	$2.50	1913	CO
(See Electro Importing Co.)	Peroxide of Lead Dry Electrolytic /SPI;OVM	$1.25	1913	CO
Philmore Mfg. Co., NY NY	Unmounted, No. 308 or 101 (later No. 7010) /O--;OHC (called "Ajax" in 1925)	$0.15	1925	CVO
	Open, No. 310 (later No. 7003) /SPS, SFM;OHC (called "Ajax" in 1925)	$0.40	1925	CVO
	Glass Enclosed, No. 309 (later clear plastic, No. 7008) /RMS;EVC (called "Ajax" in 1925)	$0.40	1925	CVO
	Philmore Fixed, No. 100 (later No. 7002) /RPM;EXM	$0.43* (*price in 1937)	1931	CVO
Powertone Radio Products Co., Minneapolis MN	Powertone Lockset /XXX;XXX	$0.50	E	O
Precision Equipment Co., Cincinnati OH (see Wireless Specialty Apparatus Co., mfr.; also Crosley Mfg. Co.; Manhattan Electrical Supply Co.; Ship Owners Radio Service, Inc.; Westmore-Savage Co.)	Detector Stand /SPI;OHC (label lists both mfr. and contract company)	$2.80	1921* (*made in 1919 by WSA, with other labels)	R
Premier Dental Mfg. Co., Philadelphia PA	Premier /O--;OHC	X	1922	R
Presto Detector Co., Denver CO	Presto /XXX;XXX	X	1925	O

MANUFACTURER	DET NAME/BASE; DETECTOR	PRICE	YEAR	REF.
(See Brach, Mfg. Co., L. S.; Radio Distributing Co.; Telephone Maintenance Co.)	Pur-a-tone /SPI;EHC	$2.75	1922	RCO
(See Nilsson, O. F., mfr.; Radio Trading Co., retailer)	Puretone Adjustable /RPS;EHC	$0.98	1934	CV
Pyramid Products Co., Chicago IL	De-Tec-Tone:			
	Type B (Panel Mount) /O--;EHM	$1.60	1924	RO
	Type C (Clip Mount) /O--;EHM	$1.50	1924	RO
(See Erisman Labs.)	Pyratek Fixed /SPM;EHM	$1.25	1924	RCVO
(See Electro-Set Co.)	Quick Change /SPI;OHC	$0.60	1916	C
(See Sears, Roebuck & Co.)	Quintuple /RWI;OHC	$1.60	1916	C
(See Edgcomb-Pyle Wireless Mfg. Co.)	R1 /SPS;OHC	$1.50	1914	C
	R2 /SPS;OHC	$2.00	1914	C
	R3 /SPI;OHP	$2.50	1914	C
	R4 /SPI;OHP	$3.00	1914	C
	R5 /SPI;OHP	$3.50	1914	C
	R6 /SPI;OHP	$3.00	1914	C
	R7 /SPI;OHC & OHP	$4.50	1914	C
	R12 /SPL;OHC & OHP	$10.00	1914	C
(See Towner Radio Mfg. Co.)	Radetec Fixed /O--;EHM	$1.00	1923	RO
Radio Apparatus Co., Philadelphia PA	Detector /SPI;OHC	$1.75	1914	O
Radio Apparatus Co., Pittsburgh PA	Speer /SPI;EHC	$2.00	1922	R
(See Sears, Roebuck & Co.)	Radio Cartridge Rectifier /O--;EHM	$1.38	1923	C
(See Electro Importing Co.)	Radiocite /SPI;OHC	$3.50	1918	CV
Radio Corp. of America (See RCA)				
Radio Distributing Co., Newark NJ (RADISCO) (see also Brach Mfg. Co., L. S.; Telephone Maintenance Co.)	RADISCO Knock-down /O--;OHC	$0.75	1922	R
	Pur-a-tone /SPI;EHC	$2.75	1922	RCO
Radio Engineering Co., New Rochelle NY	Detector /XXX;XXX	X	1925	O
(See Ford Co., K. N.)	Radiogrand /XXX;XXX	X	1925	O

MANUFACTURER	DET NAME/BASE; DETECTOR	PRICE	YEAR	REF.
(See Tuerk Mfg. Co.)	Radiophan 2 in 1 /O--;EHC or EHP	$1.50	1925	R
(See Harris Lab.; see also Liberty Mail Order House)	Radio/Radjo Permanent /SPS;EHP (appears same as Harco)	X	1923	RV
Radio Receptor Co., NY NY	Detector /XXX;XXX	X	1925	O
Radio Sales Co., San Francisco CA	Detector /SPX;XXC	$0.75	1919	R
Radio Service & Mfg. Co., NY NY	Type S-20 /SPS;OHC	$1.00	1922	R
Radio Service Co., Charleston WV, distributor (see Hedden Place Machine Co.)	BD /XXX;XXX	$1.50	1923	QO
	HPMCO /XXX;XXX	$1.50	1923	QO
Radio Shop of Newark, Newark NJ	Condenser-Detector /SPI;OHC	$3.00* (*$1.50 w/o condenser)	1922	R
(See Electro Importing Co.)	Radioson /SPI;EVE	$4.50	1914	RMCO
Radio Specialty Co., NY NY (RASCO)	Crystodyne (Zincite) /SPS;OHC	$1.75	1924	RMO
	Knocked-down /O--;OHC	$0.35	1923	R
	RASCO Baby /SPM;OHC	$0.50	1921	RQMV
Radio Supply Co., Los Angeles CA, distributor	Mexican Pyrites Tubular Fixed /O--;EHM	$0.60	1932	C
	Tom Mack Fixed /XXX;XXM	$1.25	1930	C
Radio Surplus Corp., Boston MA, distributor	NAECO No. 101 /SPS;EVC	X	E	VO
	NAECO Fixed /XXX;XXM	$0.50	1926	Q
(See Gilbert Co., A. C.; also Tuska Co., C. D., mfr.)	Radiotector /SWL;OHC	X	1920	CVO
Radio Telephone & Mfg. Co., NY NY	3-Crystal /SPI;OVC	$2.50	1922	RO
	Detector /SPS;OHC	$0.85	1922	R
Radio Trading Co., NY NY, (see Nilsson, O. F., mfr.)	Puretone Adjustable /RPS;EHC	$0.98	1934	CV
(See Radio Distributing Co.)	RADISCO Knock-down /O--;OHC	$0.75	1922	R
(See Star Mfg. Co.)	Radium Jewe*ll* /O--;EHC (later: Radium Jewe*l*)	$1.00	1922	R
(See Elec. City Novelty & Mfg. Co.)	Radjo /SPS;OHC	X	1924	RO
(See Harris Lab.)	Radjo (Radio) Permanent /SPS;EHP	X	1923	RV

MANUFACTURER	DET NAME/BASE; DETECTOR	PRICE	YEAR	REF.
Randel Wireless Co., Newark NJ, sole distributor (see also Grewol Mfg. Co., mfr.)	Grewol Fixed /ROM;EVC	$2.00	1922	RMCVO
(See Radio Specialty Co.)	RASCO Baby /SPM;OHC	$0.50	1921	RQMV
(See Davidson Radio Corp.)	Rasla Fixed /O--;EHM	$1.25	1923	CO
(See Wernes & Patch)	R. B. /XOX;XXC	$0.75	1923	C
RCA, NY NY (mfr., Wireless Specialty Apparatus Co.)	Cleartone Radio Concert Model UD 1432 /SPI;OHC	X	E	V
Redden, A. H., Irvington NJ	Redden's /SPI;OHC	$1.50	E	O
(See American Specialty Co.)	Regal No. 174 /SPI;OHC	$1.25	1922	RC
Remler Radio Mfg. Co., San Francisco CA	Remler No. 50 /SPI;OHC	X	E	V
	Turret Head /SPI;OHC (also 3 other models in 1919—unknown features)	$4.00	1919	RQ
Riley-Klotz Mfg. Co. Newark NJ	Arkay /SPS;OHC	$1.00	1922	R
Roll-O-Radio Corp., Cincinnati OH	5-in-1 /XXX;XXX	X	1924	O
(See Grebe & Co., A. H.)	RPDA /SPI;EVC & EVC	$18.00	1919	RQC
	RPDB /SPI;EVC	$5.50	1919	RQCV
R-U-F Products, Brooklyn NY	R-U-F Vernier Type 200 /O--;EHC	X	1925	O
Rusonite Products Corp., NY NY	Rusonite Fixed /SPS;EHM	$1.25	1924	QO
R-W Mfg. Co., Chicago IL	R-W Self-adjusting /SPI;EHC	$1.50	1923	RCVO
(See Radio Service & Mfg. Co.)	S-20 /SPS;OHC	$1.00	1922	R
S. A. M. Radio Co., Omaha NE	S. A. M. Perfect /O--;EHM	X	1925	R
Schenectady Radio Corp., Schenectady NY	Scyrad /XXX;XXX	X	1926	O
Schwartz & Son, Maurice, Schenectady NY (see also Federal Radio Products Co.)	Schwartz Adjustable /SFS;EHP (appears same as Whiz Bang!)	X	1925	RV
Scientific Research Labs., Baltimore MD (see Electrical Research Labs.)	Scientific Fixed /MFM;EHM (appears same as ERLA Fixed)	$1.50	1924	RVO
Scott Combination Crystal Detector Co., NY NY	Scott Combination /SPS;EHP	X	1924	R

MANUFACTURER	DET NAME/BASE; DETECTOR	PRICE	YEAR	REF.
(See Schenectady Radio Corp.)	Scyrad /XXX;XXX	X	1926	O
Sears, Roebuck & Co., Chicago IL (see also Bowman Co., A. H.; Murdock Co., Wm. J.; Welch Scientific Co.)	Acme Galena /SPI;OHC	$1.00	1916	C
	Great Lakes /SPI;OHC (appears same as Welch Scientific Co., 2615ff)	$0.95	1917	RC
	Marine Mineral /SPL;OHC (called "Bowman" by mfr.)	$2.70	1921	CV
	Midget /SPS;OHC (same as Murdock No. 324)	$0.45	1916	C
	Peerless /SPS;OHC	$2.60	1916	C
	Quintuple /RWI;OHC	$1.60	1916	C
	Radio Cartridge Rectifier /O--;EHM	$1.38	1923	C
	Standard Galena /SPI;OVC	$0.90	1917	C
	Superior Silicon /SPI;OVC	$1.25	1916	C
	Universal /SPL;OVC	$1.65	1917	C
(See Harvey & Walter Mfg. Co.)	Semi-permanent /SXS;EXC	X	1924	R
(See Electro-Set Co.)	Senior /SPI;OHC	$1.50	1916	C
(unknown mfr.)	SERCO /SPI;OHC	X	E	O
Shamrock Mfg. Co., Newark NJ (see Harkness Radio Corp.)	Shamrock-Harkness (panel mount) /O--;EHC	$2.00	1924	RM
Ship Owners Radio Service, Inc., NY NY (see Wireless Specialty Apparatus Co., mfr.; also Crosley Mfg. Co.; Manhattan Electrical Supply Co.; Precision Equipment Co.; Westmore-Savage Co.)	SORSINC /SPI;OHC (label lists both mfr. and contract company)	$1.60	1919	CVO
Signal Elec. Mfg. Co., Menominee MI (see also Andrae & Sons Co., Julius)	Signal Silicon No. R40 /SPI;OHC	$2.00	1920	RVO
	Signal Standard Galena No. 41 /SPI;OHC	$1.50	1921	RQCV
(See Doron Bros. Electrical Co.)	(Doron) Silicon No. 110 /SPI;OVC	$4.20	1916	C
(See Electro-Set Co.)	Silicon /SPL;OVC	$4.50	1916	C
(See Murdock Co., Wm. J.)	Silicon /SPL;OVC	$4.50	1913	CVO
(See Signal Elec. Mfg. Co.)	(Signal) Silicon No. R40 /SPI;OVC	$2.00	1920	RVO
(See Wireless Specialty Apparatus Co.)	Silicon Type IP200 /SPL;OHP	$10.00	1919	C

MANUFACTURER	DET NAME/BASE; DETECTOR	PRICE	YEAR	REF.
(See G & S Radio Research Lab.)	(G & S) Silkcore /O--;EHC	$1.50	1925	RO
(See Hargraves, C. E. & H. T.)	Silver Dome /XXX;XXX	X	1925	O
(See American Radio & Research Corp./AMRAD)	(AMRAD) Single, Type C-1 /SPS;EHC	$2.50	1919	RQC
(See Ship Owners Radio Service, Inc.)	SORSINC /SPI;OHC	$1.60	1919	CVO
(See Tool & Device Corp.)	Spade /XXX;XXX	X	1925	O
(See Chambers & Co., F. B.)	Special Pick /SPS;OHC	$1.50	1914	C
(See Buscher Co., C. A.)	SpeeD /RPS;EVC	$0.85	1923	C
(See Radio Apparatus Co. [Pittsburgh])	Speer /SPI;EHC	$2.00	1922	R
Stafford Radio Co., Medford Hillside MA	Stafford /SPS;EHM	$1.00	1925	O
(See McCloy Elec. Co.)	Sta-Just /SPS;OHC	$0.65	1922	R
(See Duck Co., J. J.; also see Duck Co., Wm. B.—1914)	Standard /SPI;OVC	$1.50	1913	CO
(See Andrae & Sons Co., Julius; Signal Elec. Mfg. Co.)	Standard Galena No. 41 /SPI;OHC	$1.50	1921	RC
(See Montgomery Ward & Co.)	Standard Galena /SPI;OVC	$1.43	1920	RQC
(See Sears, Roebuck & Co.)	Standard Galena /SPI;OVC	$0.90	1917	C
Star Crystal Co., Detroit MI	Crystalstat /XXX;XXX	X	1925	O
	Star Fixed /MFM;EHM (similar to ERLA Fixed)	$1.25	E	V
Star-King Co., San Francisco CA	James Blue Bullet Fixed /O--;EHM	$1.00	1925	R
Star Mfg. Co., Newark NJ	Radium Jewe*ll* /O--;EHC (later: Radium Jewe*l*)	$1.00	1922	R
	Star Semi-fixed /RPM;EHP	$1.00	1922	R
(See Novelty Radio Mfg. Co.)	Stayput /XXX;XXX	X	1925	O
Steinite Labs., Atchison KS	Steinite Rectifier /O--;EVM	$1.00	1925	RVO
Steinmetz Wireless Mfg. Co., Pittsburgh PA	Steinmetz /XXX;XXX	X	1925	O
Stellar Mfg. Co., Kansas City MO	Stellar Fixed /RPM;EXM	X	E	V
(See Sears, Roebuck & Co.)	Superior Silicon /SPI;OVC	$1.25	1916	C

MANUFACTURER	DET NAME/BASE; DETECTOR	PRICE	YEAR	REF.
(See Bisby Mfg. Co.)	Supertone /XXX;XXX	X	1925	O
Sypher Mfg. Co., Toledo OH	Sypher /SPI;OHC	$0.75	1922	R
(See Newman-Stern Co., Teagle Radio Division)	T-105 /SPI;OHC	$1.00	1922	Q
Talley, Wm. A., Beaudry AR	World Beater /SPS;OHC	X	1929	R
Tanner Radio Co., C. D., Los Angeles CA	Tanner /SPM;OHC	X	1925	RO
Telephone Maintenance Co., Chicago IL (TELMACO) (see also Brach Mfg. Co., L. S.; Radio Distributing Co./RADISCO)	Hot Spot Semi-permanent /O--;EHC	$1.50	1923	C
	Pur-a-tone /SPI;EHC	$2.75	1922	RCO
Teleradio Engineering Corp., NY NY	Mounted Detectors /XXX;XXX	X	1922	R
(See Kresge Co., S. S.)	Tellurium & Zincite /SPS;EHP	$0.25	1925	C
Tel-Radion Co., NY NY	Duplex Tel-Radion /SWI;XXX	$5.00	1916	M
	Tel-Radion Jr. /SWI;XXX	$2.00	1916	M
(See Firth & Co., John A.)	Three Mineral /SPI;OHP	X	E	VO
Tillman Radio Products Co., Brooklyn NY	Tillman /SPI;OHC	X	E	V
(See Radio Supply Co.)	Tom Mack Fixed /XXX;XXM	$1.25	1930	C
Tool & Device Corp., Troy NY	Zincite & Tellurium /SPS;EHP	$0.50	1925	RCV
	Spade /XXX;XXX	X	1925	O
Towner Radio Mfg. Co., Kansas City MO	Radetec Fixed /O--;EHM	$1.00	1923	RO
Tribune Airfone Co., Owego NY	Detector /XXX;XXX	X	1922	R
(See Lowenstein Radio Co.; Wireless Specialty Apparatus Co. /WSA)	Triple Detector Type IP-203 (SE183-A) /SPL;OHC, OHC & OHP	$35.00* (*price in 1919 by WSA)	1917	CVO
Tritector Co., St. Paul MN	Tritector /XXX;XXX	X	1925	O
Tuerk Mfg. Co., Chicago IL	Radiophan 2 in 1 /O--;EHC or EHP	$1.50	1925	R
Turney Co., Eugene T., NY NY (see also Connecticut Telephone & Elec. Co., mfr. after 1920)	Crystaloi:			
	Type A* /RPS;EVM	$3.50	1916	RCVO
	Type AA /RPI;EVM	$6.00	1916	QMCO
	Type BB /SWL;EVM	$12.00	1916	MC
	Type O* /RPS;EVM	$3.50	1916	RQMCV
	(*Type A is very similar to or same as Type O)			

MANUFACTURER	DET NAME/BASE; DETECTOR	PRICE	YEAR	REF.
(See Duck Co., Wm. B.)	Turret Cup /SPI;OHC	$2.00	1914	C
(See Remler Radio Mfg. Co.)	Turret Head /SPI;OHC	$4.00	1919	RQ
(See Elec. Mfg. & Sales Co.)	(EMSCO) Turret-Top /SPI;OHC	X	1922	RQ
Tuska Co., C. D., Hartford CT (see also Gilbert Co., A. C.)	Radiotector /SWL;OHC	X	1920	CVO
	Tuska Type 90 /SPI;OHC	$0.65	1920	C
(See Atwater Kent Mfg. Co.)	(Model 2A) Twin Unit (Part No. 4200) /RPI;OHC & OHC	X	1923	O
(See DeForest Radio Tel. & Tel. Co.)	UD-100 /O--;EVC	$4.00	1919	C
(See Buscher Co., C. A.)	UMCO /SPI;OHC	$0.85	1923	C
United Metal Stamping & Radio Co., Cincinnati OH	Adjustable /SMI;OHC	X	1925	RO
(See Clapp-Eastham Co.)	Universal:			
	Ferron /SML;OVC	$3.25	1913	CVO
	Ferron /SML;OHC or OHP	$3.25	1914	CVO
(See Doron Bros. Electrical Co.)	(Doron) Universal /SPI;OHC or OHP	$4.70	1916	RCO
(See Duck Co., Wm. B; see also Clapp-Eastham Co., re: Universal Type E Ferron detector)	Universal (Type B) /SPI;OHC	$5.00	1914	C
	Univeral (Type E) Ferron /SML; OHC or OHP	$5.00	1914	C
(See Electro Importing Co.)	Universal /SPI;OHC	$1.50	1913	CV
(See Manhattan Electrical Supply Co.)	(MESCO) Universal /SPI;OHC	$3.00	1916	RCMO
(See Montgomery Ward Co.)	Universal /SPI;OHC	$1.88	1920	RQ
(See New England Radio Co.)	Universal /SPI;OHC	$2.25	1921	Q
(See Nichols Elec. Co.)	Universal /SPI;OHC	$0.75	1915	C
(See Sears, Roebuck & Co.)	Universal /SPL;OVC	$1.65	1917	C
(See Forman & Co.)	UN-X-LD /XXX;XXX	X	1923	RO
(See Foote Mineral Co. / Foote Radio Corp.)	Variotector /O--;EHM	$1.50	1924	RVO
(See Brownlie & Co., Roland)	(Brownlie) Vernier /O--;EHM	$2.00	1924	RMCVO
Walthom Radio Products Co., NY NY	Detector /XXX;XXX	X	1925	O
(See Cheever co., Wm. E.)	Wecco /XXX;XXX	X	1926	O

MANUFACTURER	DET NAME/BASE; DETECTOR	PRICE	YEAR	REF.
Welch Scientific Co., W. M., Chicago IL (see also Sears, Roebuck & Co./S. R. & Co.)	Detector 2615ff /SPI;OHC (appears same as Great Lake, of S. R. & Co.)	$2.20	1919	C
Welty & Co., Wm. A., Chicago IL	Crystector /MPS;EVC	$4.00	1925	RO
Wernes & Patch, Chicago IL, distributor (see also Wholesale Radio Equipment Co.)	No. 325 Deluxe Adjustable /RMM;EVC	$1.25	1923	C
	R. B. /XOX;XXC	$0.75	1923	C
	Yellowtip /RPS;EVC	$2.00	1923	RCO
Western Wireless, San Francisco CA	Western Wireless /SPI;OHC	X	E	V
Westfield Machine Co., Westfield MA (see also Montgomery Ward Co.)	WMC /SPS;OVC (called "Wizard" by Montgomery Ward Co.)	$1.00	1922	RV
Westinghouse Elec. & Mfg. Co., E. Pittsburgh PA	Type DB /SPL;OVC & OVP	X	1921?	C
	Type DB /SPL;OHC & OHP	$5.00	1921	RQMCVO
Westmore-Savage Co., Boston MA (see Wireless Specialty Apparatus Co., mfr.; also Crosley Mfg. Co.; Manhattan Electrical Supply Co.; Precision Equipment Co.; Ship Owners Radio Service, Inc.)	Westmore-Savage /SPI;OHC (label lists both mfr. and contract co.)	X* (*price $1.60 in 1919, by other contract companies and by mfr., WSA)	1919	V
Westwyre Co., Westfield MA	Detector /XXX;XXX	X	1925	O
(See Federal Radio Products Co.; also Schwartz & Son, Maurice)	Whiz Bang! /SPS;EHP (appears same as Schwartz Adjustable)	$0.75	1925	RV
Wholesale Radio Equipment Co., Newark NJ, distributor (see also Wernes & Patch)	Yellowtip /RPS;EVC	$2.00	1923	RCO
Williams, J. C., Oakland CA	Williams /SPI;OHC	$3.00	1922	R
Winn Radio & Elec. Mfg. Co., Chicago IL	Winn No. 400 /SPI;OHC	$5.00	1920	RQ
Wireless Corp. of America Chicago IL	Detector /SPI;EHC	$3.00	1922	R
Wireless Improvement Co., NY NY	Double Detector, Type WI-101-A /SPI;OHC,OHC	X	1917	VO

MANUFACTURER	DET NAME/BASE; DETECTOR	PRICE	YEAR	REF.
Wireless Specialty Apparatus Co., Boston MA (see RCA)	Cleartone Radio Concert Model UD 1432 /SPI;OHC	X	E	V
(see also Crosley Mfg. Co.; Manhattan Electrical Supply Co.; Precision Equipment Co.; Ship Owners Radio Service, Inc.; Westmore-Savage Co.)	Cleartone /SPI;OHC (WSA is mfr.; five contract companies listed to the left)	$1.60	1919	CVO
(for Triple Detector, see also Lowenstein Radio Co.)	Combined Silicon & Galena Type IP202 /SPL;OHP & OHC	$25.00	1919	C
NOTE: Before 1912, WSA was located in NY NY.	Galena Type IP201 /SPL;OHC	$10.00	1919	C
	Perikon Type IP162A /SPI;OHP	X	E	V
	Perikon P520 /SPL;OHP	X	E	V
	Silicon Type IP200 /SPL;OHP	$10.00	1919	C
	Triple Detector IP203 (also called "SE183-A") /SPL;OHC,OHC & OHP	$35.00* (*price in 1919)	1917	CVO
(See Montgomery Ward Co.; Westfield Machine Co.)	Wizard /SPS;OVC (called "WMC" by mfr.)	$0.89	1922	C
	Wizard /SPS;OHC (differs from the 1922 "Wizard")	$0.89	1923	C
(See Westfield Machine Co.)	WMC /SPS;OVC (called "Wizard" by Montgomery Ward & Co.)	$1.00	1922	RV
(See Lego Corp.)	(Lego) Wonder, Fixed /O--;EHM	$1.00	1924	RM
Wonder State Crystal Co., Little Rock AR	Wonder State /XXX;XXX	X	1925	O
(See Talley, Wm. A)	World Beater /SPS;OHC	X	1929	R
Wright Radio Co., NY NY	Wright, Fixed /O--;EHM	$0.25	1922	R
Yale Radio Elec. Co., Los Angeles CA	Detector /XXX;XXX	X	1925	O
(See Wholesale Radio Equipment Co.; Wernes & Patch)	Yellowtip /RPS;EVC	$2.00	1923	RO
(See Krause, D. B.)	Z & T /SPS;EHP	X	E	V
(See Tool & Device Corp.)	Zincite & Tellurium /SPS;EHP	$0.50	1925	RCV

ADDENDUM: RECENT DETECTORS (1940–1955)

MANUFACTURER	DET NAME/BASE; DETECTOR	PRICE	YEAR	REF.
Allen, Alva F., Clinton MO	Distite Midget, Semi-fixed /MPM;OHC	$0.50	1952	O
	FD500, Fixed /O--;EHM	$1.00	1952	O
	Melomite /SFS;OHC	$0.25	1952	O

MANUFACTURER	DET NAME/BASE; DETECTOR	PRICE	YEAR	REF.
Modern Radio Labs. Garden Groves CA	MRL (various types of detectors) /XXX;XXX	X	R	O
Radio-Ore Labs., Lynn MA	Peppy Pal /SWI;OHC	X	R	V
	Radio-Ore Midget /XXM;OHC	X	R	O

CRYSTAL DETECTORS ADVERTISEMENTS

AMRAD DETECTOR STANDS

DUPLEX

$4.50

SINGLE

$2.50

TESTED AND APPROVED BY NAVY DEPARTMENT

Without question the leading instrument of its kind. Duplex type equipped with selector switch having open circuit point. Moulded base ½ inch thick. Size 4" x 3". All metal parts nickel-plated, satin finish. Crystal mountings enclosed and dust proof. Ball and socket adjustment made by tip of forefinger. All points of the crystal accessible. Adjustment remarkably proof against shock. Postage, 10c, both types.

Single type is the most compact and rugged Detector Stand built. Moulded base 2¼" x 1⅞". Equal in all other respects to our Duplex type. Substantial binding posts with non-removable tops. Tension of wisker instantly adjustable. In appearance and design it harmonizes with the finest panel construction. Especially suitable for portable sets where weight and space is a factor. Bulletin T describes both types.

Upon request we will enter on our monthly mailing list for one year the names of all active radio amateurs. To these we will send our cotalog, now in preparation, and all our latest bulletins as rapidly as printed for insertion therein. An extensive line of Amrad apparatus is under development. If there is no radio dealer in your town, ask for our list of Amrad retailers who stock our goods.

AMERICAN RADIO AND RESEARCH CORPORATION

15A Park Row New York

Address all Communications to New York Office

Factory and Laboratory Medford Hillside, Mass.

Radio News (Apr. 1920), p. 581

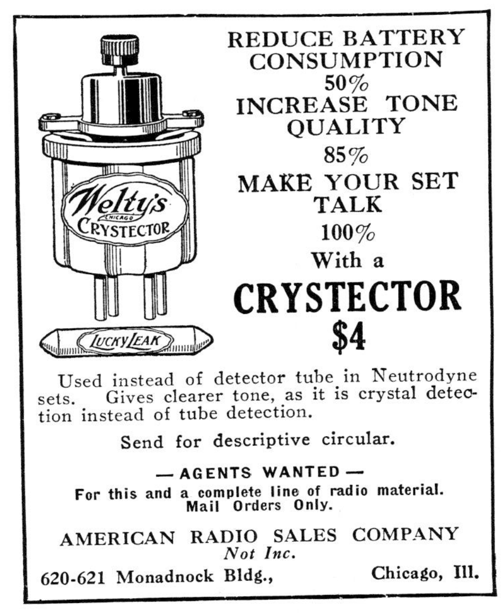

Radio News (Nov. 1924), p. 848

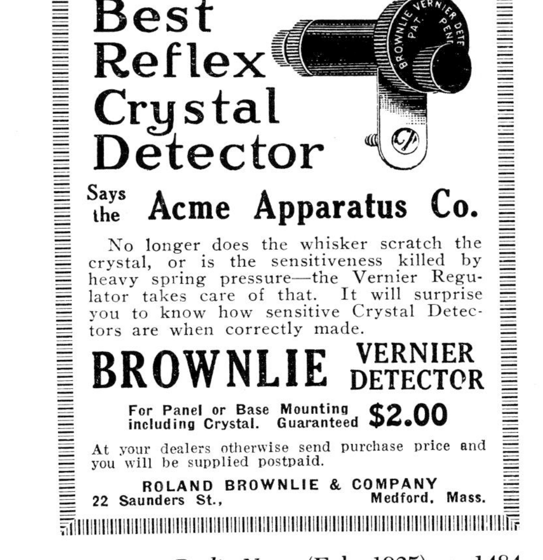

Radio News (Feb. 1925), p. 1484

Radio News (Jan. 1924), p. 1012

Radio News (Mar. 1926), p. 1374

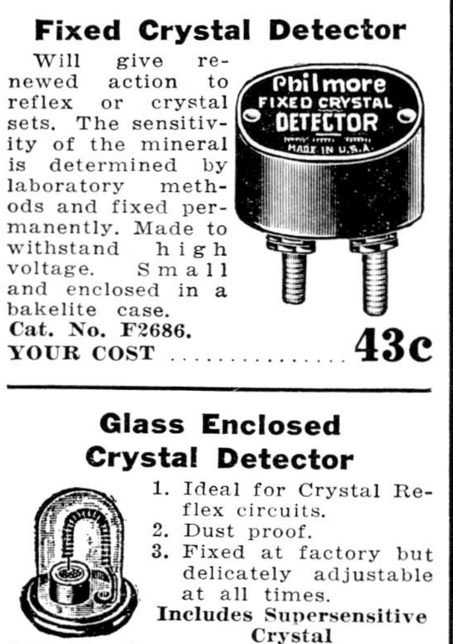

Burstein-Applebee, *Radio* Catalog No. 52 (1937), p. 105

Detectors

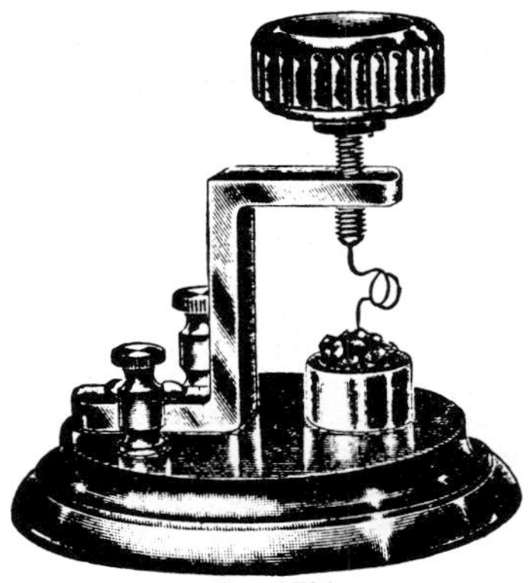

No. A504.

JUNIOR DETECTOR STAND

Contract with the mineral is established by a fine wire spring, the tension of which may be accurately adjusted. This provides a much more sensitive arrangement than the spring strip used on many detectors. Our detector will maintain its adjustment and is not easily disarranged by jars and jolts. The base is finished in dark mahogany, carefully polished. All metal fittings are polished brass, gold lacquered. The crystal is fastened in the cup by means of a small set-screw. The adjusting screw is provided with a large knurled composition knob.

No. A504 Junior Vertical Detector Stand, diameter 3 inches, height 3 inches. $0.95
Weight packed, 6 ounces.

See testimonials between pages 49-57.

No. 60X12 Detector Stand is very simple in construction, and very sensitive. Any of the substances used with detectors may be used in this detector stand.

No. A6012 Detector Stand only. . . . $0.55
Postage 11c.

CRYSTAL DETECTOR STAND.

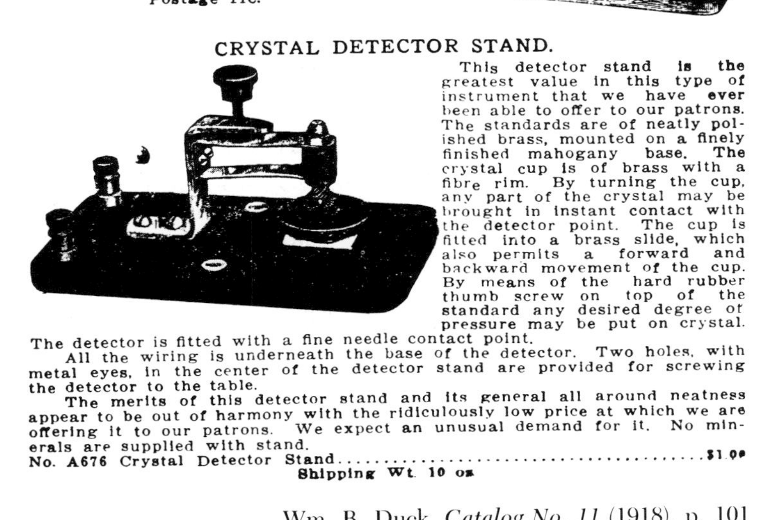

This detector stand is the greatest value in this type of instrument that we have ever been able to offer to our patrons. The standards are of neatly polished brass, mounted on a finely finished mahogany base. The crystal cup is of brass with a fibre rim. By turning the cup, any part of the crystal may be brought in instant contact with the detector point. The cup is fitted into a brass slide, which also permits a forward and backward movement of the cup. By means of the hard rubber thumb screw on top of the standard any desired degree of pressure may be put on crystal. The detector is fitted with a fine needle contact point.

All the wiring is underneath the base of the detector. Two holes, with metal eyes, in the center of the detector stand are provided for screwing the detector to the table.

The merits of this detector stand and its general all around neatness appear to be out of harmony with the ridiculously low price at which we are offering it to our patrons. We expect an unusual demand for it. No minerals are supplied with stand.

No. A676 Crystal Detector Stand. $1.00
Shipping Wt. 10 oz

Wm. B. Duck, *Catalog No. 11* (1918), p. 101

38 WILLIAM ST., NEWARK, N. J.

Dealers and Jobbers write for attractive proposition.

Radio News (Sep. 1923), p. 342

CROSLEY CRYSTAL DETECTOR STAND

This unit is especially well constructed, neatly mounted on black base covered on the bottom with green felt. All parts are bright nickel finish, complete with mounted crystal, binding posts, etc., manufactured under the following patents: "Patented January 21, 1908; November 17, 1908; June 15, 1909; September 7, 1909; July 21, 1914; September 8, 1914; November 24, 1914; April 27, 1915; January 23, 1917. Licensed for amateur, experimental or entertainment purposes only. Any other use will constitute an infringement.—$2.50.

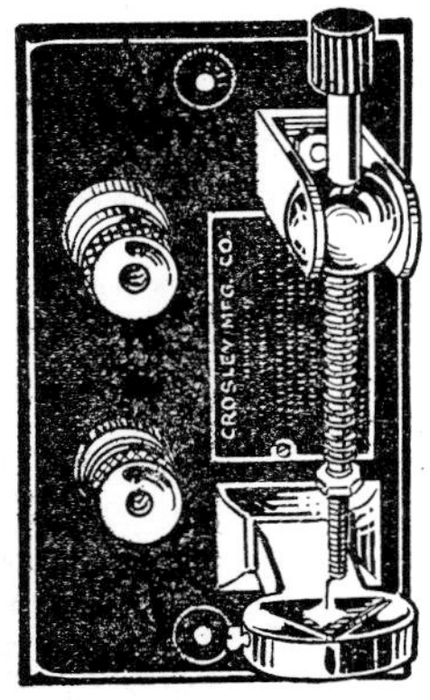
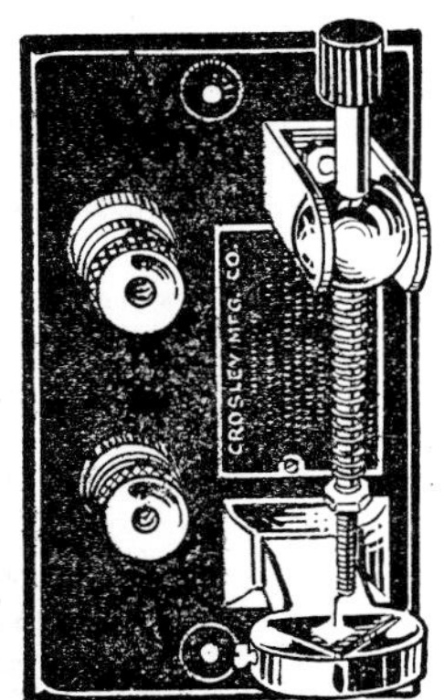

CROSLEY MANUFACTURING CO.
DEPT. Q.S.T.1 **CINCINNATI, OHIO**

ALWAYS MENTION QST WHEN WRITING TO ADVERTISERS

QST (Sep. 1922), p. 119

Allied Radio Corp., *Radio* Catalog (1933), p. 22

"You have solved the Reflex problem!"

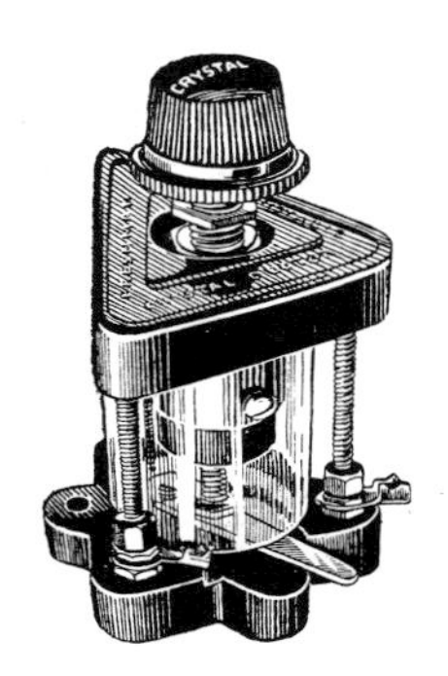

—SAID EDITOR OF

N.Y. EVE. WORLD'S RADIO MAGAZINE

"The new Freshman 'double adjustable' crystal detector 'stayed put' even when the set was deliberately shaken, stood up to 130 volts on the plate circuit without noise or distortion.

"This detector meets every requirement of the reflex circuit.

"It is enclosed and provided with two adjustments, one varying the position of the crystal, and the other regulating the brush contact adjustment.

"The crystal is a pure natural ore and is imbedded in an insulated housing, thus eliminating short circuits and consequent loud noises resulting from the cat whisker touching the metal housing.

"The Freshman detector can be panel mounted with only a small knob showing. All around it is the best crystal detector unit found for reflex work."

—Statement of Editor of N. Y. Eve. World's Radio Magazine (March 29th, 1924).

—World's Most Efficient Crystal

Every experimenter can feel confident that when the radio authority of one of the greatest newspapers in America says the Freshman is the best detector—it must be true!

Note the exclusive Freshman features:

Loop-end contact!
Non-metallic housing!
Double-adjustable!
Mounts neatly on Panel!
Stays set when adjusted!
Withstands high voltages!

FRESHMAN Double Adjustable Crystal Detector **$1.50**

For Panel or Base Use, complete with crystal

Merely turn the knob as you would a dial—No more searching for the sensitive spot!

At your dealers, otherwise send purchase price and you will be supplied postpaid.

Freshman Special Crystal with Non-Metallic Housing $ .50

Chas. Freshman Co. Inc.
Radio Condenser Products
106 Seventh Avenue, New York City

Write for free diagrams of Neutrodyne, Polydyne, Tri-Flex, Super-Heterodyne and other popular circuits.

Radio News (June 1924), p. 1865

Whiz Bang! Detector

(SEMI-FIXED)

Ideal for REFLEX and CRYSTAL Sets.

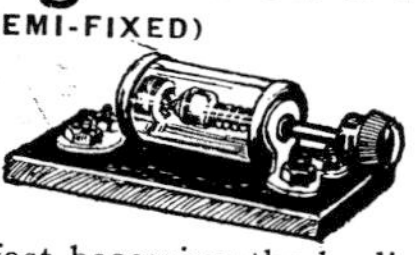

A detector that is fast becoming the leading instrument in the Radio World. Every spot is sensitive; and produces clear, mellow and loud reception. Guaranteed. Sent C.O.D. to any part of U. S.—or money orders accepted in advance. We pay postage.

75c Each

FEDERAL RADIO PRODUCTS CO.
712 Broadway Schenectady, N. Y.

Radio News (Jan. 1925), p. 1242

REJUVENATE reflex or crystal set with **THIS DETECTOR** trouble eliminator—**ALWAYS LIVE**

CRYSTAL DETEX DETECTOR $1

"ROLL THE PERFECT" Unmounted Crystal **on** 368 **GOLD** catswhiskers—**100% CONTACT $1, COMPLETE** with Crystal—Dealers or Direct. **G. FITZPATRICK,** 5560 Jackson Blvd., **Chicago**

Radio News (Apr. 1926), p. 1509

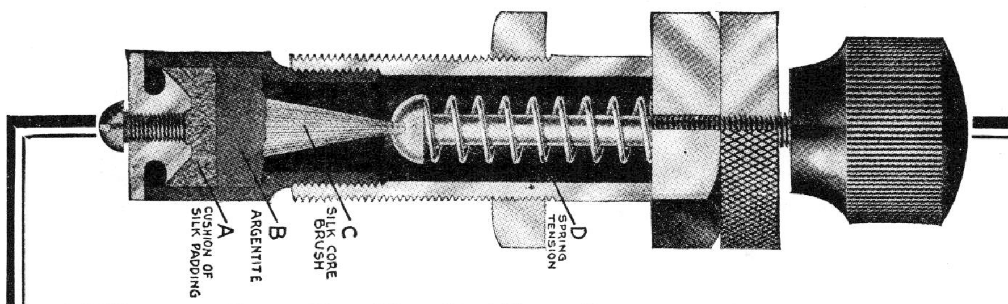

The G. & S. Silk Core Detector

A Greatly Improved Crystal Detector

TO give this detector a thorough and gruelling test, we sent samples to the leading testing laboratories of the United States. We asked the engineers to do their worst in attempting to burn out or buff the G. & S. Silk Core Detector. NOT ONE WAS ABLE TO HURT IT! They reported it to be unexcelled in volume, tone and selectivity. The distinctive feature of this detector is a brush, made of silk cord, covered with tinsel, which allows more heat expansion and provides a greater capacity and larger surface for picking up the waves.

MONEY BACK GUARANTEE

Try the G. & S. Silk Core Detector at our risk. If not satisfied, return it and your purchase price will cheerfully be refunded. Sent C. O. D. or on receipt of money order for $1.50.

DEALERS WRITE

G. & S. RESEARCH LABORATORY

1269 Cochran Avenue : : : : Los Angeles, Calif.

Radio News (Oct. 1925, p. 530

Radio News (Feb. 1924), p. 1192

Wireless MESCO Manual

TRADE MARK

Mesco Detector Stand

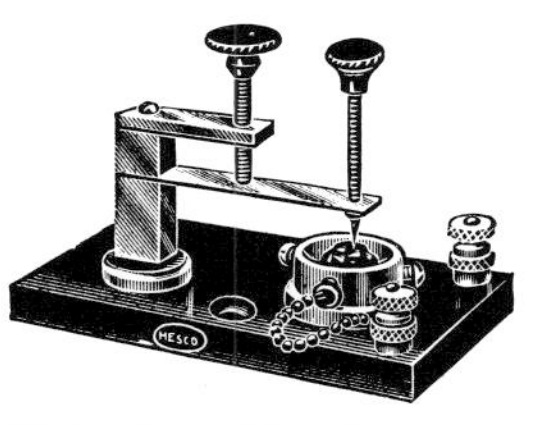

The new No. 343 Detector Stand presents many advantages in the way of wide range of adjustment, close regulation and excellent service. A double compound action is used which is not affected by any ordinary vibration. Either a heavy or a light pressure of the vertical German Silver point may be obtained quickly at will.

All metal parts are nickel plated and the base is genuine hard rubber. This Stand is a necessity to every progressive experimenter.

List No. 343 Detector Stand .... Price. **$8.00**

Manhattan Electrical Supply Co./MESCO, *Manual of Wireless Telegraphy and Catalog of Radio Telegraph Apparatus* (1916), p. 103

CRYSTAL DETECTOR

This crystal detector, submitted by the C. D. Tanner Co., 528 W. Washington Street, Los Angeles, Calif., to the RADIO NEWS LABORATORIES for test, is very well made and offers the advantage of very sensitive adjustment of the crystal contact. The detector is shown in the illustration.

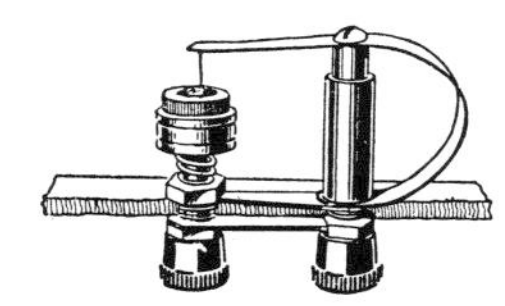

AWARDED THE RADIO NEWS LABORATORIES CERTIFICATE OF MERIT NO. 847.

Radio News (Sep. 1925), p. 327

LEGO FIXED DETECTOR

The Lego fixed crystal detector is enclosed in a glass tube fitted with metal end caps and binding post. The detector is small in size and can easily be connected in any part

of the receiver. This detector is very good for reflex receivers and works well in the ordinary crystal set. The three samples submitted by the Lego Corporation, 607 West 43rd Street, New York City, were all very sensitive and uniform as regards sensitivity. The resistance of this rectifier is about four times as great with the current passing through one direction as the other.
AWARDED THE RADIO NEWS LABORATORIES CERTIFICATE OF MERIT NO. 554.

Radio News (Dec. 1924), p. 949

MULTI-POINT PANEL CRYSTAL

The crystal shown in the illustration was submitted to the RADIO NEWS LABORATORIES for test, by the

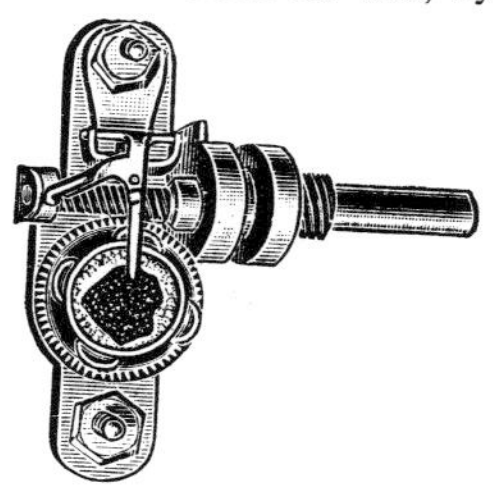

F. H. Noble Co., 29 East Madison Street, Chicago, Ill. This multi-point panel crystal detector was tested in conjunction with an ordinary crystal set and reflex receiver, and found to be satisfactory. It is of solid and efficient construction as the illustration shows and has a multi-point catwhisker.
AWARDED THE RADIO NEWS LABORATORIES CERTIFICATE OF MERIT NO. 1123.

Radio News (Apr. 1926), p. 1453

What makes possible such amazing clarity?

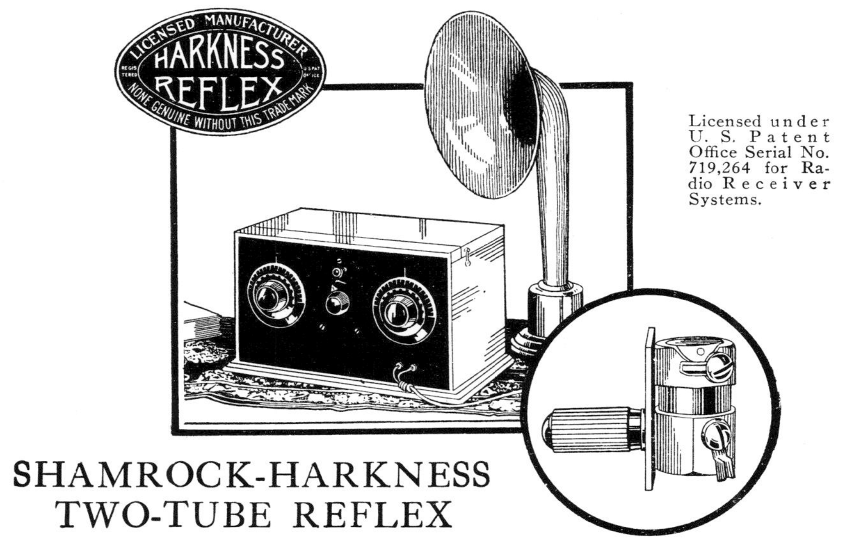

Licensed under U. S. Patent Office Serial No. 719,264 for Radio Receiver Systems.

SHAMROCK-HARKNESS TWO-TUBE REFLEX

LISTEN to it! Hear the Shamrock-Harkness for five minutes any time—even when receiving conditions are not most favorable. Above all, you'll be amazed at the way every sound comes in—clean-cut, life-like, clear as crystal! You can get *clear, consistent* loud speaker reception on this powerful little set within a radius of 1000 miles —and sometimes more!

Notice the unit pictured in the insert above. It's the crystal detector—the little wonder part that is responsible for this remarkable clarity.

Amazing Performance

The Shamrock-Harkness is astonishing the country. It combines the best features of the most powerful present day circuits—five-tube volume and distance combined with unequalled clarity. What a combination!

Only the genuine Shamrock parts will give you maximum results. Shamrock engineers have spent months in laboratory experiment designing these specially balanced parts. They are backed by our unconditional guarantee. Avoid all imitations—be sure to get a licensed Shamrock-Harkness Kit.

Clip the coupon below and send for this Instructive and Interesting booklet. Every Radio fan should have one.

SHAMROCK MANUFACTURING CO.,
Dept. A 24, Market Street, Newark, N. J.

features

- **Operates a loud speaker.**
- **Two tubes do the work of five.**
- **Cuts battery cost 60%.**
- **Does not squeal, howl or radiate.**
- **Stations can be logged.**
- **Amazing clarity and volume.**

PRICE
$35

Radio News (Mar. 1925), p. 1725

The "Electro" Commercial Detector Stand

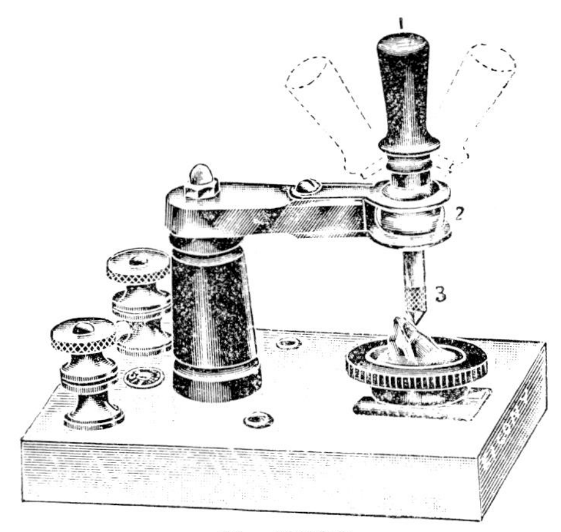

No. 9500

With ⅞ inch Italian Marble Base and Rotary Sliding Cup.

This instrument lately perfected by us stands in a class by itself. It is the highest grade Detector Stand ever introduced in the United States. It will appeal to all who wish a high grade instrument at a reasonable price.

Specification: Heavy marble base ⅞ inch thick, size 4½x3 inches Hard Rubber Pillar, on which is supported a nickel plated casting, which holds the ball swivel 2. As will be noted, the handle, 1, may be moved sideways in any direction on account of the swivel-ball arrangement; a feature not found in any other detector. This is quite an important feature, as it is often necessary with certain substances to "feel" over the surface in order to find the most sensitive spot.

The handle, 1, **can also be pulled up vertically**, as it is held back by a spring inside of the ball. Any amount of tension of the spring may be had by adjusting knurl, 3. This gives to the contact point on the detector substance any required tension. Therefore this detector is capable of the greatest variations, not alone in its free movement, but it can be adjusted **from the lightest contact to the heaviest.**

Only the best materials are used in the construction of this detector, all metal parts being finely nickel plated and highly polished.

There are two heavy nickel binding posts and there is a **FELT COVERING ON THE BOTTOM OF THE MARBLE BASE.**

Size over all 4½x3x5 inches.

No. 9500 "Electro" Commercial Detector Stand, as described. Price **$3.75**

Dear Sirs: **Cassopolis, Mich.**

I have bought goods at Chicago from your branch office, which were VERY SATISFACTORY. I intend now to purchase a mineral detector and a head band for my receiver.

Yours respectfully,

CHAS. A. MACDONALD.

Electro Importing Co., *Wireless* Catalog (1913), p. 12

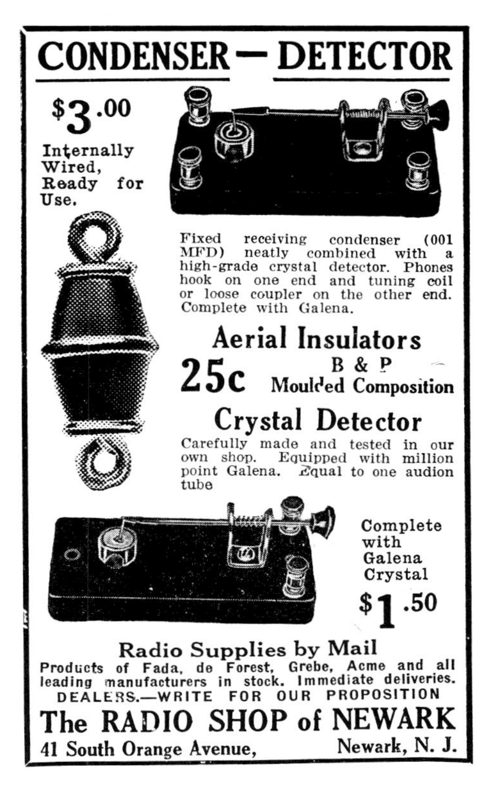

Radio News (Apr.-May 1922), p. 1050

Standard Galena Detector Stand

Improved Model

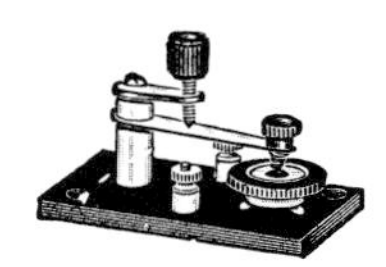

A popular detector stand. Tested piece of galena is mounted in cup which can be rotated. Crystal contact of phosphor bronze wire coiled and pointed and soldered on flat spring. Very fine adjustment obtainable with screw. Molded base and adjustment knob. Base size, 3 by 3 inches. Shipping weight, 1¼ pounds.

63 J 5305 .... **$1.15**

Wizard Detector Stand

This detector is an excellent value for the money. Carefully made of highest grade materials. Uses any detector mineral. Adjustment can be made to any position. Black polished composition base. Metal parts brass, nickel finish. Shipping weight, 4 ounces.

63 J 6535 .... **89¢**

Complete Set of Crystal Detector Parts

Complete set of metal parts to make up a high grade crystal detector. Made of brass, nickel finish. No base or crystal included. Shipping weight, per set, 3 ounces.

63 J 6534 .... **29¢**

Montgomery Ward & Co., *Radio* Catalog (1923), p. 20

Radio News (Nov. 1922), p. 956

High Grade DETECTORS and DIALS

Highly nickelplated Detector with ball and socket joint. List price 85c

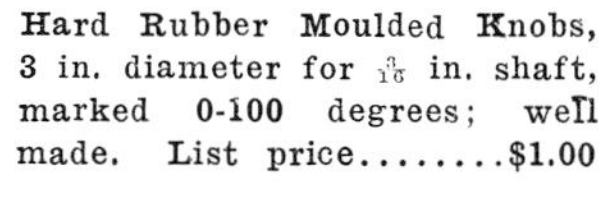

Hard Rubber Moulded Knobs, 3 in. diameter for 3/16 in. shaft, marked 0-100 degrees; well made. List price........$1.00

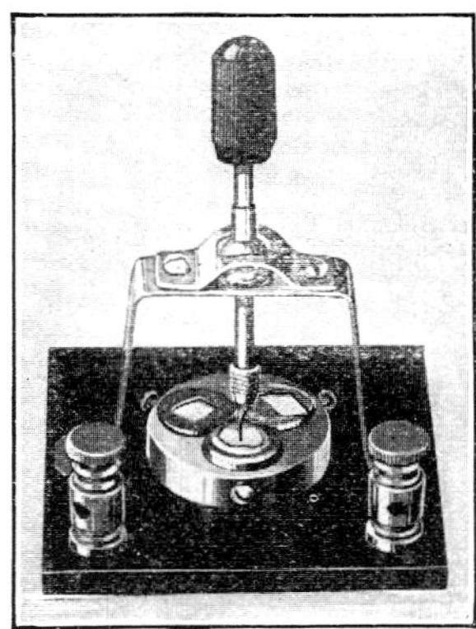

A handy Detector with a ball joint. This detector has a three-crystal cup so that galena, silicon and radiocite can be used. List price (without crystals) $2.00
Complete $2.50

Send for Dealers' proposition

RADIO TELEPHONE & MFG. CO.

150-152 Chambers St. New York, N. Y.

Radio News (Apr.-May 1922), p. 1015

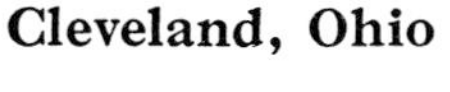

Teagle Radio Division

Newman-Stern Bldg.

Cleveland, Ohio

Teagle Crystal Detector

Fully adjustable. A *better* detector.

No. T-105. Price.......**$1.00**

Radio News (Oct. 1922), p. 685

HERE is the Crystal Detector exactly as used on the Kolster Decremeter. The most easily adjusted detector ever designed. Satisfaction guaranteed in direct comparison with any other type of Crystal Detector. Price **$2.50.** Silicon Crystal **25c.** Super-sensitive Galena Crystal, **40c.** Both mounted in Woods' Metal.

Examine Firco products at all leading radio dealers

JOHN FIRTH & CO., INC., 18 BROADWAY, NEW YORK

QST (Feb. 1922), p. 101

EMSCO RADIO PRODUCTS - MADE BY EXPERIENCED ELECTRICAL MANUFACTURERS

TURRET-TOP DETECTOR

Decidedly the best detector so far produced. Sturdy and well made of the best materials. "Turret-Top" swivel adjustment permits the greatest range of adjustment and holds the phosphor-bronze "cat whisker contactor" steadily in any position. Thumblock assures permanent adjustment. Locking-notch feature of binding posts exclusive with this detector. Unbreakable base. A fair example of EMSCO quality and value.

You take no chances when you buy EMSCO products. Behind each article is the skill of many years experience making wireless and electrical equipment. That is why the trademark, EMSCO, always signifies outstanding merit. Reliable dealers handle EMSCO products. If your dealer hasn't them write us.

JUST A FEW EMSCO PRODUCTS

Single and Double
Wound Rotors
3"—3½"
Single Slide Tuners
6"—8"
Binding Posts
Double Slide Tuners
6"—8"

Square Rods Cut and Drilled
3/16x7 ¼x7
x8 x8
x10 x10
x12 x12
3/16x7 Extension Threaded
Enamel Wound Coils
6x3" 8x3" 8x3½"

Weatherproof Strain Insulators
Vario Couplers
Sliders
3/16" and ¼"

ELECTRIC MFG. & SALES CO., 92 Academy St., Newark, N. J.

QST (June 1922), p. 78

B-METAL REFLEX DETECTOR

This is a permanent crystal detector especially designed for Reflex circuits. The crystal is securely sealed in a glass container and mounted on a base as shown in the illustration. The detector was found exceptionally sensitive and gave excellent results. It is manufactured by the B-Metal Refining Co., 3134 Trumbull Ave., Detroit, Mich., and is known as "Type C Reflex Tube Detector."

Arrived in excellent packing.

AWARDED THE RADIO NEWS LABORATORIES CERTIFICATE OF MERIT NO. 312.

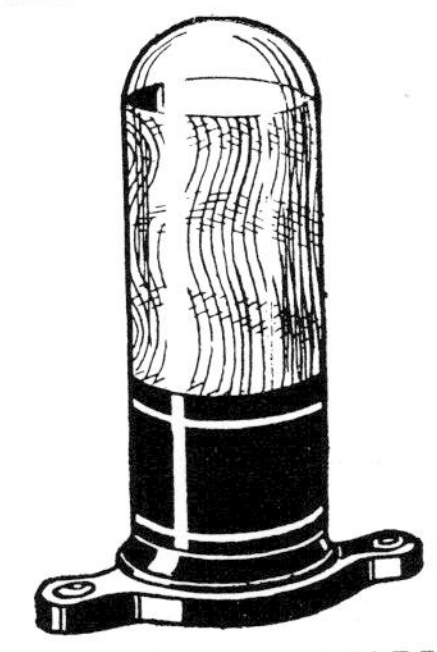

B-METAL ADJUSTABLE DETECTOR

The B-Metal Refining Co. also manufactures the adjustable type D crystal detector. This detector, shown in the illustration, is noted for its exceptionally small size, ease of adjustment and ability to hold its adjustment. It is very sensitive, the sensitivity, of course, depending upon the crystal used.

Arrived in excellent packing.

AWARDED THE RADIO NEWS LABORATORIES CERTIFICATE OF MERIT NO. 293.

Radio News (Feb. 1924), p. 1089

RADJO CRYSTAL DETECTOR

This is a very neat crystal detector that may be panel or base mounted and is constructed of two parts so that the crystal holder can be easily exchanged. The novel features of this detector are the vernier or micrometer adjustment provided and the use of an insulated metal screen in front of the crystal. The purpose of this screen is to hold the catwhisker in a fixed position so that its pressure on the crystal can be regulated without

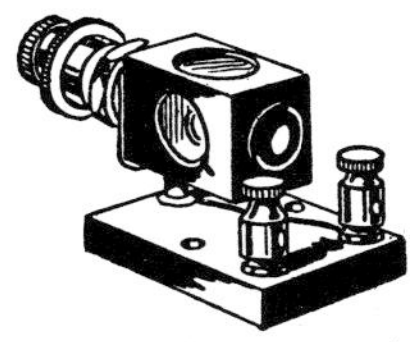

having it slip off the sensitive spot. This detector is manufactured by the Electric City Novelty and Manufacturing Co., 126 Odell Street, Schenectady, N. Y.

AWARDED THE RADIO NEWS LABORATORIES CERTIFICATE OF MERIT NO. 569.

Radio News (Dec. 1924), p. 949

AIRADER CRYSTAL DETECTOR

This detector was submitted to the Radio News Laboratories for test, by Bernard's Radio Co., 11 Twelfth Street, Providence, R. I. The crystal holder is shown in the illustration

tion and operates very satisfactorily with any of the ordinary types of crystals.

AWARDED THE RADIO NEWS LABORATORIES CERTIFICATE OF MERIT NO. 827.

Radio News (Aug. 1925), p. 203

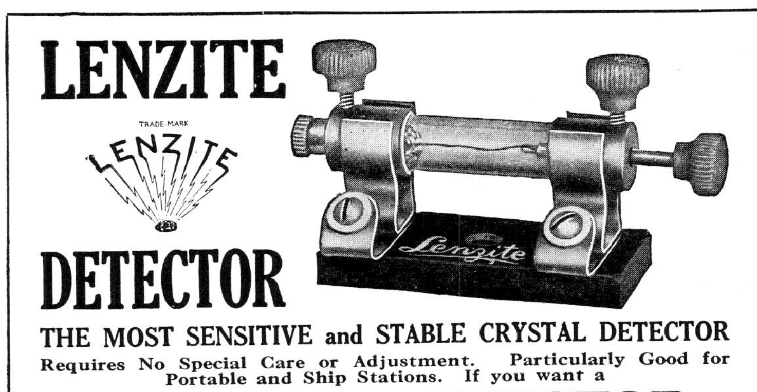

THE MOST SENSITIVE and STABLE CRYSTAL DETECTOR

Requires No Special Care or Adjustment. Particularly Good for Portable and Ship Stations. If you want a

PERFECT DETECTOR

BUY A LENZITE DETECTOR

If at the end of 30 days you want your $5.00 back you will get it

Lenzite Detector, $5.00 — LENZITE CRYSTAL CORP. Pasadena, Cal.

Everyday Mechanics (Jan. 1917), p. XIX

The Jones Patented Crystal Detector

presents six different surfaces of the Crystal to the action of the needle, without changing its position. It keeps you on the LIVE SPOT at all times and thereby gives you the loudest and clearest signals.

Four Dollars net, Postpaid.

Send for new Illustrated descriptive Bulletin describing these and other High Grade Radio Apparatus.

THE JONES RADIO COMPANY, 384 Monroe Street, Brooklyn, N. Y.

The Wireless Age (Apr. 1920), p. 47

CRYSTALS TABLE

GUIDE TO ABBREVIATIONS USED IN TABLE FOR CRYSTALS:

MINERAL:

A Alloy
B Bornite
Cd Cadmium
C Carborundum
Cp Copper Pyrites
Ip Iron Pyrites (Ferron)
G Galena
Gn Germanium
M Molybdenum
O Other
P Pyrite
S Silicon
Sg Steel Galena
Sc Synthetic
T Tellurium
X Unknown
Z Zincite

PRICE: Earliest retail price found; X unknown.

YEAR:*

E Early (Before or by 1929)
L Later (1930–1939)
R Recent (1940–1955)

*If year is unknown or uncertain, one of the above letters is used for an approximation, based on available evidence; when year is listed, it is for the earliest references found.

REFERENCES:

R *Radio News*
Q *QST*
M Other Magazine(s)
C Catalogs
V Viewed by Author
O Other Source(s)

Note: The list includes only minerals that are packaged (in boxes, tins, envelopes, or other containers; or attached to cards) and labelled, based on the best information available. If both mounted and unmounted minerals were marketed, only the prices for the mounted crystals are given.

MANUFACTURER	TRADE NAME/MINERAL	PRICE	YEAR	REF.
(See California Radio Minerals)	A-1 Wonder /G	$0.60	1924	RVO
(See Electro-Set Co.; Sears, Roebuck & Co.)	AAA (Triple A) /B	$0.30	1917	C
	AAA /C,Cp,Ip,G,M,S	$0.25	1917	C
	AAA /Z	$0.35	1917	C
Adams-Morgan Co. (AMCO), Upper Montclair NJ	Paragon No. 46 /X	$0.50	1922	C

MANUFACTURER	TRADE NAME/MINERAL	PRICE	YEAR	REF.
Adbrin Labs., Newark NJ	Adbrin /G,S	X	1925	O
(See Continental Sales Agency)	Aero Master /X	X	1925	O
Aerovox Wireless Corp., NY NY	Aerovox /X	X	1925	O
Airia Crystal Co., Moline IL	Airia /X	X	1925	O
(See Philmore Mfg. Co.)	Ajax /Sg	X	1925	VO
(See Howe & Co.)	Alderson Airzone /X	X	1925	O
Allen, Alva F., Clinton MO (see also Burstein-Applebee Co.; also see Addendum)	Melomite /Sg	$0.10	1937	CO
Allen, Co., Harry G., Seattle WA	Cascade /G	$0.50	1922	M
American Radio Mfg. Co., Kansas City MO	American Reflex /X	$0.35	1924	RV
Anchor Co., Pittsburgh PA	Anc-hor-ite /X	$0.25	1923	R
Andrae & Sons Co., Julius, Milwaukee WI (JASCO)	JASCO /G	$0.30	1923	C
Andrea Co., Frank A. D., NY NY (FADA)	FADA Galena /G	$0.50	1921	R
Appliance Radio Co., (address unknown)	Syn-tec /Sc	X	1926	M
Argentite Radio Corp., Los Angeles CA	Argentite /Sg	$0.30* (*price in 1930)	1925	CO
ARJO Radio Products Co. (address unknown)	Mul-ti-tec /X	X	E	O
(See Sears, Roebuck & Co.)	Arlington Tested /G,S	$0.20	1917	C
	Arlington Tested /B & Z	$0.40	1917	C
(See Newman-Stern Co.)	Arlington Tested NAA /G,S	$0.35	1920	RQMCVO
(See Keystone Products Co.)	(Keystone) Atomite /X	$0.50	1924	RO
(See Taylor Elec. Co.)	Audion /X	X	1925	O
(See Foote Radio Corp.)	Baby Giant /G,Ip	$0.35	1925	O
Bangert Elec. Co., Jamaica NY	Bangert /X	$0.30	1922	R
Barawik Co., Chicago IL, retailer	Broadcast Tested /X	$0.10	1929	C
	Detector Crystals /G,S	$0.20	1922	RC
	Meteorite /Sc	$0.12	1927	C
	Silver Clay (Fused) /Sc	$0.17	1927	C
(See Hyman & Co., Henry)	Bestone /X	X	1925	O

MANUFACTURER	TRADE NAME/MINERAL	PRICE	YEAR	REF.
(See Classy Specialty Co.)	Black Diamond /X	X	1925	O
B-Metal Refining Co., Detroit MI (see also Miller, A. H. Radio Co.)	B-Metal Loud Talking /Sc	$0.50	1923	RQMCVO
Brach Mfg. Co., Newark NJ	Brach /X	X	1925	O
Brady, J. J., San Francisco CA	Brady /X	X	1925	O
(See Bronx Radio Equip. Co.)	BRECO /X	X	1925	O
(See Barawik Co.)	Broadcast Tested /X	$0.10	1929	C
Bronx Radio Equipment Co., NY NY (BRECO)	BRECO /X	X	1925	O
Brownlie, R., Everett MA (later, Medford MA)	Brownlie Quick Contact /G,S	$0.50	1920	RQMCVO
Bunnell & Co., J. H., NY NY	Bunnell Galena /G	$0.25	1920	RQ
	Ennes /G,S	$0.15	1923	C
Burstein-Applebee Co., Kansas City MO, retailer (see also Allen, Alva T., mfr.; also see Addendum)	Melomite /Sg	$0.10	1937	C
Buscher Co., C. A., Kansas City MO, retailer	Buscher's /G	$0.15	1923	C
	Buscher's /P,S	$0.20	1923	C
	Hot Point /X	$0.35	1923	C
California Radio Minerals, Burlingame CA	A-1 Wonder /G	$0.60	1924	RVO
California Radio Supplies Co., Los Angeles CA	California Silvertone /X	X	1923	Q
(unknown mfr.)	Canadian /Sg	X	E	O
(See Allen, Harry G.)	Cascade /G	$0.50	1922	M
Case Toy & Sales Co., Pittsburgh PA	Case /X	X	1925	O
Celerundum Radio Products Co., Boston MA	Celerundum /X	$0.50	1924	RO
Chapman Radio Co., San Francisco CA	Chapman Supernatural Radio Crystal /X	X	1926	R
Cheever Co., Wm. E., Providence RI	Wecco /XXX;XXX	X	1926	O
Chemical Research Co., Chicago IL, mfr. (see also Everett Radio Co.)	DX-Alena /Sc	$0.50	1924	R

MANUFACTURER	TRADE NAME/MINERAL	PRICE	YEAR	REF.
Cherington Radio Products Co., Waukegan IL	C-R-P /XXX;XXX	X	1926	O
Chief Radio Crystal Co., Los Angeles CA	Chief /X	X	1925	O
Christy Radio Crystal Co., Cincinnati OH	Christy /X	X	1925	O
(unknown mfr.)	Clara Bell /X	X	E	V
Classy Specialty Co., Cincinnati OH	Black Diamond /X	X	1925	O
Clearco Crystal Co., Idaho Springs CO & Milwaukee WI	Clearco:			
	Straight Rim /Sg	$0.15	1925	RCVO
	Flange Rim /Sg	$0.18	1925	RCO
(See RCA; Wireless Specialty Apparatus Co.)	Cleartone /G	X	1919	O
(See Radiall Elec. Co.)	Coldmount /X	X	1922	R
Continental Sales Agency, Detroit MI	Aero Master /X	X	1925	O
Cook Co., East Orange NJ	Cook /X	$0.35	1922	R
(See Cherington Radio Products Co.)	C-R-P /XXX;XXX	X	1926	O
(See Montgomery Ward Co.)	Culina /X	$0.26	1925	C
(See RCA; Westinghouse Elec. & Mfg. Co.)	DD (Catwhisker Type) /X	$1.00	1922	C
	DE (Pressure Type) /X	$1.00	1922	C
(See Pacific Radio Specialty Co.)	Death Valley /X	$0.30	1925	O
Delta Co., Detroit MI	Delta /X	X	1925	O
(See Barawik Co.)	Detector Crystals /G,S	$0.20	1922	R
Detectron Sales Co., St. Louis MO	Detectron /X	X	1924	R
(See Keystone Products Co.)	Diamond Mounts /A	$0.25	1922	RO
Drefuss Co., P. M., NY NY	Permanite /Sg	$0.40	1923	RO
Dubilier Condenser & Radio Corp., NY NY	Du-Tec /Sc	$0.30	1923	RCO
Duck Co., Wm. B., Toledo OH	(Duck) Galena /G	$0.20	1921	C
Durban Elec. Co., Chicago IL	Durban /X	X	1925	O

MANUFACTURER	TRADE NAME/MINERAL	PRICE	YEAR	REF.
(See Skinner, W. E.)	D-V (Deep Vein) /X	X	1923	R
(See Chemical Research Co.; Everett Radio Co.)	DX-Alena /Sc	$0.50	1924	R
(See Specialty Service Co.)	D-X (Sure-Fire) /Sc	$0.50	1924	RO
Edelman, P. E., NY NY	Edelman's Multitone /G	$0.35	1922	R
	Edelmite /Sc	$0.65	1922	R
Electro Importing Co., NY NY (EICO)	Radiocite /Ip	$0.50	1916	MCO
	E. I. Galena /G	$0.15	1918	C
	Radio "DeLuxe" Set /G, Ip, & S	$2.00	1918	C
Electro-Set Co., Cleveland OH (see also Sears, Roebuck & Co.)	AAA /B	$0.30	1917	C
	AAA /C,Cp,Ip,G,S	$0.25	1917	C
	AAA /Z	$0.35	1917	C
	Arlington Tested /C,S	$0.20	1917	C
	Arlington Tested /B & Z	$0.40	1917	C
El Piacho Mining Co., Las Vegas NV	El Piacho /X	X	1925	O
Empire Elec. Machinery Co., Joplin MO	REX (*R*adio *E*mpire *X*tal) /G	$0.25	1922	RQ
(See Bunnell & Co., J. H.)	Ennes /G,S	$0.15	1923	C
ERTCO Lab., Houston TX	Pep-ite /X	X	1923	R
Everett Radio Co., Chicago IL—agent (see Chemical Research Co., mfr.)	DX-Alena /Sc	$0.50	1924	R
E-Z Crystal Co., Philadelphia PA	E-Z /X	X	E	V
(unknown mfr.)	Ferron /Ip	X	E	O
Firth & Co., John, NY NY (FIRCO)	FIRCO /S	$0.25	1922	R
	FIRCO /G	$0.40	1922	R
(See Smith, Fred G.)	F. G. S. /X	X	1923	R
Foote Mineral Co. (by 1925, Foote Radio Corp.), Philadelphia PA	Foote's:			
	Radio Twins (Hexagon) /Ip	$0.50	1922	RO
	Single /G,Ip, or S	$0.75	1923	RO
	Twin /G & Ip	$0.90	1923	RO
	Triplet /G,Ip & S	$1.00	1923	RO
	Baby Giant /G,Ip	$0.35	1925	O
	Little Giant /G,Ip	$0.50	1925	O
	Giant /G,Ip	$0.50	1925	O
	Giant Triplet /G,Ip & S	$1.00	1925	O

MANUFACTURER	TRADE NAME/MINERAL	PRICE	YEAR	REF.
Ford Co., K. N., Los Angeles CA	Radiogrand /X	X	1925	O
Forman & Co., NY NY	UN-X-LD /G,P,S	$0.10	1923	RO
Freed-Eisemann Radio Corp., NY NY	Freed-Eisemann /G	$0.20	1923	M
Freshman Co., Chas., NY NY	Freshman /X	$0.50	1924	MO
Galena Crystal Mfg. Co., Brooklyn NY	Galena /G	$0.35	1922	RM
(See Radio Shop of Newark)	Galena & Radiocite Millionpoint /Ip & Sg	$0.40	1922	RQ
(See Harris Lab.)	Ganaerite /X	$0.50	1922	R
(See Martin-Hewitt Lab.)	Garsite /X	X	1922	R
(See Foote Radio Corp.)	Giant /G,Ip	$0.50	1925	O
	Giant Triplet /G,Ip & S	$1.00	1925	O
Gibbons-Dustin Radio Mfg. Co., Los Angeles CA	Magnetite /Sg	$0.50	1923	RO
(See Hamburg Bros.)	Goldite /Ip	X	1925	O
(See Morris, Walter C.)	Gold Medal /X	X	1925	O
Gordon Radio & Elec. Mfg. Co., Seattle WA	Gordon /X	X	1925	O
Gotham Radio Corp., NY NY	Gotham Galena /G	X	E	O
(See Walthom Radio Products)	Gracite /X	X	1925	O
Great Lakes Radio Co., Chicago IL	Great Lakes /G,S	$0.12	1922	R
Grewol Mfg. Co., Newark NJ	Grewol 2 in 1 /X	$0.50	1924	RO
(unknown mfr.)	Gun Set /X	X	E	V
Haller, W. B., Pittsburgh PA	Hallerio /X	X	1925	O
Hamburg Bros., Pittsburgh PA	Goldite /X	X	1925	O
HARCO, (address unknown) (Harcourt Radio Co., Toronto, Canada?) (see Montgomery Ward & Co.)	HARCO Culina /X	$0.26	1925	CO
	HARCO Silvertone /X	X	E	V
Hargraves C. E. & H. T., Lakewood RI	Silgo /X	X	1925	O
	Silver Dome /X	X	1926	O

MANUFACTURER	TRADE NAME/MINERAL	PRICE	YEAR	REF.
Harris Lab., NY NY	Ganaerite /X Harris Galena /G	$0.50 X	1922 E	R VO
Hatfield & Son, H. D., Hollywood CA	Mexican Placerite Silver Galena /Sg	$0.40	1925	RO
Hearwell Elec. Co., Boston MA	Hearwell /X	X	1925	O
(unknown mfr.)	Hertzite /Sg	X	E	O
(See T. N. T. Products Co.)	(T. N. T.) High Power /Sg	$0.50	1926	O
(See Philmore Mfg. Co.)	Hi-Volt /Sc	X	E	V
(See Milton, E. J.)	Hi-Wave /X	X	1926	R
(See Buscher Co., C. A.)	Hot Point /X	$0.35	1923	C
(See Telephone Maint. Co.)	Hot Spot /X	$0.25	1923	C
Howe & Co., Chicago IL	Alderson Airzone /X	X	1925	O
Hyman & Co., Henry, NY NY	Bestone /X	X	1925	O
Illinois Radio Engineering Co., Chicago IL	Klear-Rite /X	X	1925	O
Indianapolis Radio Supply Co., Indianapolis IN	Galena /G	$0.20	1921	Q
(See Radio Supply Co.)	Iron Pyrites /Ip	$0.25	1930	C
(See U.C. Battery & Elec. Co.)	IXL Radio /X	$0.50	1925	M
(See Star-King Co.)	James 2-Side /X	$0.60	1925	RO
(See Andrae & Co., Julius)	JASCO /G	$0.30	1923	C
Johnson Radio Co., Newark NJ	Johnson /X	X	1925	O
Karlowa Radio Co., Rock Island IL	Karlowa /G,S	$0.25	1921	Q
Kelso National Mining Co., Georgetown CO	Kelso /Sg	X	1925	O
Keystone Products Co., Royal Oak MI	Diamond Mounts /A Keystone Atomite /X	$0.25 $0.50	1922 1924	RO R
(See Illinois Radio Engineering Co.)	Klear-Rite /X	X	1925	O
Krystal-Kleer Co., NY NY	Krystal-Kleer /X	$0.50	1922	R
Lee Elec. & Mfg. Co., San Francisco CA	LEMCO /XXX;XXX	X	1925	O

MANUFACTURER	TRADE NAME/MINERAL	PRICE	YEAR	REF.
Lenzite Crystal Corp., Pasadena CA	Lenzite /Sg	X	E	O
Leumas Labs., NY NY	Leumite /X	$0.25	1919	RQ
(See Foote Radio Corp.)	Little Giant /G,Ip	$0.50	1925	O
Los Angeles Radio Supply Co., Los Angeles CA	Mineral /X	X	1925	O
(See B-Metal Refining Co.)	(B-Metal) Loud Talking /Sc	$0.50	1923	RQMCVO
(See T. N. T. Products Co.)	(T. N. T.) Low Loss /Sg	$0.50	1926	RV
McCarthy Radio Products Co., Tacoma WA	Mount Tacoma /X	X	1925	O
(unknown mfr.)	Magic /Sg	X	E	O
(See Gibbons-Dustin Radio Mfg. Co.)	Magnetite /Sg	$0.50	1923	RO
Magnus Elec. Co., NY NY (by 1925, Magnus Elec. & Radio Co.)	Radiosite /X	$0.30	1923	CO
Martin-Hewitt Lab., Bronx NY (by 1925, Maspeth, L. I. NY)	Garsite /X	X	1922	R
	"X"-Ray /X	X	1925	O
(See Radio Owners Service Co., agent for Mathison Silver Ores Co. of Alaska)	Mathison's Polar /Sg	$0.85	1925	M
Melodian Labs., Independence MO	Melody King /X	X	1925	O
(See Allen, Alva F.; Burstein-Applebee Co.; also see Addendum)	Melomite /Sg	$0.10	1937	CO
(See Barawik Co.)	Meteorite /Sc	$0.12	1927	C
(See Hatfield & Son, H. D.)	Mexican Placerite Silver Galena /Sg	$0.40	1925	R
(See Radio Supply Co.)	Mexican Pyrites /P	$0.15	1932	C
(See Wolverine Radio Co.)	Mexite /X	X	1925	O
(unknown mfr.)	Midite /Sg	X	E	O
Midland Elec. Co., Indianapolis IN	Midland /G	X	1924	R
Midwest Mineral Co., Kansas City MO	Midwest Mineral /G	$0.35	1922	R

MANUFACTURER	TRADE NAME/MINERAL	PRICE	YEAR	REF.
Miller Radio Co., A. H., Detroit MI (see also B-Metal Refining Co.)	Miller B-Metal /Sc	$0.50	1924	M
(See M. P. M. Sales Co.)	Million Point Mineral (MPM) /Sg	$0.25	1922	RQMCV
(See Radio Shop of Newark)	(Radio Shop Galena & Radiocite) Million Point /Ip & Sg	$0.40	1922	RQ
Milton, E. J., Denver CO	Hi-Wave /X	X	1926	R
Mineral Novelty Co., Joplin MO	Mineral /X	X	1926	R
Mineral Products Sales Corp., NY NY	Radio Minerals /Cd,G,P,S,T	X	1923	R
Minerals & Insulation Supply Co., NY NY	Mineral /X	X	1925	O
(unknown mfr.)	Missourite /X	X	E	O
(See Radio Supply Co.)	Model A /X	$0.25	1930	C
Modern Radio Labs., Garden Groves CA	MRL /C,Ip,S,Sg	$0.50* (*price in 1984)	1938	CO
(See Phoenix Radio Lab.)	Molybdic Galena /G	$0.50	1923	MO
Montgomery Ward & Co., Chicago IL (see also HARCO; Tip Top Crystal Co.)	HARCO Culina /X	$0.26	1925	CO
	Tip-Top /X	$0.26	1925	CO
Morris, Walter C., Yonkers NY	Gold Medal /X	X	1925	O
	Silver Medal /X	X	1925	O
(See McCarthy Radio Products Co.)	Mount Tacoma /X	X	1925	O
M. P. M. Sales Co., Los Angeles CA	Million Point Mineral (MPM) /Sg	$0.25	1922	RQMCVO
(See Modern Radio Labs.)	MRL /C,Ip,S,Sg	$0.50* (*price in 1984)	1938	CO
(See Rusonite Products Corp.)	(Rusonite) Multipoint /Sc	$0.50	1923	RQMCO
(See ARJO Radio Products Co.)	Mul-ti-tec /X	X	E	O
(See Edelman, P. E.)	(Edelman's) Multitone /G	$0.35	1922	R
(See Newman-Stern Co.)	(Arlington Tested) NAA /G,S	$0.35	1920	RQMCVO
Nassau Engineering Works, Baldwin NY	N-E-W /G	$0.30	1922	R
	N-E-W /S	$0.35	1922	R
(See Radio Efficiency Co.)	National Efficiency /X	X	1926	R

MANUFACTURER	TRADE NAME/MINERAL	PRICE	YEAR	REF.
(unknown mfr.)	Nevadium /G	X	E	O
(See Post Trading Co.)	Neverfail /X	X	1925	O
Newman-Stern Co., Cleveland OH	Arlington Tested NAA /G,S	$0.35	1920	RQMCVO
Non-Skid Crystal Mfg. Co., Kansas City MO	Shur-Hot /X	$0.75	1925	VO
Novelty Radio Mfg. Co., St. Louis MO	Stayput /X	X	1925	O
Ozark Crystal Co., Morelton MO	Ozark /G	$0.25	1922	Q
Pacific Radio Specialty Co., Philadelphia PA	Death Valley /X	$0.30	1925	O
(See Adams-Morgan Co.)	Paragon No. 46 /X	$0.50	1922	C
Parkin Mfg. Co., San Rafael CA	Parkin /G	$0.25	1925	CVO
Passera, P. F., Amsterdam NY	Rival /X	X	E	O
(See ERTCO Lab.)	Pep-ite /X	X	1923	R
(See Dreyfuss Co., P. M.)	Permanite /Sg	$0.40	1923	RO
Philmore Mfg. Co., NY NY	Ajax /Sg	X	1925	VO
	Hi-Volt /Sc	X	E	V
	Philmore /G	$0.10	1931	C
	Philmore /Sg	$0.20	1931	C
	"X"TRA Loud /P	X	E	O
Phoenix Radio Lab., Phoenix AZ	Molybdic Galena /G	$0.50	1923	MO
(See Hatfield & Sons, H. D.)	(Mexican) Placerite Silver Galena /Sg	$0.40	1925	R
Post Trading Co., NY NY	Neverfail /X	X	1925	O
Powertone Radio Products Co., Minneapolis MN	Powertone /X	$0.25	E	O
(See Radio Trading Co.)	Puretone /Sc	$0.10	1934	C
(unknown mfr.)	Pyron /Ip	X	E	O
(See Western Radio Mfg. Co.)	QRK /G,S	$0.20	1922	R
(See Brownlie, R.)	(Brownlie) Quick Contact /G,S	$0.50	1920	RQMCV
Radiall Elec. Co., Passaic NJ	Coldmount /X	X	1922	R
	Radiall /G,S, & X	$0.25	1922	R
(See Electro Importing Co.)	Radiocite /Ip	$0.50	1916	MCO

MANUFACTURER	TRADE NAME/MINERAL	PRICE	YEAR	REF.
(See Radio Shop of Newark)	(Radio Shop Galena &) Radiocite Million Point /Sg & Ip	$0.50	1916	RQO
Radio Corp. of America (See RCA)				
(See Electro Importing Co.)	Radio "DeLuxe" Set /G,Ip, & S	$2.00	1918	C
Radio Efficiency Co., Kansas City MO	National Efficiency /X	X	1926	R
Radio Engineering Co., New Rochelle NY	Radio Minerals /X	X	1925	O
(See Ford Co., K. N.)	Radiogrand /X	X	1925	O
Radio Minerals Co., Detroit MI	Radio Minerals /X	X	1925	O
(See Mineral Products Sales Corp.)	Radio Minerals /Cd,G,P,S,T	X	1923	R
Radio Owners Service Co., Oakland CA (agent for Mathison Silver Ores Co. of Alaska)	Mathison's Polar /Sg	$0.85	1925	M
Radio Shop of Newark, Newark NJ	Radio Shop Galena & Radiocite Million Point /Sg & Ip	$0.40	1922	RQ
(See Magnus Elec. Co.)	Radiosite /Sg	$0.30	1923	CO
Radio Specialties Co., San Francisco CA	Radio Minerals /X	X	1925	O
Radio Supply Co., Los Angeles CA (see Argentite Radio Corp., re: *1925* listing of Argentite)	Argentite /Sg	$0.30	1930	CO
	Iron Pyrites /Ip	$0.25	1930	C
	Mexican Pyrites /P	$0.15	1932	C
	Model A /X	$0.25	1930	C
Radio Testing Station, Binghamton NY	RTS /G,S	$0.25	1920	RQ
Radio Trading Co., NY NY, distributor	Puretone /Sc	$0.10	1934	C
(See Foote Mineral Co.)	(Foote's) Radio Twins (Hexagon) /Ip	$0.50	1922	RO
Ray-Di-Co Organization, Chicago IL	Ray-Di-Co /X	$0.25	1922	Q

MANUFACTURER	TRADE NAME/MINERAL	PRICE	YEAR	REF.
RCA, NY NY (mfrs.: *for Cleartone*—Wireless Specialty Apparatus Co.; *for Type DD & DE*—Westinghouse Elec. & Mfg. Co.)	Cleartone /G Type DD ("Catwhisker Type") /X Type DE ("Pressure Type") /X	X $1.00 $1.00	1919 1922 1922	O CV CV
(See American Radio Mfg. Co.)	(American) Reflex /X	$0.35	1924	RV
Reinhardt & Co., St. Louis MO	Silverblend /X	X	1925	O
(See Empire Elec. Machinery Co.)	REX (*R*adio *E*mpire *X*tal) /G	$0.25	1922	RQ
Reynolds Radio Specialty Co., Colorado Springs CO	Reynolds /X	X	1925	R
(See Passera, P. F.)	Rival /X	X	E	O
Rocky Mountain Radio Products, NY NY	Rocky Mountain /X	$0.35	1922	Q
Roll-O Crystal Co. /Roll-O Radio Corp., Cincinnati OH	Roll-O /X	$1.00* (*price for 3)	1924	RV
Rosenthal Lab., Camden NJ	Rosite /X	$0.50	E	V
(See Radio Testing Station)	RTS /G,S	$0.25	1920	RQ
(unknown mfr.)	Rubyite /Sg	X	E	O
R-U-F Products Co., Brooklyn NY	Wonder /X	X	1925	O
Rusonite Products Corp., NY NY	Rusonite Multipoint /Sc	$0.50	1923	RQMCO
San Co., NY NY	San /G	X	E	V
Schaffer, Geo. W., Allentown PA	Schaffer /X	X	1925	R
Sears, Roebuck & Co., Chicago IL (see also Electro-Set Co.)	AAA (Triple A) /B AAA /C,Cp,Ip,G,M,S AAA /Z Arlington Tested /G,S Arlington Tested /B & Z	$0.30 $0.25 $0.35 $0.20 $0.40	1917 1917 1917 1917 1917	C C C C C
(See Wilmat Co.)	"7/11" /X	X	1925	O
Shamrock Mfg. Co., Newark NJ	Shamrock-Harkness /X	X	1925	O
Shaw-Root Radio Minerals Co., San Francisco CA	T. N. T. Steel Galena /Sg	$0.50	1925	V
(See Non-Skid Crystal Mfg. Co.)	Shur-Hot /X	$0.75	1925	VO

MANUFACTURER	TRADE NAME/MINERAL	PRICE	YEAR	REF.
(See Hargraves, C. E. & H. T.)	Silgo /X	X	1925	O
(unknown mfr.)	Silver Bell /G	X	E	O
(See Reinhardt & Co.)	Silverblend /X	X	1925	O
(See Hargraves, C. E. & H. T.)	Silver Dome /X	X	1926	O
(See Barawik Co.)	Silver Clay (Fused) /Sc	$0.17	1927	C
(unknown mfr.)	Silverite /S	X	E	O
(See Morris, Walter C.)	Silver Medal /X	X	1925	O
(See California Radio Supplies Co.)	(California) Silvertone /X	X	1923	Q
(See HARCO)	(HARCO) Silvertone /X	X	E	V
(See Foote Mineral Co.)	(Foote's) Single/G,Ip,S	$0.75	1923	RO
Skinner, W. E., Minneapolis MN	D-V (Deep Vein) /X	X	1923	R
	Skinner /G	X	1923	R
Smith, Fred G., Huron MI	F. G. S. /X	X	1923	R
(unknown mfr.)	Sparkite /Sg	X	E	O
Specialty Service Co., Brooklyn NY	D-X (Sure-Fire) /Sc	$0.50	1924	RO
Standard Crystal Co., Newark NJ (called Standard Mineral Co. by 1923)	Standard /X	$0.50	1922	QM
Star Crystal Co., Detroit MI	Star /G,S,P	$0.10	1925	CV
(see B-Metal Refining Co.)	B-Metal Loud Talking /Sc	$0.50	1925	O
Star-King Co., San Francisco CA	James 2-Side /X	$0.60	1925	RO
(See Novelty Radio Co.)	Stayput /X	X	1925	O
Steinite Labs., Atchison KS	Steinite /X	$0.50	1924	RO
Steinmetz Wireless Mfg. Co., Pittsburgh PA	Steinmetz /G	$0.15	1922	R
(See U. S. Mfg. & Distributing Co.)	Sunset /G	$0.50	1923	RO
	Sunset /P	$0.25	1923	RO
(See Chapman Radio Co.)	(Chapman) Supernatural Radio Crystal /X	X	1926	R
(See Specialty Service Co.)	(D-X) Sure-Fire /Sc	$0.50	1924	RO
Sydney Specialty Co., NY NY	Sydney /G,Ip,S,Z	$0.75	1921	R

MANUFACTURER	TRADE NAME/MINERAL	PRICE	YEAR	REF.
(See Appliance Radio Co.)	Syn-tec /Sc	X	1926	M
Taylor Elec. Co., Providence RI	Audion /X	X	1925	O
Telephone Maintenance Co., Chicago IL (TELMACO)	Hot Spot /X	$0.25	1923	C
Teleradio Engineering Corp., NY NY	Teleradium /X	X	1922	R
(unknown mfr.)	Texas /Sg	X	E	O
Tip-Top Crystal Co., Newark NJ (see Montgomery Ward Co.)	Tip-Top /X	$0.26	1925	CO
T. N. T. Products Co., San Francisco CA	T. N. T. High Power /Sg	$0.50	1925	O
	T. N. T. Low Loss /Sg	$0.50	1925	RVO
(See Shaw-Root Radio Minerals Co.)	T. N. T. Steel Galena /Sg	$0.50	1925	VO
Tri-City Radio Elec. Supply Co. (TRESCO), Davenport IA	TRESCO /G	X	1922	R
Trimount Trading Co., Boston MA	Trimount /X	X	1925	O
(See Electro-Set Co.; Sears, Roebuck & Co.)	Triple-A (AAA):			
	/B	$0.30	1917	C
	/C,Cp,Ip,G,M,S	$0.25	1917	C
	/Z	$0.35	1917	C
(See Foote Mineral Co.)	(Foote's:)			
	Triplet /G,Ip & S	$1.00	1923	RO
	Twin /G & Ip	$0.90	1923	RO
	(Radio) Twins /Ip	$0.50	1922	RO
U. C. Battery & Elec. Co., Berkeley CA	IXL Radio /X	$0.50	1925	M
United Metal Stamping & Radio Co., Cincinnati OH	United /X	X	1925	O
(See Forman & Co.)	UN-X-LD /G,P,S	$0.10	1923	RO
U. S. Mfg. & Distributing Co., Newark NJ	Sunset /G	$0.50	1923	R
	Sunset /P	$0.25	1923	R
U. S. Radio Co. of Penna., Pittsburgh PA	U. S. Eagle Galena /G	$0.25	1922	RM
	U. S. Eagle Goldena /Ip	$0.35	1922	RM
Walthom Radio Products Co., NY NY	Gracite /X	X	1925	O
(See Westville Mfg. Co.)	W. C. /X	X	1925	O

MANUFACTURER	TRADE NAME/MINERAL	PRICE	YEAR	REF.
(See Cheever Co., Wm. E.)	Wecco /X	X	1926	O
Western Radio Mfg. Co., Chicago IL	QRK Detector /G,S	$0.20	1922	R
Westinghouse Elec. & Mfg. Co., E. Pittsburgh PA (see also RCA)	Type DD ("Catwhisker Type") /X Type DE ("Pressure Type") /X	$1.00 $1.00	1922 1922	C C
Westville Mfg. Co., Philadelphia PA	W. C. /X	X	1925	O
Whitall Elec. Co., Westerly RI	Whitall /G,S	$0.25	1921	Q
Wilmat Co., Los Angeles CA	"7/11" /X	X	1925	O
Wireless Specialty Apparatus Co., Boston MA (see also RCA)	Cleartone /G	X	1919	O
Wolverine Radio Co., Detroit MI	Mexite /X	X	1925	O
(See California Radio Minerals)	(A-1) Wonder /G	$0.60	1924	RVO
(See R-U-F Products Co.)	Wonder /X	X	1925	O
Wonder State Crystal Co., Little Rock AR	Wonder State /Sg	$0.50	1922	RO
(See Martin-Hewitt Lab.)	"X"-Ray /X	X	1925	O
(See Philmore Mfg. Co.)	"X"TRA Loud /P	X	E	O

ADDENDUM: RECENT CRYSTALS (1940–1955)

MANUFACTURER	TRADE NAME/MINERAL	PRICE	YEAR	REF.
Allen, Alva F., Clinton MO	AAA (Triple-A) /X Midget /X	$0.50 $0.35	1952 1952	C C
(various mfrs.)	Diodes /Gn,S	X	R	VO
Radio-Ore Labs., Lynn MA	Radio-Ore /X	X	R	O
(various mfrs.)	Transistors /Gn	X	R	VO

Eye Test Beats Ear Guess

WE have been selling tested commercial crystals for ten years. Our new method of **METER** testing is a big step forward. Ear tests, while valuable, are mere indicators. Eye tests with sensitive measuring instruments, are absolutely exact.

N A A Meter-Tested Crystals

Every NAA Crystal is **METER-TESTED** in ten positions to a minimum of 76 microamperes of rectified current, by impulses picked up from a specially designed oscillating circuit. The average crystal is considered good if it rectifies half this current.

Perfect for Reflex

Cold mounted in Newman-Stern improved holder cup (patent pending). No damage to crystal from hot alloy. Recessed for protection. A vital necessity in every reflex receiver—the heart of the set.

NAA perfected METER-TESTED CRYSTALS are sold by best jobbers and dealers, or sent postpaid by mail for 60 cents. Send name of your dealer.

1742 East 12th Street Cleveland, Ohio

Originators of tested crystals in 1914. Oldest and largest producers. Pioneers in Radio Equipment in Ohio.

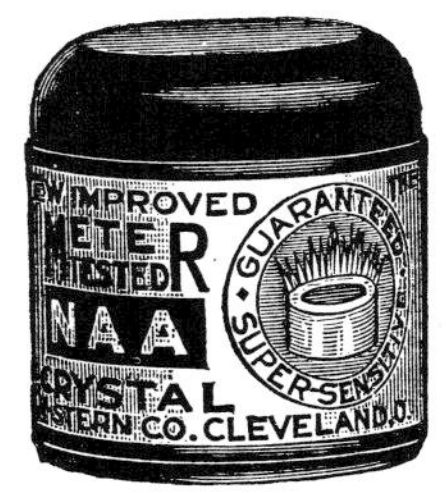

Radio News (Jan. 1925), p. 1284

Radio News (Sep. 1925), p. 378

Reception Louder Clearer More Natural

(Reg. U. S. Patent Office)

Functions with any conductor.
No special Cat-Whisker necessary.
Will not burn out.................. **50c**

Celerundum "DE-TEX-IT"

A perfect fixed detector......... **$1.25**

Celerundum Products carry a MONEY BACK GUARANTEE

Celerundum Radio Products Co.

170 Summer Street, Boston — Standard Radio Corp., Chicago

Radio News (Apr. 1924), p. 1446

CRYSTAL FANS

Are you satisfied with your crystal, or do you want better results?

MOLYBDIC GALENA

A new detector just discovered is sensitive even when ground to dust.

Send 50c for one of these wonder crystals. We will refund your money if you are not entirely satisfied.

Phoenix Radio Laboratory

Box 842, Phoenix, Arizona

Radio (June 1923), p. 62

Meter Tested Crystals

Produce loud and clear signals. These are exceptionally fine quality crystals. Laboratory tested on sensitive meters. Shpg. wt., 2 oz.
3B3820—Silicon.
3B3821—Galena.
List price, 60c.
Net price.............. **39c**

Du-tec

3B3822—
List price, 30c.
Net price... **18c**
A Dubilier chemically made radio rectifier. Replace crystals. Every point sensitive and remains so. No adjustments necessary. Mounted in cup ready for use. Shpg. wt., 2 oz.
3B3823—Genuine million point crystal. Sensitive all over. Mounted. Each................ **21c**
3B3824—Meteorite. A synthetic crystal mounted in metal cup. Very sensitive. Each............ **12c**
3B3826—Silver clay crystal, a fused crystal of great volume and clarity. Each.................. **17c**

Test Buzzer

3B3828—
Each... **85c**
Watch case buzzer. Polished nickel finish. Diam., 2⅝ in. Shpg. wt., 4 oz.

Test Buzzer Push Button

3B3829—
Each.......... **21c**
A small compact button with pearl center. Held in ⅝-inch hole by small spring clips. Shpg. wt., 2 oz.

The Barawik Co., *Radio Catalog* (1927-1928), p. 110

76 The Electro Importing Co., N. Y., Mfrs.

Minerals and Crystals

When you buy a mineral or wireless crystal you are interested in only a very few things. First you want to know value. EVERY CRYSTAL SOLD BY US IS TESTED FOR SENSITIVITY. Don't pay more for so-called **"special"** and **"extra"** grades. **Our competitors' "special" and "extra" grades are** OUR REGULAR STOCK QUALITY. Now for quantity. Note that we sell by weight wherever possible. When we say you get an ounce, you get an ounce, **not a piece.** This means a big saving to you. Now on delivery. We carry more wireless minerals in stock than any other concern in the world. **We guarantee prompt delivery.** Being the largest buyers and sellers of this class of material we are naturally offered the pick of the world. In that way by buying your Crystals and Minerals from the E. I. Co., you buy the best tested goods that is found at the lowest possible prices. Your first order will convince you of our claims.

BORNITE

Used a great deal abroad. Can be used with a phosphor bronze contact wire, or with zincite. Marvelously sensitive.

No. CE2416 Bornite, per oz. **$0.35**
Shipping weight 2 oz.

GALENA

NO. AE2504

This mineral is thought by many to be one of the most sensitive discovered so far. Used to best advantage by having a fine phosphor bronze or brass wire spring, size about No. 26 B. & S., press very lightly on the Galena. We carry only a specially selected cubic crystal grade.

No. AE2504 Galena, per ounce **$0.15**
Shipping weight 2 oz.

COPPER PYRITES

Very sensitive and very stable. Even sensitiveness along whole surface. Not easily jarred out. Use phosphor bronze contact wire. **GUARANTEED 100 PER CENT. PURE.**

No. CE2419 Copper Pyrites, per oz. **$0.35**
Shipping weight 2 oz.

MOLYBDENITE

This new substance is the only one discovered so far which does not get out of adjustment, when used in a sensitive Detector, and when placed near a sending gap. Most substitutes suffer a great deal from strong sending currents, but it is impossible to damage the adjustment of the Molybdenite Detector, and a heavy discharge does not affect it. Molybdenite proves quite sensitive when distant stations are to be picked up.

No. EK9210 Molybdenite, per oz. **$0.50**
Shipping weight 2 oz.

CARBORUNDUM

Specially selected for experimenting with the Carborundum Detector. Quite sensitive. Used by commercial companies for many years.

No. CK9308 Carborundum, per oz. **$0.30**
Shipping weight 2 oz.

ZINCITE

The aristocrat of all wireless minerals. Too well known and too far famed to praise it here. **Undoubtedly the most sensitive of all crystals. GUARANTEED 100 PER CENT. PURE.**

No. ABE2417 Zincite, per oz **$1.25**
Shipping weight 2 oz.

No. CE2418 Zincite, ¼ oz. **$0.35**
Shipping weight 1 oz.

SILICON

There are two kinds of this material: Silicon crystals and fused Silicon. The former, manufactured in this country, is absolutely unfit to use; the latter, imported by us, is the only kind that should be used. It comes in chunks and somewhat resembles graphite. It is very hard and extremely brittle.

No. CE9209 Silicon, per oz. **$0.35**
Shipping weight 2 oz.

No. AE9209a Silicon, ¼ oz. **$0.15**
Shipping weight 1 oz.

IRON PYRITES

Our iron pyrites is all imported Spanish stock that may be used for years without deterioration. Very sensitive.

No. CK2505 Iron Pyrites (Ferron), extremely sensitive, per oz. **$0.30**
Shipping weight 2 oz.

PEROXIDE OF LEAD

No. CK2506 Peroxide of Lead, compressed tablets, each **$0.30**
Shipping weight 2 oz.

MINERAL SETS

No. GK2502 Zincite and Copper Pyrites (Perikon), per set **$0.70**
Shipping weight per set 4 oz.

MINERAL ASSORTMENT

Consisting of generous pieces each of the nine minerals and crystals shown on these pages. An excellent assortment for the wireless experimenter. Each mineral in separate box. No Radiocite supplied.

No. AEK2346 Mineral Assortment (9 minerals) **$1.50**
Shipping weight 1 lb.

Electro Importing Co., *Wireless* Catalog (1918), pp. 76, 77

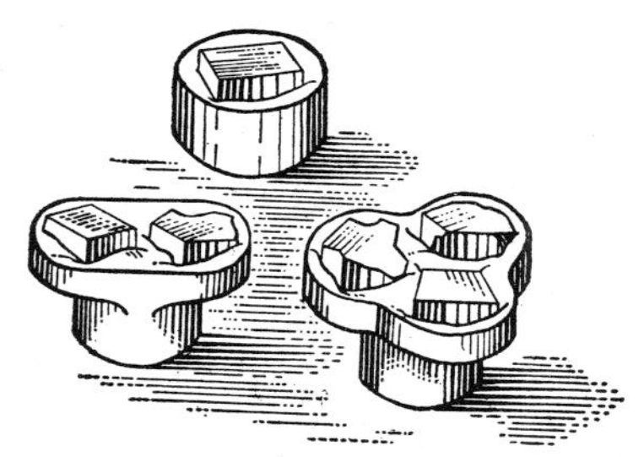

100,000 CRYSTAL FANS

Use Foote's Triple-Test Multi-Crystals

And they are continually breaking long distance records. Guaranteed QSA. (Your signals are strong.) Order by Mail and Learn by Experimenting.

Triplet Galena & Pyrite with Silicon, Twin Galena & Pyrite, Single Galena, Single Pyrite, Single Silicon, Tripletone Solid Gold Cat-Whisker, Combined value $2.90.

All supersensitive or money back (or replacement) by return mail.

Special Postpaid $2.00 for all six Cash with order

FOOTE MINERAL CO., INC., MFRS.
107 N. 19th St., Philadelphia
Estab. 47 Yrs. Ref. Your Banker (Dun-Bradstreet)
Distributors Wanted.

Radio News (May 1923), p. 2056

FOOTE DETECTOR CRYSTALS

The Foote Mineral Company, Inc., 107 N. 19th Street, Philadelphia, Pa., submitted a box containing 50 small tin boxes of various radio crystals. There were three types of

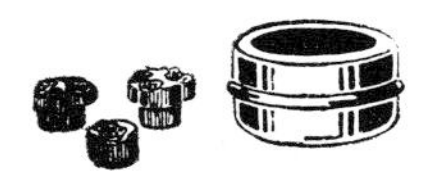

crystals submitted, viz., galena, pyrite and silicon. The crystals are mounted in an alloy having a ½" diameter projection, so that they can be clamped in a standard ½" detector cup. Single, double and triple-mounted crystals were submitted. There were in all 15 single galena crystals, 10 single pyrite crystals, five single silicon crystals, 10 double mountings of galena and pyrites, and 10 triple mountings containing galena, pyrites and silicon. All were tested and found to be very sensitive, but the galena, as was expected, was the most sensitive. Instruction sheet and diagram were furnished in each small metal box of crystals. The crystals are wrapped in tinfoil.

AWARDED THE RADIO NEWS LABORATORIES CERTIFICATE OF MERIT NO. 178.

Radio News (Sep. 1923), p. 286

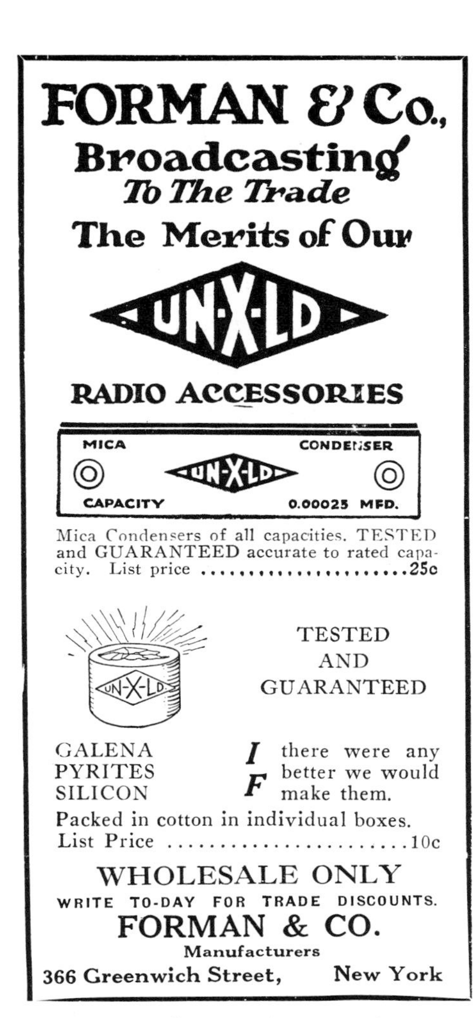

Radio News (Nov. 1923), p. 644

Make Your Reflex Set 100% Efficient by Using

MPM

Million Point Mineral CRYSTAL

APPROVED RADIO NEWS LABORATORIES 1922

PRICE

35c Mounted 25c Unmounted

M. P. M.

Million Point Mineral CRYSTALS

This remarkable crystal is super-sensitive—reproducing from every point on its surface. Greatly increases both audibility and radius—records having been reported up to 1200 miles. Will not burn out or corrode. M. P. M. crystal is unsurpassed for reception both with or without amplification.

PRICE
$4.50

M. P. M.

Reflex Radio Frequency TRANSFORMER

Absolute precision in the assembling of these transformers assures remarkable range and volume with positive elimination of distortion.

Special Offer

Detailed working diagram of a successful reflex circuit will be sent free with all orders for either crystals or transformers.

M. P. M. Sales Co.

Dept. N, 247 South Central Ave.,
Los Angeles, California

Radio News (Jan. 1924), p. 986

Radio News (Jan. 1925), p. 1232

ANNOUNCEMENT

NOW $1.00

NOW $1.50

SINCE we placed the NATIONAL AIRPHONE "GOLD-GRAIN" DETECTOR on the market, six months ago, it has taken the country by storm. Thousands upon thousands of radio fans, amateurs and experimenters are now using "GOLD-GRAIN" DETECTORS for Crystal outfits, for Reflex Sets and for many other radio purposes.

The demand for "GOLD-GRAIN" DETECTORS has been so tremendous that we have been enabled to effect vast economies in our manufacturing processes. This has made it possible for us to make sweeping price reductions in these detectors.

NATIONAL AIRPHONE "GOLD-GRAIN" DETECTORS

Now $1.00

Actual Size

FOR PANEL MOUNTING

After you have fussed with catwhiskers, springs, balls and adjustment handles, and after you have almost become a nervous wreck, hunting for "the elusive sensitive spot"—you will welcome with open arms our 100 per cent. **GOLD-GRAIN DETECTOR.**

This Detector is not a fixed Detector, but is foolproof; it has no catwhiskers, no springs, no balls, no adjusting handles; no fussing. The Detector is Entirely enclosed in hard rubber composition cartridge.

A special crystal is used, while contact elements are made of **pure gold.** There is always a multiplicity of contacts. The Detector is sealed hermetically. The contact with the crystal is always perfect.

This detector has been pronounced by experts as the greatest detector in existence. It reproduces voice, and music in natural color of tone, without distortion. You will be surprised at the wonderful results and satisfaction it gives.

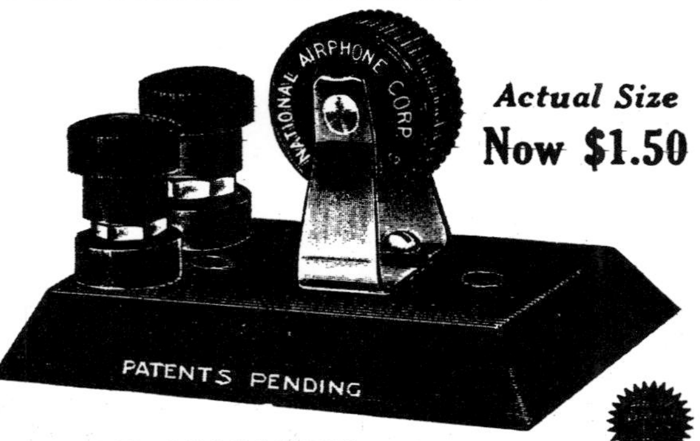

Actual Size

Now $1.50

YOU ARE PROTECTED BY THIS GUARANTEE

Should any National "Gold-Grain" Detector not be in first-class condition when purchased and within 10 days you return it to us unopened, or in unbroken condition, we will repair it or send you a new one free of charge. **Order from your Dealer—or direct from us.**

HOW TO MAKE A REFLEX SET

With the reflex circuits illustrated, and with the values as given, it is now possible, with a single tube and a NATIONAL AIRPHONE "GOLD-GRAIN" DETECTOR, to receive distances over 1500 miles on a small aerial.

The price of the parts as shown in the illustrations should not come higher than from $20.00 to $22.00 (excluding Vacuum tube and phones).

The results are really remarkable, and by using a WD-11 Tube it is not even necessary to use a storage battery. A small "B" Battery and a dry cell can be used.

An ideal portable outfit can be constructed quite readily with the Reflex, and for local stations, within a radius of 50 miles, an outdoor aerial is not required. A small two-foot loop may be used, and it becomes then possible to obtain a moderate volume of sound on a loud speaker.

The Reflex outfit as shown in the circuit herewith has been constructed by our engineering department and we shall be glad to demonstrate it to the radio fraternity. The extraordinary results obtained with this circuit are in part due to the NATIONAL AIRPHONE "GOLD-GRAIN" DETECTOR. Recent changes made in this Detector have improved it to such an extent that it is now entirely automatic and will stay put with an occasional adjustment.

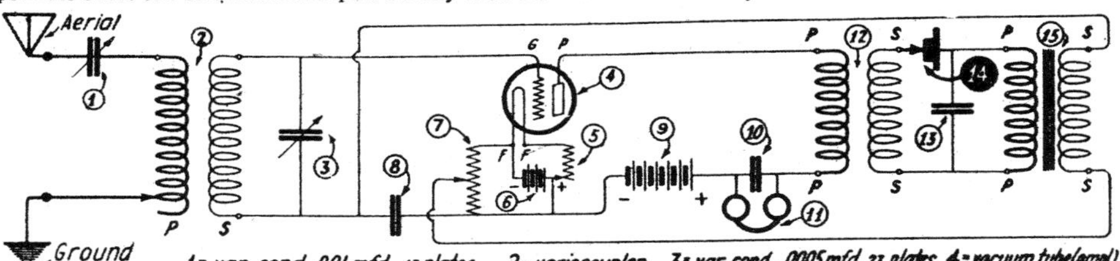

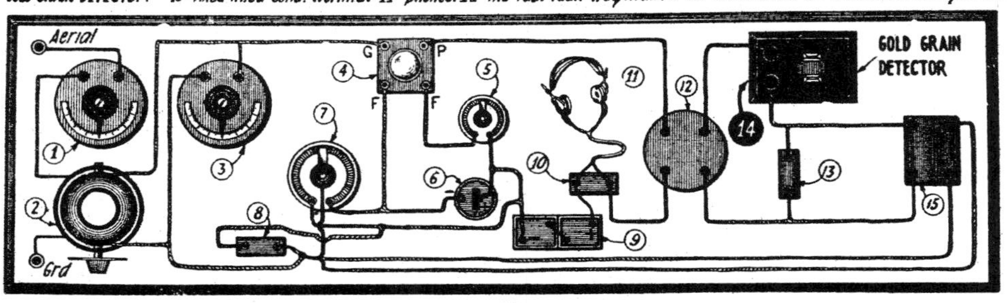

NATIONAL AIRPHONE CORPORATION

Distributors—Write for Territory.

Dealers—Write for Discounts.

18 HUDSON ST. NEW YORK

Radio News (June 1923), p. 2136

Index

The Index consists of two parts. General index items are in *Part 1*. *Part 2* contains the listings for all crystal sets, crystal detectors, and crystals (by trade names), and their manufacturers. Each of these four categories is grouped under a separate subheading in *Part 2*.
Italicized page numbers indicate items in advertisements, illustrations, or their captions.

Index Part 1: *General*

(See *Index Part 2* for specific item listings by trade name and for the manufacturers.)

Index Part 2

CRYSTAL DETECTORS BY TRADE NAMES

CRYSTALS BY TRADE NAMES

CRYSTAL SETS BY TRADE NAMES

MANUFACTURERS (ORIGINATING COMPANIES)

Crystal Clear

Vintage American Crystal Sets,
Crystal Detectors,
and Crystals

Composed by Eastern Graphics in
11 point Caledonia,
2 points leaded, with display type
set in Caledonia Bold Italic.

Title page by Katha Fauty
Cover photo by Stephen J. Appel